Lecture Notes Editorial Policies

Lecture Notes in Statistics provides a format for the informal and quick publication of monographs, case studies, and workshops of theoretical or applied importance. Thus, in some instances, proofs may be merely outlined and results presented which will later be published in a different form.

Publication of the Lecture Notes is intended as a service to the international statistical community, in that a commercial publisher, Springer-Verlag, can provide efficient distribution of documents that would otherwise have a restricted readership. Once published and copyrighted, they can be documented and discussed in the scientific literature.

Lecture Notes are reprinted photographically from the copy delivered in camera-ready form by the author or editor. Springer-Verlag provides technical instructions for the preparation of manuscripts. Volumes should be no less than 100 pages and preferably no more than 400 pages. A subject index is expected for authored but not edited volumes. Proposals for volumes should be sent to one of the series editors or addressed to "Statistics Editor" at Springer-Verlag in New York.

Authors of monographs receive 50 free copies of their book. Editors receive 50 free copies and are responsible for distributing them to contributors. Authors, editors, and contributors may purchase additional copies at the publisher's discount. No reprints of individual contributions will be supplied and no royalties are paid on Lecture Notes volumes. Springer-Verlag secures the copyright for each volume.

Series Editors:

Professor P. Bickel
Department of Statistics
University of California
Berkeley, California 94720
USA

Professor P. Diggle
Department of Mathematics
Lancaster University
Lancaster LA1 4YL
England

Professor S. Fienberg
Department of Statistics
Carnegie Mellon University
Pittsburgh, Pennsylvania 15213
USA

Professor K. Krickeberg
3 Rue de L'Estrapade
75005 Paris
France

Professor I. Olkin
Department of Statistics
Stanford University
Stanford, California 94305
USA

Professor N. Wermuth
Department of Psychology
Johannes Gutenberg University
Postfach 3980
D-6500 Mainz
Germany

Professor S. Zeger
Department of Biostatistics
The Johns Hopkins University
615 N. Wolfe Street
Baltimore, Maryland 21205-2103
USA

Lecture Notes in Statistics 136

Edited by P. Bickel, P. Diggle, S. Fienberg, K. Krickeberg, I. Olkin, N. Wermuth, S. Zeger

Springer
New York
Berlin
Heidelberg
Barcelona
Budapest
Hong Kong
London
Milan
Paris
Singapore
Tokyo

Gregory C. Reinsel
Raja P. Velu

Multivariate Reduced-Rank Regression

Theory and Applications

Springer

Gregory C. Reinsel
Department of Statistics
University of Wisconsin, Madison
Madison, WI 53706

Raja P. Velu
School of Management
Syracuse University
Syracuse, NY 13244

CIP data available.
Printed on acid-free paper.

Camera ready copy provided by the authors.
Printed and bound by Braun-Brumfield, Ann Arbor, MI.
Printed in the United States of America.

9 8 7 6 5 4 3 2 1

ISBN 0-387-98601-4 Springer-Verlag New York Berlin Heidelberg SPIN 10688907

Preface

In the area of multivariate analysis, there are two broad themes that have emerged over time. The analysis typically involves exploring the variations in a set of interrelated variables or investigating the simultaneous relationships between two or more sets of variables. In either case, the themes involve explicit modeling of the relationships or dimension-reduction of the sets of variables. The multivariate regression methodology and its variants are the preferred tools for the parametric modeling and descriptive tools such as principal components or canonical correlations are the tools used for addressing the dimension-reduction issues. Both act as complementary to each other and data analysts typically want to make use of these tools for a thorough analysis of multivariate data. A technique that combines the two broad themes in a natural fashion is the method of reduced-rank regression. This method starts with the classical multivariate regression model framework but recognizes the possibility for the reduction in the number of parameters through a restriction on the rank of the regression coefficient matrix. This feature is attractive because regression methods, whether they are in the context of a single response variable or in the context of several response variables, are popular statistical tools. The technique of reduced-rank regression and its encompassing features are the primary focus of this book.

The book develops the method of reduced-rank regression starting from the classical multivariate linear regression model. The major developments in the area are presented along with the historical connections to concepts and topics such as latent variables, errors-in-variables, and discriminant analysis that arise in various science areas and a variety of statistical areas. Applications of reduced-rank regression methods in fields such as economics, psychology, and biometrics are reviewed. Illustrative examples are presented with details sufficient for readers to construct their own applications. The computational algorithms are described so that readers can potentially use them for their analysis, estimation, and inference.

The book is broadly organized into three areas: regression modeling for general multi-response data, for multivariate time series data, and for longitudinal data. Chapter 1 gives basic results pertaining to analysis for the classical multivariate linear regression model. Chapters 2 and 3 contain the important developments and results for analysis and applications for the multivariate reduced-rank regression model. In addition to giving the fundamental methods and technical results, we also provide relationships between these methods and other familiar statistical topics such as linear structural models, linear discriminant analysis, canonical correlation analysis, and principal components. Practical numerical examples are also presented to illustrate the methods. Together, Chapters 1 through 3 cover the basic theory and applications related to modeling of multi-response data, such as multivariate cross-sectional data that may occur commonly in business and economics.

Chapters 4 and 5 focus on modeling of multivariate time series data. The fourth chapter deals with multivariate regression models where the errors are autocorrelated, a situation that is commonly considered in econometrics and other areas involving time series data. Chapter 5 is devoted to multiple autoregressive time series modeling. In addition to natural extensions of the reduced-rank models introduced in Chapters 1 through 3 to the time series setting, we also elaborate on nested reduced-rank models for multiple autoregressive time series. A particular feature that may occur in time series is unit-root nonstationarity, and we show how issues about nonstationarity can be investigated through reduced-rank regression methods and associated canonical correlation methods. We also devote a section to the related topic of cointegration among multiple nonstationary time series, the associated modeling, estimation, and testing. Numerical examples using multivariate time series data are provided to illustrate several of the main topics in Chapters 4 and 5.

Chapter 6 deals with the analysis of balanced longitudinal data, especially through use of the popular growth curve or generalized multivariate analysis of variance (GMANOVA) model. Results for the basic growth curve model are first developed and presented. Then we illustrate how the model can be modified and extended to accommodate reduced-rank features in the multivariate regression structure. As a result, we obtain an expanded collection of flexible models that can be useful for more parsimonious modeling of growth curve data. The important results on estimation and hypothesis testing are developed for these different forms of growth curve models with reduced-rank features, and a numerical example is presented to illustrate the methods. The use of reduced-rank growth curve methods in linear discriminant analysis is also briefly discussed.

In Chapter 7, we consider seemingly unrelated regression (SUR) equations models, which are widely used in econometrics and other fields. These models are more general than the traditional multivariate regression models because they allow for the values of the predictor variables associated with different response variables to be different. After a brief review of estimation and inference aspects of SUR models, we investigate the possibility of reduced rank for the coefficient matrix pooled from the SUR equations. Although the predictor variables take on different values for the different response variables, in practical applications the predictors generally represent similar variables in nature. Therefore we may expect certain commonality among the vectors of regression coefficients that arise from the different equations. The reduced-rank regression approach also provides a type of canonical analysis in the context of the SUR equations models. Estimation and inference results are derived for the SUR model with reduced-rank structure, including details on the computational algorithms, and an application in marketing is presented for numerical illustration.

In Chapter 8, the use of reduced-rank methods in the area of financial economics is discussed. It is seen that, in this setting, several applications of the reduced-rank regression model arise in a fairly natural way from economic theories. The context in which these models arise, particularly in the area of asset pricing, the particular form of the models and some empirical results are briefly surveyed. This topic provides an illustration of the use of reduced-rank models not mainly as an empirical statistical modeling tool for dimension-reduction but more as arising somewhat naturally in relation to certain economic theories.

The book concludes with a final chapter that highlights some other techniques associated with recent developments in multivariate regression modeling, many of which represent areas and topics for further research. This includes brief surveys and discussion of other related multivariate regression modeling methodologies that have similar parameter reduction objectives as reduced-rank regression, such as multivariate ridge regression, partial least squares, joint continuum regression, and other shrinkage and regularization techniques.

Because of the specialized nature of the topic of multivariate reduced-rank regression, a certain level of background knowledge in mathematical and statistical concepts is assumed. The reader will need to be familiar with basic matrix theory and have some limited exposure to multivariate methods. As noted above, the classical multivariate regression model is reviewed in Chapter 1 and many basic estimation and hypothesis testing results pertaining to that model are derived and presented. However, it is still anticipated that the reader will have some previous experience with the fundamentals of univariate linear regression modeling.

We want to acknowledge the useful comments by reviewers of an initial draft of this book, which were especially helpful in revising the earlier chapters. In particular, they provided perspectives for the material in Chapter 2 on historical developments of the topic. We are grateful to T. W. Anderson for his constructive comments on many general aspects of the subject matter contained in the book. We also want to thank Guofu Zhou for his helpful comments, in particular on material in Chapter 8 devoted to applications in financial economics. The scanner data used for illustration in Chapter 7 was provided by A. C. Nielsen and we want to thank David Bohl for his assistance. Raja Velu would also like to thank the University of Wisconsin-Whitewater for granting him a sabbatical during the 1996-97 academic year and the University of Wisconsin-Madison for hosting him; the sabbatical was partly devoted to writing this book.

Finally, we want to express our indebtedness to our wives, Sandy and Yasodha, and our children, for their support and love which were essential to complete this book.

Gregory C. Reinsel and Raja P. Velu

Contents

1
Multivariate Linear Regression

1.1 Introduction

Regression methods are perhaps the most widely used statistical tools in data analysis. When several response variables are studied simultaneously, we are in the sphere of multivariate regression. The usual description of the multivariate regression model, that relates the set of m multiple responses to a set of n predictor variables, assumes implicitly that the $m \times n$ regression coefficient matrix is of full rank. It can then be demonstrated that the simultaneous estimation of the elements of the coefficient matrix, by least squares or maximum likelihood estimation methods, yields the same results as a set of m multiple regressions, where each of the m individual response variables is regressed separately on the predictor variables. Hence, the fact that the multiple responses are likely to be related is not involved in the estimation of the regression coefficients as no information about the correlations among the response variables is taken into account. Any new knowledge gained by recognizing the multivariate nature of the problem and the fact that the responses are related is not incorporated when estimating the regression parameters jointly. There are two practical concerns regarding this general multivariate regression model. First, the accurate estimation of all the regression coefficients may require a relatively large number of observations, which might involve some practical limitations. Second, even if the data are available, interpreting simultaneously a large number of the regression coefficients can become unwieldy. Achieving parsimony in the number of unknown parameters may be desirable both from the estimation and interpretation points of view of multivariate regression analysis. We will address these concerns by recognizing a feature that enters the multivariate regression model.

The special feature that can enter into the multivariate linear regression model case is that we admit the possibility that the rank of the regression coefficient matrix can be deficient. This implies that there are linear restrictions on the coefficient matrix, and these restrictions themselves are often not known a priori. Alternately, the coefficient matrix can be written as a product of two component matrices of lower dimension. Such a model is called a *reduced-rank* regression model, and the model structure and estimation method for this model will be described in detail later. It follows from this estimation method that the assumption of lower rank for the regression coefficient matrix leads to estimation results which take into account the interrelations among the multiple responses and which use the entire set of predictor variables in a systematic fashion, compared to some ad hoc procedures which may use only some portion of the predictor variables in an attempt to achieve parsimony. The above mentioned features both have practical implications. When we model with a large number of response and predictor variables, the implication in terms of restrictions serves a useful purpose. Certain linear combinations of response variables can, eventually, be ignored for the regression modeling purposes, since these combinations will be found to be unrelated to the predictor variables. The alternate implication indicates that only certain linear combinations of the predictor variables need to be used in the regression model, since any remaining linear combinations will be found to have no influence on the response variables given the first set of linear combinations. Thus, the reduced-rank regression procedure takes care of the dimension reduction aspect of multivariate regression model building through the assumption of lower rank for the regression coefficient matrix.

In the case of multivariate linear regression, where the dependence of response variables among each other is utilized through the assumption of lower rank, the connection between the regression coefficients and the descriptive tools of principal components and canonical variates can be demonstrated. There is a correspondence between the rank of the regression coefficient matrix and a specified number of principal components or canonical correlations. Therefore, the use of these descriptive tools in multivariate model building is explicitly demonstrated through the reduced-rank assumption. Before we proceed with the consideration of various aspects of reduced-rank models in detail in subsequent chapters, we present some basic results concerning the classical full-rank multivariate linear regression model in the remainder of this chapter. More detailed and in-depth presentations of results for the full-rank multivariate linear regression model may be found in the books by Anderson (1984a, Chap. 8), Muirhead (1982, Chap. 10), and Srivastava and Khatri (1979, Chaps. 5 and 6), among others.

1.2 Multivariate Linear Regression Model and Least Squares Estimator

We consider the general multivariate linear model,

$$Y_k = CX_k + \epsilon_k, \qquad k = 1, \ldots, T, \tag{1.1}$$

where $Y_k = (y_{1k}, \ldots, y_{mk})'$ is an $m \times 1$ vector of response variables, $X_k = (x_{1k}, \ldots, x_{nk})'$ is an $n \times 1$ vector of predictor variables, C is an $m \times n$ regression coefficient matrix, and $\epsilon_k = (\epsilon_{1k}, \ldots, \epsilon_{mk})'$ is the $m \times 1$ vector of random errors, with mean vector $E(\epsilon_k) = 0$ and covariance matrix $\text{Cov}(\epsilon_k) = \Sigma_{\epsilon\epsilon}$, an $m \times m$ positive-definite matrix. The ϵ_k are assumed to be independent for different k. We shall assume that T vector observations are available, and define the $m \times T$ and $n \times T$ data matrices, respectively, as

$$\boldsymbol{Y} = [Y_1, \ldots, Y_T] \qquad \text{and} \qquad \boldsymbol{X} = [X_1, \ldots, X_T]. \tag{1.2}$$

We assume that $m+n \leq T$ and that $\boldsymbol{X}$ is of full rank $n = \text{rank}(\boldsymbol{X}) < T$. This last condition is required to obtain a unique (least squares) solution to the first order equations. We arrange the error vectors ϵ_k, $k = 1, \ldots, T$, into an $m \times T$ matrix $\boldsymbol{\epsilon} = [\epsilon_1, \epsilon_2, \ldots, \epsilon_T]$, similar to the data matrices given by (1.2). We also will consider the errors arranged into an $mT \times 1$ vector $e = (\boldsymbol{\epsilon}'_{(1)}, \ldots, \boldsymbol{\epsilon}'_{(m)})'$, where $\boldsymbol{\epsilon}_{(j)} = (\epsilon_{j1}, \ldots, \epsilon_{jT})'$ is the vector of errors corresponding to the jth response variable in the regression equation (1.1). Then we have that

$$E(e) = 0, \qquad \text{Cov}(e) = \Sigma_{\epsilon\epsilon} \otimes I_T, \tag{1.3}$$

where $A \otimes B$ denotes the Kronecker product of two matrices A and B. That is, if A is an $m \times n$ matrix and B is a $p \times q$ matrix, then the Kronecker product $A \otimes B$ is the $mp \times nq$ matrix with block elements $A \otimes B = (a_{ij}B)$, where a_{ij} is the (i,j)th element of A.

The unknown parameters in model (1.1) are the elements of the regression coefficient matrix C and the error covariance matrix $\Sigma_{\epsilon\epsilon}$. These can be estimated by the method of least squares, or equivalently, the method of maximum likelihood under the assumption of normality of the ϵ_k. Before we present the details on the estimation method, we rewrite the model (1.1) in terms of the complete data matrices $\boldsymbol{Y}$ and $\boldsymbol{X}$ as

$$\boldsymbol{Y} = C\boldsymbol{X} + \boldsymbol{\epsilon}. \tag{1.4}$$

The least squares criterion is given as

$$e'e = \text{tr}(\boldsymbol{\epsilon}\boldsymbol{\epsilon}') = \text{tr}[(\boldsymbol{Y} - C\boldsymbol{X})(\boldsymbol{Y} - C\boldsymbol{X})'] = \text{tr}\Big[\, \textstyle\sum_{k=1}^{T} \epsilon_k \, \epsilon_k' \,\Big], \tag{1.5}$$

where $\text{tr}(A)$ denotes the trace of a square matrix A, that is, the sum of the diagonal elements of A. The least squares (LS) estimate of C is the value

$\tilde{C}$ that minimizes the criterion in (1.5). Using results on matrix derivatives (e.g., Rao, 1973, p. 72), in particular $\partial \text{tr}(CZ)/\partial C = Z'$, we have the first order equations to obtain a minimum as

$$\frac{\partial \text{tr}(\boldsymbol{\epsilon}\boldsymbol{\epsilon}')}{\partial C} = 2\left[-\boldsymbol{Y}\boldsymbol{X}' + C(\boldsymbol{X}\boldsymbol{X}')\right] = 0 \tag{1.6}$$

which yields a unique solution for C as

$$\tilde{C} = \boldsymbol{Y}\boldsymbol{X}'(\boldsymbol{X}\boldsymbol{X}')^{-1} = \left(\tfrac{1}{T}\boldsymbol{Y}\boldsymbol{X}'\right)\left(\tfrac{1}{T}\boldsymbol{X}\boldsymbol{X}'\right)^{-1}. \tag{1.7}$$

If we let $C'_{(j)}$ denote the jth row of the matrix C, it is directly seen from (1.7) that the corresponding jth row of the LS estimate $\tilde{C}$ is

$$\tilde{C}'_{(j)} = \boldsymbol{Y}'_{(j)}\boldsymbol{X}'(\boldsymbol{X}\boldsymbol{X}')^{-1} = \left(\tfrac{1}{T}\boldsymbol{Y}'_{(j)}\boldsymbol{X}'\right)\left(\tfrac{1}{T}\boldsymbol{X}\boldsymbol{X}'\right)^{-1}, \tag{1.8}$$

for $j = 1, \ldots, m$, where $\boldsymbol{Y}'_{(j)} = (y_{j1}, \ldots, y_{jT})$ denotes the jth row of the matrix $\boldsymbol{Y}$, that is, $\boldsymbol{Y}_{(j)}$ is the $T \times 1$ vector of observations on the jth response variable. Thus, it is noted here that $\boldsymbol{Y}'_{(j)}\boldsymbol{X}' = \sum_{k=1}^{T} y_{jk}X'_k$ refers to the cross products between the jth response variable and the n predictor variables, summed over all T observations. Hence, the rows of the matrix C are estimated in (1.7) by the least squares regression of each response variable on the predictor variables and therefore the covariances among the response variables do not enter into the estimation.

From the presentation of model (1.1), it follows that our focus here is on the regression coefficient matrix C of the response variables $\boldsymbol{Y}$ on the predictor variables $\boldsymbol{X}$; the intercept term typically will not be explicitly represented in the model. With a constant term explicitly included, the model would take the form $Y_k = d + CX_k + \epsilon_k$, where d is an $m \times 1$ vector of unknown parameters. For convenience, in such a model it will be assumed that the predictor variables have been centered by subtraction of appropriate sample means, so that the resulting data matrix $\boldsymbol{X}$ has zero means, i.e., $\bar{X} = \frac{1}{T}\sum_{k=1}^{T} X_k = 0$. These mean-corrected observations would typically be the basis of all the calculations that follow. Under this setup, the basic results concerning the LS estimate of C would continue to hold, while the LS estimate of the intercept term is simply $\tilde{d} = \bar{Y} = \frac{1}{T}\sum_{k=1}^{T} Y_k$. For practical applications, an intercept is almost always needed in modeling, and in terms of estimation of C and $\Sigma_{\epsilon\epsilon}$ the inclusion of a constant term is equivalent to performing calculations with response variables Y_k that have (also) been corrected for the sample mean $\bar{Y}$.

The maximum likelihood method of estimation, under the assumption that values of the predictor variables X_k are known constant vectors and the assumption of normality of the error terms ϵ_k in addition to the properties given in (1.3), yields the same results as least squares. The likelihood is equal to

$$f(\boldsymbol{\epsilon}) = (2\pi)^{-mT/2}|\Sigma_{\epsilon\epsilon}|^{-T/2}\exp\left[\,-(1/2)\,\text{tr}\left(\Sigma_{\epsilon\epsilon}^{-1}\boldsymbol{\epsilon}\boldsymbol{\epsilon}'\right)\,\right], \tag{1.9}$$

where $\epsilon = \boldsymbol{Y} - C\boldsymbol{X}$. For maximum likelihood (ML) estimation, the criterion that must be minimized to estimate C is seen from (1.9) to be

$$\operatorname{tr}\left(\Sigma_{\epsilon\epsilon}^{-1} \epsilon\epsilon' \right) = \operatorname{tr}\left[\Sigma_{\epsilon\epsilon}^{-1/2} (\boldsymbol{Y} - C\boldsymbol{X})(\boldsymbol{Y} - C\boldsymbol{X})' \Sigma_{\epsilon\epsilon}^{-1/2} \right], \tag{1.10}$$

which can also be expressed as $e'(\Sigma_{\epsilon\epsilon}^{-1} \otimes I_T)e$. From (1.10), the first order equations to obtain a minimum are $\partial \operatorname{tr}(\Sigma_{\epsilon\epsilon}^{-1} \epsilon\epsilon')/\partial C = 2\, \Sigma_{\epsilon\epsilon}^{-1} [-\boldsymbol{Y}\boldsymbol{X}' + C(\boldsymbol{X}\boldsymbol{X}')] = 0$, which yields the same solution for C as given in (1.7). In later chapters we shall consider a more general criterion,

$$\operatorname{tr}(\Gamma \epsilon\epsilon') = \operatorname{tr}[\Gamma^{1/2} (\boldsymbol{Y} - C\boldsymbol{X})(\boldsymbol{Y} - C\boldsymbol{X})' \Gamma^{1/2}], \tag{1.11}$$

where Γ is a positive-definite matrix. The criteria (1.5) and (1.10) follow as special cases of (1.11) by appropriate choices of the matrix Γ. The ML estimator of the error covariance matrix $\Sigma_{\epsilon\epsilon}$ is obtained from the least squares residuals $\hat{\epsilon} = \boldsymbol{Y} - \tilde{C}\boldsymbol{X}$ as $\tilde{\Sigma}_{\epsilon\epsilon} = \frac{1}{T}(\boldsymbol{Y} - \tilde{C}\boldsymbol{X})(\boldsymbol{Y} - \tilde{C}\boldsymbol{X})' \equiv \frac{1}{T}S$, where $S = (\boldsymbol{Y} - \tilde{C}\boldsymbol{X})(\boldsymbol{Y} - \tilde{C}\boldsymbol{X})' = \sum_{k=1}^{T} \hat{\epsilon}_k \, \hat{\epsilon}_k'$ is the error sum of squares matrix, and $\hat{\epsilon}_k = Y_k - \tilde{C}X_k$, $k = 1, \ldots, T$, are the least squares residual vectors. An unbiased estimator of $\Sigma_{\epsilon\epsilon}$ is obtained as

$$\overline{\Sigma}_{\epsilon\epsilon} = \frac{1}{T-n} \hat{\epsilon}\hat{\epsilon}' = \frac{1}{T-n} \sum_{k=1}^{T} \hat{\epsilon}_k \, \hat{\epsilon}_k'. \tag{1.12}$$

1.3 Further Inference Properties in the Multivariate Regression Model

Under the assumption that the matrix of predictor variables $\boldsymbol{X}$ is fixed and nonstochastic, properties of the least squares estimator $\tilde{C}$ in (1.7) can readily be derived. In practice, the X_k in (1.1) may consist of observations on a stochastic vector, in which case the (finite sample) distributional results to be presented for the LS estimator $\tilde{C}$ can be viewed as conditional on fixed observed values of the X_k. First, $\tilde{C}$ is an unbiased estimator of C since from (1.4)

$$E(\tilde{C}) = E[\boldsymbol{Y}\boldsymbol{X}'(\boldsymbol{X}\boldsymbol{X}')^{-1}] = E[\boldsymbol{Y}]\boldsymbol{X}'(\boldsymbol{X}\boldsymbol{X}')^{-1} = C\,\boldsymbol{X}\boldsymbol{X}'(\boldsymbol{X}\boldsymbol{X}')^{-1} = C.$$

From (1.8), we find that

$$\operatorname{Cov}(\tilde{C}_{(i)}, \tilde{C}_{(j)}) = (\boldsymbol{X}\boldsymbol{X}')^{-1}\boldsymbol{X} \operatorname{Cov}(\boldsymbol{Y}_{(i)}, \boldsymbol{Y}_{(j)})\, \boldsymbol{X}'(\boldsymbol{X}\boldsymbol{X}')^{-1} = \sigma_{ij}(\boldsymbol{X}\boldsymbol{X}')^{-1}$$

for $i, j = 1, \ldots, m$, where σ_{ij} denotes the (i,j)th element of the covariance matrix $\Sigma_{\epsilon\epsilon}$, since $\operatorname{Cov}(\boldsymbol{Y}_{(i)}, \boldsymbol{Y}_{(j)}) = \sigma_{ij} I_T$. The distributional properties of $\tilde{C}$ follow easily from multivariate normality of the error terms ϵ_k. Specifically, we consider the distribution of $\operatorname{vec}(\tilde{C}')$, where the "vec" operator transforms an $n \times m$ matrix into an nm-dimensional column vector by stacking the columns of the matrix below each other. Thus, for an $m \times n$ matrix $\tilde{C}$, $\text{vec}(\tilde{C}') = (\tilde{C}'_{(1)}, \ldots, \tilde{C}'_{(m)})'$, where $\tilde{C}'_{(j)}$ refers to the jth row of the matrix $\tilde{C}$. A useful property that relates the vec operation with the Kronecker product of matrices is that if A, B, and C are matrices of appropriate dimensions so that the product ABC is defined, then $\text{vec}(ABC) = (C' \otimes A)\text{vec}(B)$ [see Neudecker (1969)]. Another useful property is $\text{tr}(ABCB') = \{\text{vec}(B)\}'(C' \otimes A)\text{vec}(B)$.

Now notice that the model (1.4) may be expressed in the vector form as

$$y = \text{vec}(\boldsymbol{Y}') = \text{vec}(\boldsymbol{X}'C') + \text{vec}(\boldsymbol{\epsilon}') = (I_m \otimes \boldsymbol{X}')\,\text{vec}(C') + e, \qquad (1.13)$$

where $y = \text{vec}(\boldsymbol{Y}') = (\boldsymbol{Y}'_{(1)}, \ldots, \boldsymbol{Y}'_{(m)})'$, and $e = \text{vec}(\boldsymbol{\epsilon}') = (\boldsymbol{\epsilon}'_{(1)}, \ldots, \boldsymbol{\epsilon}'_{(m)})'$ is similar in form to y with $\text{Cov}(e) = \Sigma_{\epsilon\epsilon} \otimes I_T$. In addition, the least squares estimator $\tilde{C}$ in (1.7) can be written in vector form as

$$\text{vec}(\tilde{C}') = \text{vec}((\boldsymbol{X}\boldsymbol{X}')^{-1}\boldsymbol{X}\boldsymbol{Y}') = (I_m \otimes (\boldsymbol{X}\boldsymbol{X}')^{-1}\boldsymbol{X})\,\text{vec}(\boldsymbol{Y}'), \qquad (1.14)$$

where $\text{vec}(\boldsymbol{Y}')$ is distributed as multivariate normal with mean vector $(I_m \otimes \boldsymbol{X}')\,\text{vec}(C')$ and covariance matrix $\Sigma_{\epsilon\epsilon} \otimes I_T$. It then follows directly that $\text{vec}(\tilde{C}')$ is also distributed as multivariate normal, with

$$E[\text{vec}(\tilde{C}')] = (I_m \otimes (\boldsymbol{X}\boldsymbol{X}')^{-1}\boldsymbol{X})E[\text{vec}(\boldsymbol{Y}')] = \text{vec}(C'),$$

$$\begin{aligned}
\text{Cov}[\text{vec}(\tilde{C}')] &= (I_m \otimes (\boldsymbol{X}\boldsymbol{X}')^{-1}\boldsymbol{X})\,\text{Cov}[\text{vec}(\boldsymbol{Y}')]\,(I_m \otimes \boldsymbol{X}'(\boldsymbol{X}\boldsymbol{X}')^{-1}) \\
&= \Sigma_{\epsilon\epsilon} \otimes (\boldsymbol{X}\boldsymbol{X}')^{-1}.
\end{aligned}$$

We summarize the main results on the LS estimate $\tilde{C}$ in the following.

Result 1.1 For the model (1.1) under the normality assumption on the ϵ_k, the least squares estimator $\tilde{C} = \boldsymbol{Y}\boldsymbol{X}'(\boldsymbol{X}\boldsymbol{X}')^{-1}$ is the same as the maximum likelihood estimator of C, and the distribution of the least squares estimator $\tilde{C}$ is such that

$$\text{vec}(\tilde{C}') \sim N(\text{vec}(C'), \Sigma_{\epsilon\epsilon} \otimes (\boldsymbol{X}\boldsymbol{X}')^{-1}). \qquad (1.15)$$

Note, in particular, that this result implies that the jth row of $\tilde{C}$, $\tilde{C}'_{(j)}$, which is the vector of least squares estimates of regression coefficients for the jth response variable, has the distribution $\tilde{C}_{(j)} \sim N(C_{(j)}, \sigma_{jj}(\boldsymbol{X}\boldsymbol{X}')^{-1})$.

The inference on the elements of the matrix C can be made using the result in (1.15). In practice, because $\Sigma_{\epsilon\epsilon}$ is unknown, a reasonable estimator such as the unbiased estimator in (1.12) is substituted for the covariance matrix in (1.15). All the necessary calculations for the above linear least squares inference procedures can be carried out using standard computer softwares, such as Statistical Programs for Social Sciences (SPSS) and Statistical Analysis System (SAS).

We shall also indicate the likelihood ratio procedure for testing simple linear hypotheses regarding the regression coefficient matrix. Consider $\boldsymbol{X}$ partitioned as $\boldsymbol{X} = [\boldsymbol{X}_1', \boldsymbol{X}_2']'$ and corresponding $C = [C_1, C_2]$, so that the model (1.4) is written as $\boldsymbol{Y} = C_1\boldsymbol{X}_1 + C_2\boldsymbol{X}_2 + \boldsymbol{\epsilon}$ where C_1 is $m \times n_1$ and C_2 is $m \times n_2$ with $n_1 + n_2 = n$. It may be of interest to test the null hypothesis $H_0 : C_2 = 0$ against the alternative $C_2 \neq 0$. The null hypothesis implies that the predictor variables $\boldsymbol{X}_2$ do not have any (additional) influence on the response variables $\boldsymbol{Y}$, given the impact of the variables $\boldsymbol{X}_1$. Using the likelihood ratio (LR) testing approach, it is easy to see from (1.9) that the LR test statistic is $\lambda = U^{T/2}$, where $U = |S|/|S_1|$, $S = (\boldsymbol{Y} - \tilde{C}\boldsymbol{X})(\boldsymbol{Y} - \tilde{C}\boldsymbol{X})'$ and $S_1 = (\boldsymbol{Y} - \hat{C}_1\boldsymbol{X}_1)(\boldsymbol{Y} - \hat{C}_1\boldsymbol{X}_1)'$. The matrix S is the residual sum of squares matrix from maximum likelihood (i.e., least squares) fitting of the full model, while S_1 is the residual sum of squares matrix obtained from fitting the reduced model with $C_2 = 0$ and $\hat{C}_1 = \boldsymbol{Y}\boldsymbol{X}_1'(\boldsymbol{X}_1\boldsymbol{X}_1')^{-1}$. This LR statistic result follows directly by noting that the value of the multivariate normal likelihood function evaluated at the maximum likelihood estimates $\tilde{C}$ and $\tilde{\Sigma}_{\epsilon\epsilon} = (1/T)S$ is equal to a constant times $|S|^{-T/2}$, because $\mathrm{tr}(\, \tilde{\Sigma}_{\epsilon\epsilon}^{-1}\hat{\boldsymbol{\epsilon}}\hat{\boldsymbol{\epsilon}}' \,) = \mathrm{tr}(\, T\,\tilde{\Sigma}_{\epsilon\epsilon}^{-1}\,\tilde{\Sigma}_{\epsilon\epsilon} \,) = Tm$ is a constant when evaluated at the ML estimate. Thus, the likelihood ratio for testing $H_0 : C_2 = 0$ is $\lambda = |S_1|^{-T/2}/|S|^{-T/2} = U^{T/2}$. It has been shown [see Anderson (1984a, Chap. 8)] that for moderate and large sample size T, the test statistic

$$\mathcal{M} = -[T - n + (n_2 - m - 1)/2]\log(U) \tag{1.16}$$

is approximately distributed as chi-squared with n_2m degrees of freedom ($\chi^2_{n_2m}$); the hypothesis is rejected when $\mathcal{M}$ is greater than a constant determined by the $\chi^2_{n_2m}$ distribution.

It is useful to express the LR statistic $\lambda = U^{T/2}$ in an alternate equivalent form. Write $\boldsymbol{X}\boldsymbol{X}' = A = [(A_{ij})]$, where $A_{ij} = \boldsymbol{X}_i\boldsymbol{X}_j'$, $i,j = 1,2$, and let $A_{22.1} = A_{22} - A_{21}A_{11}^{-1}A_{12}$. Then the LS estimate of C_1 under the reduced model with $C_2 = 0$ can be expressed as $\hat{C}_1 = \boldsymbol{Y}\boldsymbol{X}_1'(\boldsymbol{X}_1\boldsymbol{X}_1')^{-1} = \tilde{C}_1 + \tilde{C}_2A_{21}A_{11}^{-1}$, because of the full model LS normal equations (1.6), $\boldsymbol{Y}\boldsymbol{X}_1' = \tilde{C}_1\boldsymbol{X}_1\boldsymbol{X}_1' + \tilde{C}_2\boldsymbol{X}_2\boldsymbol{X}_1'$. Hence

$$\boldsymbol{Y} - \hat{C}_1\boldsymbol{X}_1 = (\boldsymbol{Y} - \tilde{C}\boldsymbol{X}) + \tilde{C}_2(\boldsymbol{X}_2 - A_{21}A_{11}^{-1}\boldsymbol{X}_1),$$

where the two terms on the right are orthogonal since $(\boldsymbol{Y} - \tilde{C}\boldsymbol{X})\boldsymbol{X}' = 0$ because $\tilde{C}$ is a solution to the first order equations (1.6). It thus follows that

$$\begin{aligned} S_1 &= (\boldsymbol{Y} - \hat{C}_1\boldsymbol{X}_1)(\boldsymbol{Y} - \hat{C}_1\boldsymbol{X}_1)' \\ &= (\boldsymbol{Y} - \tilde{C}\boldsymbol{X})(\boldsymbol{Y} - \tilde{C}\boldsymbol{X})' + \tilde{C}_2(A_{22} - A_{21}A_{11}^{-1}A_{12})\tilde{C}_2' \\ &= S + \tilde{C}_2A_{22.1}\tilde{C}_2', \end{aligned} \tag{1.17}$$

and also note that $A_{22.1}^{-1}$ is the $n_2 \times n_2$ lower diagonal block of the matrix $(\boldsymbol{X}\boldsymbol{X}')^{-1}$ [e.g., see Rao (1973, p. 33)]. Hence we see that

$$U = |S|/|S_1| = |S|/|S + H| = 1/|I + S^{-1}H|$$

where $H = \tilde{C}_2 A_{22.1} \tilde{C}_2'$ is referred to as the hypothesis sum of squares matrix (associated with the null hypothesis $C_2 = 0$). Therefore, we have $-\log(U) = \sum_{i=1}^{m} \log(1+\hat{\lambda}_i^2)$ where $\hat{\lambda}_i^2$ are the eigenvalues of $S^{-1}H$, and the LR statistic $\mathcal{M}$ in (1.16) is directly expressible in terms of these eigenvalues. In addition, it can be shown [e.g., see Reinsel (1997, Sec. A4.3)] that the $\hat{\lambda}_i^2$ are related to the sample partial canonical correlations $\hat{\rho}_i$ between $\boldsymbol{Y}$ and $\boldsymbol{X}_2$, eliminating $\boldsymbol{X}_1$, as $\hat{\lambda}_i^2 = \hat{\rho}_i^2/(1-\hat{\rho}_i^2)$, $i = 1, \ldots, m$. We summarize the above in the following result.

Result 1.2 For the model (1.1) under the normality assumption on the ϵ_k, the likelihood ratio for testing $H_0: C_2 = 0$ is $\lambda = U^{T/2}$, where $U = |S|/|S_1| = 1/|I + S^{-1}H|$ with $H = \tilde{C}_2 A_{22.1} \tilde{C}_2'$. Therefore, $-\log(U) = \sum_{i=1}^{m} \log(1 + \hat{\lambda}_i^2)$ where $\hat{\lambda}_i^2$ are the eigenvalues of $S^{-1}H$. For large sample size T, the asymptotic distribution of the test statistic $\mathcal{M} = -[T-n+(n_2-m-1)/2]\log(U)$ is chi-squared with $n_2 m$ degrees of freedom.

Finally, we briefly consider testing the more general null hypothesis of the form $H_0: F_1 C G_2 = 0$ where F_1 and G_2 are known full-rank matrices of appropriate dimensions $m_1 \times m$ and $n \times n_2$, respectively. In the hypothesis $F_1 C G_2 = 0$, the matrix G_2 allows for restrictions among the coefficients in C across (subsets of) the predictor variables, whereas F_1 provides for restrictions in the coefficients of C across the different response variables. For instance, the hypothesis $C_2 = 0$ considered earlier is represented in this form with $G_2 = [0', I_{n_2}']'$ and $F_1 = I_m$, and the hypothesis of the form $F_1 C G_2 = 0$ with the same matrix G_2 and $F_1 = [I_{m_1}, 0]$ would postulate that the predictor variables $\boldsymbol{X}_2$ do not enter into the linear regression relationship *only* for the first $m_1 < m$ components of the response variables $\boldsymbol{Y}$. Using the LR approach, the likelihood ratio is given by $\lambda = U^{T/2}$ where $U = |S|/|S_1|$ and $S_1 = (\boldsymbol{Y} - \hat{C}\boldsymbol{X})(\boldsymbol{Y} - \hat{C}\boldsymbol{X})'$ with $\hat{C}$ denoting the ML estimator of C obtained subject to the constraint that $F_1 C G_2 = 0$. In fact, using Lagrange multiplier methods, it can be shown that the constrained ML estimator under $F_1 C G_2 = 0$ can be expressed as

$$\hat{C} = \tilde{C} - SF_1'(F_1 S F_1')^{-1} F_1 \tilde{C} G_2 [G_2'(\boldsymbol{X}\boldsymbol{X}')^{-1} G_2]^{-1} G_2'(\boldsymbol{X}\boldsymbol{X}')^{-1}. \quad (1.18)$$

An alternate way to develop this last result is to consider matrices $F = [F_1', F_2']'$ and $G = [G_1, G_2]$ such that $F_1 F_2' = 0$ and $G_1' G_2 = 0$ (for convenience), and apply the transformation to the model (1.4) as $\boldsymbol{Y}^* = F\boldsymbol{Y} = FCGG^{-1}\boldsymbol{X} + F\boldsymbol{\epsilon} \equiv C^*\boldsymbol{X}^* + \boldsymbol{\epsilon}^*$, where $C^* = FCG$, $\boldsymbol{Y}^* = F\boldsymbol{Y}$, and $\boldsymbol{X}^* = G^{-1}\boldsymbol{X}$. With C^* partitioned into submatrices C_{ij}^*, $i, j = 1, 2$, the hypothesis $F_1CG_2 = 0$ is equivalent to $C_{12}^* = 0$. Maximum likelihood estimation of the parameters C^* in this transformed "canonical form" model, subject to the restriction $C_{12}^* = 0$, and transformation of the resulting ML estimate $\hat{C}^*$ back to the parameters C of the original model as $\hat{C} = F^{-1}\hat{C}^*G^{-1}$ then leads to the result in (1.18). Using either method, from (1.18) we therefore have

$$\begin{aligned} \boldsymbol{Y} - \hat{C}\boldsymbol{X} &= (\boldsymbol{Y} - \tilde{C}\boldsymbol{X}) \\ &\quad + SF_1'(F_1SF_1')^{-1}F_1\tilde{C}G_2[G_2'(\boldsymbol{X}\boldsymbol{X}')^{-1}G_2]^{-1}G_2'(\boldsymbol{X}\boldsymbol{X}')^{-1}\boldsymbol{X} \end{aligned}$$

where the two terms on the right are orthogonal. Following the above developments as in (1.17), we thus readily find that

$$\begin{aligned} S_1 = S \,+\, & SF_1'(F_1SF_1')^{-1}(F_1\tilde{C}G_2)[G_2'(\boldsymbol{X}\boldsymbol{X}')^{-1}G_2]^{-1} \\ & \times (F_1\tilde{C}G_2)'(F_1SF_1')^{-1}F_1S \equiv S + H, \end{aligned} \tag{1.19}$$

and $U = 1/|I + S^{-1}H|$ so that the LR test can be based on the eigenvalues of $S^{-1}H$.

The nonzero eigenvalues of the $m \times m$ matrix $S^{-1}H$ are the same as those of the $m_1 \times m_1$ matrix $S_*^{-1}H_*$, where $S_* = F_1SF_1'$ and $H_* = F_1HF_1' = F_1\tilde{C}G_2[G_2'(\boldsymbol{X}\boldsymbol{X}')^{-1}G_2]^{-1}(F_1\tilde{C}G_2)'$, so that $U^{-1} = |I + S_*^{-1}H_*| = \prod_{i=1}^{m_1}(1 + \hat{\lambda}_i^2)$ where the $\hat{\lambda}_i^2$ are the eigenvalues of $S_*^{-1}H_*$. Analogous to (1.16), the test statistic used is $\mathcal{M} = -[T - n + \frac{1}{2}(n_2 - m_1 - 1)]\log(U)$ which has an approximate $\chi_{n_2m_1}^2$ distribution under H_0. Somewhat more accurate approximation of the null distribution of $\log(U)$ is available based on the F-distribution, and in some cases (notably when $n_2 = 1$ or 2, or $m_1 = 1$ or 2) exact F-distribution results for U are available [see Anderson (1984a, Chap. 8)]. Values relating to the (exact) distribution of the statistic U for various values of m_1, n_2, and T–n have been tabulated by Schatzoff (1966), Pillai and Gupta (1969), and Lee (1972). However, for most of our applications we will simply use the large sample $\chi_{n_2m_1}^2$ approximation for the distribution of the LR statistic $\mathcal{M}$.

Concerning the exact distribution theory associated with the test procedure, under normality of the ϵ_k, it is established (e.g., Srivastava and Khatri, 1979, Chap. 6) that S_* has the $W(F_1\Sigma_{\epsilon\epsilon}F_1', T - n)$ distribution, the Wishart distribution with matrix $F_1\Sigma_{\epsilon\epsilon}F_1'$ and $T - n$ degrees of freedom, and H_* is distributed as Wishart under H_0 with matrix $F_1\Sigma_{\epsilon\epsilon}F_1'$ and n_2 degrees of freedom. The sum of squares matrices S_* and H_* are also independently distributed, which follows from the well known fact that the residual sum of squares matrix S and the least squares estimator $\tilde{C}$ are independent, and H_* is a function of $\tilde{C}$ alone. Thus, in particular, in the special case $m_1 = 1$ we see that $S_*^{-1}H_* = H_*/S_*$ is distributed as $n_2/(T$–$n)$ times the F-distribution with n_2 and $T - n$ degrees of freedom.

Tests other than the LR procedure may also be considered for the multivariate linear hypothesis $F_1CG_2 = 0$. The most notable of these is the test developed by Roy (1953) based on the union-intersection principle of test construction. Roy's test rejects for large values of the test statistic which is equal to the largest eigenvalue of the matrix $S_*^{-1}H_*$. Charts and tables of the upper critical values for the largest-root distribution have been computed by Heck (1960) and Pillai (1967).

Notice that the above hypotheses concerning C can be viewed as an assumption that the coefficient matrix C has a *reduced* rank, but of a particular *known* or specified form (i.e., the matrices F_1 and G_2 in the constraint $F_1CG_2 = 0$ are known). In the next chapter, we consider further investigating the possibility of a reduced-rank structure for the regression coefficient matrix C, but where the form of the reduced rank is not known. In particular, in such cases of reduced-rank structure for C, linear constraints of the form $F_1C = 0$ will hold but with the $m_1 \times m$ matrix F_1 unknown $(m_1 < m)$. Notice from (1.18) that the appropriate ML estimator of C under (known) constraints on C of such a form is given by

$$\hat{C} = (I_m - SF_1'(F_1SF_1')^{-1}F_1)\tilde{C} \equiv P\tilde{C}, \tag{1.20}$$

where $P = I_m - SF_1'(F_1SF_1')^{-1}F_1$ is an idempotent matrix of rank $r = m - m_1$. In Sections 2.3 and 2.6, we will find that the ML estimator $\hat{C}$ of C subject to the constraint of having reduced rank r (but with the linear constraints $F_1C = 0$ unknown) can be expressed in a similar form as in (1.20), but in terms of an estimate $\hat{F}_1$ whose rows are determined from examination of the eigenvectors of the matrix $S^{-1/2}\tilde{C}(\boldsymbol{X}\boldsymbol{X}')\tilde{C}'S^{-1/2}$ associated with its m_1 smallest eigenvalues.

1.4 Prediction in the Multivariate Linear Regression Model

Another problem of interest in regard to the multivariate linear regression model (1.1) is confidence regions for the mean responses corresponding to a fixed value X_0 of the predictor variables. The mean vector of responses at X_0 is CX_0 and its estimate is $\tilde{C}X_0$, where $\tilde{C} = \boldsymbol{Y}\boldsymbol{X}'(\boldsymbol{X}\boldsymbol{X}')^{-1}$ is the LS estimator of C. Under the assumption of normality of the ϵ_k, noting that $\tilde{C}X_0 = \mathrm{vec}(X_0'\tilde{C}') = (I_m \otimes X_0')\mathrm{vec}(\tilde{C}')$, we see that $\tilde{C}X_0$ is distributed as multivariate normal $N(CX_0, X_0'(\boldsymbol{X}\boldsymbol{X}')^{-1}X_0\,\Sigma_{\epsilon\epsilon})$ from Result 1.1. Since $\tilde{C}$ is distributed independently of $\overline{\Sigma}_{\epsilon\epsilon} = \frac{1}{T-n}S$ with $(T-n)\overline{\Sigma}_{\epsilon\epsilon}$ distributed as Wishart with matrix $\Sigma_{\epsilon\epsilon}$ and $T-n$ degrees of freedom, it follows that

$$\mathcal{T}^2 = (\tilde{C}X_0 - CX_0)'\overline{\Sigma}_{\epsilon\epsilon}^{-1}(\tilde{C}X_0 - CX_0)/\{X_0'(\boldsymbol{X}\boldsymbol{X}')^{-1}X_0\} \tag{1.21}$$

has the Hotelling's T^2-distribution with T–n degrees of freedom (Anderson, 1984a, Chap. 5). Thus, $\{(T-n-m+1)/m(T-n)\}\mathcal{T}^2$ has the F-distribution with m and $T-n-m+1$ degrees of freedom, so that a $100(1-\alpha)\%$ confidence ellipsoid for the true mean response CX_0 at X_0 is given by

$$(\tilde{C}X_0 - CX_0)'\overline{\Sigma}_{\epsilon\epsilon}^{-1}(\tilde{C}X_0 - CX_0) \leq \{X_0'(\boldsymbol{X}\boldsymbol{X}')^{-1}X_0\}\left[\frac{m(T-n)}{T-n-m+1}F_{m,T-n-m+1}(\alpha)\right] \quad (1.22)$$

where $F_{m,T-n-m+1}(\alpha)$ is the upper 100αth percentile of the F-distribution with m and $T-n-m+1$ degrees of freedom. Because

$$(a'\tilde{C}X_0 - a'CX_0)^2/(a'\overline{\Sigma}_{\epsilon\epsilon}a) \leq (\tilde{C}X_0 - CX_0)'\overline{\Sigma}_{\epsilon\epsilon}^{-1}(\tilde{C}X_0 - CX_0)$$

for every nonzero $m\times 1$ vector a, by the Cauchy–Schwarz inequality, $100(1-\alpha)\%$ *simultaneous* confidence intervals for all linear combinations $a'CX_0$ of the individual mean responses are given by

$$a'\tilde{C}X_0 \pm \{(a'\overline{\Sigma}_{\epsilon\epsilon}a)\,X_0'(\boldsymbol{X}\boldsymbol{X}')^{-1}X_0\}^{1/2}\left[\frac{m(T-n)}{T-n-m+1}F_{m,T-n-m+1}(\alpha)\right]^{1/2}.$$

Notice that the mean response vector CX_0 of interest represents a particular example of the general parametric function F_1CG_2 discussed earlier in the hypothesis testing context of Section 1.3, corresponding to $F_1 = I_m$, $G_2 = X_0$, and hence $n_2 = 1$.

A related problem of interest is the prediction of values of a new response vector $Y_0 = CX_0 + \epsilon_0$ corresponding to the predictor variables X_0, where it is assumed that Y_0 is independent of the values $Y_1, \ldots, Y_T$. The predictor of Y_0 is given by $\hat{Y}_0 = \tilde{C}X_0$, and it follows similarly to the above result (1.22) that a $100(1-\alpha)\%$ prediction ellipsoid for Y_0 is

$$(Y_0 - \tilde{C}X_0)'\overline{\Sigma}_{\epsilon\epsilon}^{-1}(Y_0 - \tilde{C}X_0) \leq \{1 + X_0'(\boldsymbol{X}\boldsymbol{X}')^{-1}X_0\}\left[\frac{m(T-n)}{T-n-m+1}F_{m,T-n-m+1}(\alpha)\right] \quad (1.23)$$

Another prediction problem that may be of interest on occasion is that of prediction of a subset of the components of Y_0 given that the remaining components have been observed. This problem would also involve the covariance structure $\Sigma_{\epsilon\epsilon}$ of Y_0, but it will not be discussed explicitly here. Besides the various dimension reduction interpretations, one additional purpose of investigating the possibility of a reduced-rank structure for C, as will be undertaken in the next chapter, is to obtain more precise predictions of future response vectors as a result of more accurate estimation of the regression coefficient matrix C in model (1.1).

1.5 A Numerical Example

In this section we present some details for a numerical example on multivariate linear regression analysis of some biochemical data, to illustrate some of the methods and results described in the earlier sections of this chapter.

EXAMPLE 1.1. We consider the basic multivariate regression analysis methods of the previous sections applied to some biochemical data taken from a study by Smith et al. (1962). The data involve biochemical measurements on several characteristics of urine specimens of 17 men classified into two weight groups, overweight and underweight. Each subject, except for one, contributed two morning samples of urine, and the data considered in this example consist of the 33 individual samples (treated as independent with no provision for the possible correlations within the two determinations of each subject) for five response variables ($m = 5$), pigment creatinine (y_1), and concentrations of phosphate (y_2), phosphorous (y_3), creatinine (y_4), and choline (y_5), and the concomitant or predictor variables, volume (x_2) and specific gravity (x_3), in addition to the predictor variable for weight of the subject (x_1). The multivariate data set for this example is presented in Table A.1 of the Appendix.

We consider a multivariate linear regression model for the kth vector of responses of the form

$$Y_k = c_0 + c_1 x_{1k} + c_2 x_{2k} + c_3 x_{3k} + \epsilon_k \equiv C X_k + \epsilon_k\,, \qquad k = 1, \ldots, T, \qquad (1.24)$$

with $T = 33$, where $Y_k = (y_{1k}, y_{2k}, y_{3k}, y_{4k}, y_{5k})'$ is a 5×1 vector, and $X_k = (1, x_{1k}, x_{2k}, x_{3k})'$ is a 4×1 vector (hence $n = 4$). With $\boldsymbol{Y}$ and $\boldsymbol{X}$ denoting 5×33 and 4×33 data matrices, respectively, the least squares estimate of the regression coefficient matrix is obtained as

$$\tilde{C} = \boldsymbol{Y}\boldsymbol{X}'(\boldsymbol{X}\boldsymbol{X}')^{-1} = \begin{bmatrix} \begin{matrix} 15.2809 \\ (4.3020) \end{matrix} & \begin{matrix} -2.9090 \\ (1.0712) \end{matrix} & \begin{matrix} 1.9631 \\ (0.6902) \end{matrix} & \begin{matrix} 0.2043 \\ (1.1690) \end{matrix} \\ \begin{matrix} 1.4159 \\ (0.8051) \end{matrix} & \begin{matrix} 0.6044 \\ (0.2005) \end{matrix} & \begin{matrix} -0.4816 \\ (0.1292) \end{matrix} & \begin{matrix} 0.2667 \\ (0.2188) \end{matrix} \\ \begin{matrix} 2.0187 \\ (0.6665) \end{matrix} & \begin{matrix} 0.5768 \\ (0.1659) \end{matrix} & \begin{matrix} -0.4245 \\ (0.1069) \end{matrix} & \begin{matrix} -0.0401 \\ (0.1811) \end{matrix} \\ \begin{matrix} 1.8717 \\ (0.5792) \end{matrix} & \begin{matrix} 0.6160 \\ (0.1442) \end{matrix} & \begin{matrix} -0.5781 \\ (0.0929) \end{matrix} & \begin{matrix} 0.3518 \\ (0.1574) \end{matrix} \\ \begin{matrix} -0.8902 \\ (7.0268) \end{matrix} & \begin{matrix} 1.3798 \\ (1.7496) \end{matrix} & \begin{matrix} -0.6289 \\ (1.1273) \end{matrix} & \begin{matrix} 2.8908 \\ (1.9095) \end{matrix} \end{bmatrix}$$

The unbiased estimate of the 5×5 covariance matrix of the errors ϵ_k (with correlations shown above the diagonal) is given by

$$\overline{\Sigma}_{\epsilon\epsilon} = \frac{1}{33-4}\hat{\epsilon}\hat{\epsilon}' = \begin{bmatrix} 11.2154 & -0.4225 & -0.3076 & -0.3261 & -0.0780 \\ -0.8867 & 0.3928 & 0.8951 & 0.6769 & 0.3418 \\ -0.5344 & 0.2911 & 0.2692 & 0.6605 & 0.2891 \\ -0.4924 & 0.1913 & 0.1545 & 0.2033 & 0.0343 \\ -1.4285 & 1.1718 & 0.8203 & 0.0845 & 29.9219 \end{bmatrix}$$

where $\hat{\epsilon} = \boldsymbol{Y} - \tilde{C}\boldsymbol{X}$. The estimated standard deviations of the individual elements of $\tilde{C}$ are obtained as the square roots of the diagonal elements of $\overline{\Sigma}_{\epsilon\epsilon} \otimes (\boldsymbol{X}\boldsymbol{X}')^{-1}$, where

$$(\boldsymbol{X}\boldsymbol{X}')^{-1} = \begin{bmatrix} 1.65016 & -0.08066 & -0.19343 & -0.37982 \\ -0.08066 & 0.10231 & -0.01457 & -0.02425 \\ -0.19343 & -0.01457 & 0.04247 & 0.04236 \\ -0.37982 & -0.02425 & 0.04236 & 0.12186 \end{bmatrix}$$

These estimated standard deviations are evaluated and displayed in parentheses below the corresponding estimates in $\tilde{C}$. We find that many of the individual estimates in $\tilde{C}$ are greater than twice their estimated standard deviations.

We illustrate the LR test procedure by testing the hypothesis that the two concomitant predictor variables x_2 and x_3 have no influence on the response variables, that is, $H_0 : C_2 \equiv [c_2, c_3] = 0$. The matrix due to this null hypothesis is obtained as

$$H = \tilde{C}_2 A_{22.1} \tilde{C}_2' = \begin{bmatrix} 129.365 & -38.712 & -28.064 & -47.170 & -106.563 \\ -38.712 & 12.469 & 8.410 & 15.266 & 40.140 \\ -28.064 & 8.410 & 6.088 & 10.248 & 23.225 \\ -47.170 & 15.266 & 10.248 & 18.695 & 49.587 \\ -106.563 & 40.140 & 23.225 & 49.587 & 164.779 \end{bmatrix}$$

where

$$A_{22.1} = \begin{bmatrix} 0.042472 & 0.042358 \\ 0.042358 & 0.121855 \end{bmatrix}^{-1} = \begin{bmatrix} 36.0395 & -12.5278 \\ -12.5278 & 12.5613 \end{bmatrix}$$

and the error matrix is $S = \hat{\epsilon}\hat{\epsilon}' = (33-4)\overline{\Sigma}_{\epsilon\epsilon}$. The nonzero eigenvalues of the matrix $S^{-1}H = S^{-1}\tilde{C}_2 A_{22.1}\tilde{C}_2'$ are the same as those of

$$\tilde{C}_2' S^{-1} \tilde{C}_2 A_{22.1} = \begin{bmatrix} 2.55873 & -1.12685 \\ -2.16833 & 1.49117 \end{bmatrix}$$

and are found to equal $\hat{\lambda}_1^2 = 3.67671$ and $\hat{\lambda}_2^2 = 0.37319$. Hence, with $T = 33$, $n = 4$, $n_2 = 2$, and $m = 5$, the value of the test statistic in (1.16) is obtained as

$$\mathcal{M} = [33 - 4 + (2-5-1)/2]\,[\log(1 + 3.6767) + \log(1 + 0.3732)] = 50.2126$$

with $n_2 m = (2)(5) = 10$ degrees of freedom. Since the critical value of the χ^2_{10} distribution at level $\alpha = 0.05$ is 18.31, H_0 is clearly rejected and the implication is that the concomitant variables (x_2, x_3) have influence on (at least some of) the response variables y_1, y_2, y_3, y_4, and y_5. From informal examination of the estimates in $\tilde{C}$ and their estimated standard deviations, it is expected that the significance of this hypothesis testing result is due mainly to the influence of the variable volume (x_2), with the variable x_3 having much less contribution. (In fact, the value of the test statistic in (1.16) for $H_0 : c_3 = 0$ yields the value $\mathcal{M} = 14.6389$ with 5 degrees of freedom and indicates only mild influence for x_3 on the response variables.)

Similarly, we can test the hypothesis that the weight variable (x_1) has no influence on the response variables, $H_0 : c_1 = 0$. For the LR test of this hypothesis, since $n_2 = 1$, the single nonzero eigenvalue of $S^{-1}H$ can be obtained as the scalar $\tilde{c}_1' S^{-1} \tilde{c}_1 / 0.10231 = 0.08184/0.10231 = 0.79995$, noting that 0.10231 represents the (2,2)-element of the matrix $(\boldsymbol{X}\boldsymbol{X}')^{-1}$. So the value of the test statistic in (1.16) is $\mathcal{M} = [33-4+(1-5-1)/2] \log(1+0.79995) = 15.5756$ with 5 degrees of freedom. The null hypothesis H_0 is rejected at the 0.05 level. Because $n_2 = 1$ for this H_0, an "exact" test is available based on the F-distribution, similar to the form presented in relation to (1.21) with $X_0 = (0,1,0,0)'$ so that $CX_0 = c_1$, using the statistic $\frac{33-4-5+1}{5(33-4)} \tilde{c}_1' \overline{\Sigma}_{\epsilon\epsilon}^{-1} \tilde{c}_1 / 0.10231 = 3.9998$ with $m = 5$ and $T - n - m + 1 = 33 - 4 - 5 + 1 = 25$ degrees of freedom. A similar conclusion is obtained as with the approximate chi-squared LR test statistic.

Concluding, we also point out that the last three columns of the LS estimate $\tilde{C}$ (excluding the column of the constant terms) give some indications of a reduced-rank structure for the 5×3 matrix $[c_1, c_2, c_3]$ of regression coefficients in model (1.24). For example, notice the similarity of coefficient estimates among the second through fourth rows of the last three columns of $\tilde{C}$. This type of reduced-rank structure will be examined formally in the subsequent chapters.

2
Reduced-Rank Regression Model

2.1 The Basic Reduced-Rank Model and Background

The classical multivariate regression model presented in Chapter 1, as noted before, does not make direct use of the fact that the response variables are likely to be correlated. A more serious practical concern is that even for a moderate number of variables whose interrelationships are to be investigated, the number of parameters in the regression matrix can be large. For example, in a multivariate analysis of economic variables (see Example 2.2), Gudmundsson (1977) uses $m = 7$ response variables and $n = 6$ predictor variables, thus totaling 42 regression coefficient parameters (excluding intercepts) to be estimated, in the classical regression setup. But the number of vector data points available for estimation is only $T = 36$; these are quarterly observations from 1948 to 1956 for the United Kingdom. Thus, in many practical situations, there is a need to reduce the number of parameters in model (1.1) and we approach this problem through the assumption of lower rank of the matrix C in model (1.1). More formally, in the model $Y_k = CX_k + \epsilon_k$ we assume that

$$\text{rank}(C) = r \leq \min(m, n). \tag{2.1}$$

The methodology and results that we will present for the multivariate regression model under this assumption of reduced rank apply equally to both the cases of dimension $m \leq n$ and $m > n$. However, for some of the initial discussion below we shall assume $m \leq n$ for convenience of notation.

The rank condition (2.1) has two related practical implications. First, with $r < m$ it implies that there are $(m-r)$ linear restrictions on the regression coefficient matrix C of the form

$$\ell_i' C = 0, \qquad i = 1, 2, \ldots, (m-r), \tag{2.2}$$

and these restrictions themselves are often not known a priori in the sense that $\ell_1, \ldots, \ell_{m-r}$ are unknown. Premultiplying (1.1) by ℓ_i', we have

$$\ell_i' Y_k = \ell_i' \epsilon_k. \tag{2.3}$$

The linear combinations, $\ell_i' Y_k$, $i = 1, 2, \ldots, (m-r)$, could be modeled without any reference to the predictor variables X_k and depend only on the distribution of the error term ϵ_k. Otherwise, these linear combinations can be isolated and can be investigated separately.

The second implication that is somewhat complementary to (2.2) is that with assumption (2.1), C can be written as a product of two lower dimensional matrices that are of full ranks. Specifically, C can be expressed as

$$C = AB \tag{2.4}$$

where A is of dimension $m \times r$ and B is of dimension $r \times n$, but both have rank r. Note that the r columns of the left matrix factor A in (2.4) can be viewed as a basis for the column space of C, while the r rows of B form a basis for the row space. The model (1.1) can then be written as

$$Y_k = A(BX_k) + \epsilon_k, \qquad k = 1, \ldots, T, \tag{2.5}$$

where BX_k is of reduced dimension with only r components. A practical use of (2.5) is that the r linear combinations of the predictor variables X_k are sufficient to model the variation in the response variables Y_k and there may not be a need for all n linear combinations or otherwise for all n variables, as would be the case when no single predictor variable can be discarded from the full-rank analysis. Hence, there is a gain in simplicity and interpretation through the reduced-rank regression modeling, although from a practitioner point of view one would still need to have measurements on all n predictor variables to perform the reduced-rank analysis leading to the construction of the smaller number of relevant linear combinations.

We shall describe now how the reduced-rank model arises in various contexts from different disciplines and this will also provide a historical background. Anderson (1951) was the first to consider in detail the reduced-rank regression problem, for the case where X_k, the set of predictor variables, is fixed. In an application to economics, he called Y_k the economic variables but X_k is taken to be the vector of noneconomic variables that can be manipulated. The linear combinations $\ell' Y_k$ are called structural relations; these are stable relationships among the macroeconomic variables that could not be manipulated. Estimating the restrictions 'ℓ' is thus important to understand the basic structure of the economy. Izenman (1975) introduced the term 'reduced-rank regression' and examined this model in detail. The reduced-rank regression model and its statistical properties were examined further by Robinson (1973, 1974), Tso (1981), Davies and Tso (1982), Zhou (1994), and Geweke (1996), among others.

Subsequent to but separate from the initial work of Anderson (1951), reduced-rank regression concepts were considered in various contexts and under different terminologies by several authors, as noted by van der Leeden (1990, Chaps. 2 and 3). Rao (1964) studied principal components and presented results that can be related to reduced-rank regression, referring to the most useful linear combinations of the predictor variables as *principal components of instrumental variables.* Fortier (1966) considered reduced-rank modeling, which he referred to as *simultaneous linear prediction* modeling, and van den Wollenberg (1977) discussed basically the same procedure as an alternative to canonical correlation analysis, which is known as *redundancy analysis.* The basic results in these works are essentially the same and are related to a particular version of reduced-rank estimation, which we note explicitly in Section 2.3. Keller and Wansbeek (1983), starting from an errors-in-variable model form, demonstrated the formulation of a reduced-rank model both in terms of linear restrictions on the regression coefficient matrix and in terms of lower-dimensional component matrices, referred to as the *principal relations* and the *principal factors* specifications, respectively.

An appealing physical interpretation of the reduced-rank model was offered by Brillinger (1969). This is to regard a situation where the n component vector X_k represents information which is to be used to send a message Y_k having m components ($m \leq n$) but such a message can only be transmitted through r channels ($r \leq m$). Thus, BX_k acts as a code and on receipt of the code, form the vector ABX_k by premultiplying the code, which it is hoped, would be as close as possible to the desired Y_k.

Aside from this attractiveness as a statistical technique for parameter reduction, it is also known that a reduced-rank regression structure has an interpretation in terms of underlying hypothetical constructs that drive the two sets of variables. The concept of unobservable 'latent' variables has a long history in economics and the social sciences [see Zellner (1970) and Goldberger (1972)]. These latent variables are hypothetical constructs and the observable variables may appear as both effects and causes of these latent variables. The latent models are extensively used in sociology and economics. To illustrate, let $Y_k = AY_k^* + U_k$ where Y_k^* is a scalar latent variable, A is a column vector, and Y_k is the set of indicators of the underlying hypothetical construct. Suppose Y_k^* is determined by a set of causal variables X_k through $Y_k^* = BX_k + V_k$, where B is a row vector. By combining the two models we have the reduced-form equation connecting the multiple causes to the multiple indicators (denoted as MIMIC model),

$$Y_k = ABX_k + (AV_k + U_k) = CX_k + \epsilon_k. \tag{2.6}$$

The above multivariate regression model falls in the reduced-rank framework and it follows that the coefficient matrix is principally constrained to have rank one. Jöreskog and Goldberger (1975) describe the above model with an application in sociology. Clearly, this situation can be extended to allow for $r > 1$ latent variables, with the resulting model (2.6) having coefficient matrix of reduced rank r.

In the context of modeling macroeconomic time series $\{Y_t\}$, Sargent and Sims (1977) discuss the notion of 'index' models, which fit into the above framework. The indexes, $X_t^* = BX_t$, which are smaller in number are constructed from a large set of time series predictor variables X_t and are found to drive the vector of observed response variables Y_t. They are further interpreted through a theory of macroeconomics. More recent developments of reduced-rank models applied to vector autoregressive time series analysis, for example, where the predictor variables X_t in (2.5) represent lagged values Y_{t-1} of the time series Y_t, have been pursued by Reinsel (1983), Velu, Reinsel, and Wichern (1986), Ahn and Reinsel (1988, 1990), and Johansen (1988, 1991), among others.

Models closely related to the reduced-rank regression structure in (2.5) have been studied and applied in various other areas. For example, in work preceding that of Anderson's, Tintner (1945, 1950) and Geary (1948) studied a similar problem within an economics setting in which the (unknown) "systematic parts" W_k of the Y_ks are not necessarily specified to be linear functions of a known set of predictor variables X_k but were still assumed to satisfy a set of $m-r$ linear relationships $\ell_i' W_k = 0$. That is, the model considered was

$$Y_k - \mu = W_k + \epsilon_k \tag{2.7}$$

where W_k is unobservable and is taken to be the systematic part and ϵ_k is the random error. It is assumed that the systematic part $\{W_k\}$ varies in a lower-dimensional linear space of dimension r $(< m)$, so that the W_k satisfy $LW_k = 0$ for all k (assuming the W_k are "centered" at zero) with L denoting an unknown $(m-r) \times m$ full-rank matrix having rows ℓ_i' as in (2.3). Typically, analysis and estimation of the parameters L in such a model requires some special assumptions on the covariance structure $\Sigma_{\epsilon\epsilon}$ of the error part ϵ_k, such as $\Sigma_{\epsilon\epsilon}$ known or $\Sigma_{\epsilon\epsilon} = \sigma^2\Psi_0$ where σ^2 is an unknown scalar parameter but Ψ_0 is a known $m \times m$ matrix. This problem falls within the framework of models for functional relationships or structural relationships in the systematic part $\{W_k\}$, that is, the relationships are represented by $LW_k = 0$. The terminology "functional" is commonly used when the W_k are taken to be fixed, and "structural" when the W_k are treated as random.

The above model can also be considered from the multivariate errors-in-variables regression model viewpoint. To give a more explicit representation in this form, write the $(m-r) \times m$ full-rank matrix L as $L = [L_1, L_2]$ where L_1 is $(m-r) \times (m-r)$ and assumed to be of full rank, and partition Y_k, W_k, and ϵ_k in a compatible fashion as $Y_k = (Y'_{1k}, Y'_{2k})'$, $W_k = (W'_{1k}, W'_{2k})'$, and $\epsilon_k = (\epsilon'_{1k}, \epsilon'_{2k})'$. Then we can multiply the linear relation $LW_k = L_1W_{1k} + L_2W_{2k} = 0$ on the left by L_1^{-1} to obtain

$$W_{1k} = -L_1^{-1}L_2W_{2k} \equiv K_2W_{2k},$$

where $K_2 = -L_1^{-1}L_2$. So we have the multivariate regression relation

$$Y_{1k} - \mu_1 = W_{1k} + \epsilon_{1k} \equiv K_2W_{2k} + \epsilon_{1k}, \tag{2.8}$$

where the "independent" variables W_{2k} in the regression model are not directly observed but are observed with error, in that $Y_{2k} - \mu_2 = W_{2k} + \epsilon_{2k}$ is observed. Analysis and estimation of the above models (2.7) and (2.8) have been studied and described by many authors including Sprent (1966), Moran (1971), Gleser and Watson (1973), Gleser (1981), Theobald (1975), Anderson (1976, 1984b), Bhargava (1979), and Amemiya and Fuller (1984). A few specific estimation results for these models will be presented briefly in relation to the topic of principal components analysis which is discussed in Section 2.4.

The model (2.7) with the linear restrictions $LW_k = 0$ can also be given a "parametric" or factor analysis representation as $Y_k - \mu = AF_k + \epsilon_k$ [for example, see Anderson (1984b)], where A is an $m \times r$ matrix of factor loadings of full rank $r < m$ (such that $LA = 0$) and F_k is an $r \times 1$ vector of unobservable common factors. (Model (2.5) can be viewed as a special case of this form with $F_k = BX_k$ being specified as r unknown linear combinations of the observable set of n predictor variables X_k, and (2.5) has been advocated for use in the factor analysis context as a special case of fixed effects factor analysis by van der Leeden (1990, Chap. 5).) Typically, for the factor analysis model it is assumed that the F_k are random vectors with mean zero and covariance matrix I_r, and $\Sigma_{\epsilon\epsilon} = \text{diag}(\sigma_1^2, \ldots, \sigma_m^2)$ is diagonal. Interest in factor analysis modeling stems originally from its applications in psychology and education, and early developments were by Thurstone (1947), Reiersøl (1950), Lawley (1940, 1943, 1953), Rao (1955), Anderson and Rubin (1956), and Jöreskog (1967, 1969).

When the error covariance matrix in model (2.7) is assumed to be unknown and arbitrary, replicated vector observations are needed for estimation. For example, we may have $N \geq 2$ observations for each k with the model $Y_{kj} - \mu = W_k + \epsilon_{kj}$, $j = 1, \ldots, N$, $k = 1, \ldots, T$. This is a multivariate one-way ANOVA model, and analysis of such a model under the reduced-rank or linear restriction assumptions for the (fixed effects) W_k has been considered by Anderson (1951, 1984b), Villegas (1961, 1982), Theobald (1975), Healy (1980), and Campbell (1984), among others, following earlier work by Fisher (1938) who considered a related problem in regard to discriminant analysis. This one-way ANOVA model situation will be discussed in more detail later in Section 3.2.

The above brief summaries indicate that research and interest in applications of models that directly involve reduced-rank regression concepts or are closely related have a long history, and that the scope is quite broad. However, because the focus of this book is directly on the (reduced-rank) multivariate regression modeling, primarily, it will not be possible to explore all of the above discussed related areas in detail. Articles by Anderson (1984b, 1991) provide excellent summaries of the developments and results in some of these areas, such as estimating linear functional and structural relationships, factor analysis models, and the relation to estimation of simultaneous equations models in econometrics. Interested readers may refer to these articles for more detailed survey of developments in these areas.

2.2 Some Examples of Application of the Reduced-Rank Model

In this section we introduce and discuss some practical examples where reduced-rank methods have been applied.

EXAMPLE 2.1. Davies and Tso (1982) have applied reduced-rank regression methods to study the experimental properties of hydrocarbon fuel mixtures in relating response to composition. The responses are the engine test ratings and they are related to the composition of hydrocarbon fuels, which is characterized by the technique of gas-liquid chromatography (g.l.c.). The responses which are correlated among themselves are known to be dependent on the same portion of g.l.c. chromatograms. The precision of the g.l.c. determination of composition is quite high, so that the resultant data vector of explanatory variables may be regarded as deterministic, but the responses exhibit a wide variation. From the description of the problem, it is possible that a multivariate regression model with rank constraints might be appropriate. A small number of certain features of the chromatogram could be used to predict fuel properties. Because the experimentation in this area is quite expensive, Davies and Tso (1982) point out the usefulness of the model in reducing the number of components.

A total of 30 vector observations, corresponding to different gasoline blends, is divided into two data sets of 14 and 16 blends, respectively, for validation purposes. On each blend, five gasoline distillation measurements $y_1, \ldots, y_5$ were considered, each representing percentage weight evaporated by a specified temperature, and ten x variables were considered to represent the corresponding blend composition by condensing the blend's low-resolution chromatogram into a 10-component vector by grouping of peak areas. Thus, the classical multivariate regression would call for estimating fifty regression coefficient parameters (excluding intercepts) with thirty vector observations. Using the mean square error criterion for model fit,

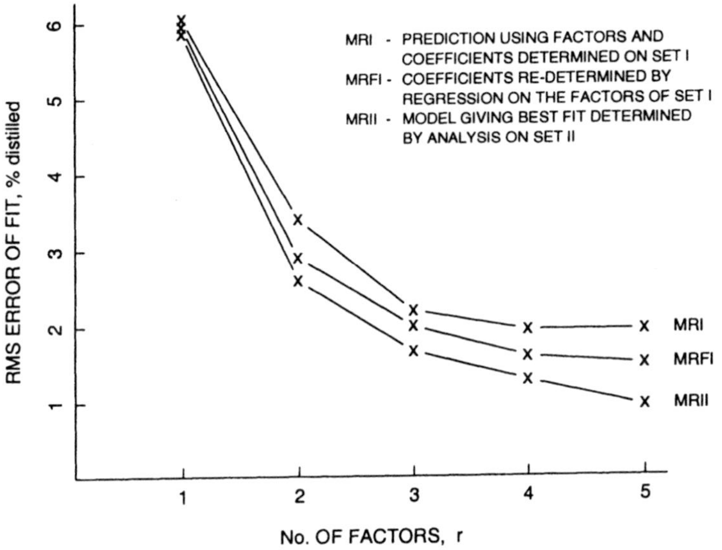

Figure 2.1. Prediction and fitting to distillation data for Set II gasolines

$\frac{1}{Tm} \parallel \boldsymbol{Y} - \hat{\boldsymbol{Y}}_{(r)} \parallel^2$, where $\hat{\boldsymbol{Y}}_{(r)}$ is the matrix of predicted values of the responses when the rank of the regression coefficient matrix is taken to be r and is estimated by minimizing the overall error sum of squares criterion, Davies and Tso demonstrate that $r = 2$ is fairly sufficient. In fact, in both data sets, over 95 percent of the variation in Y_k is explained by taking a model with $r = 2$ linear combinations of the X_k. Thus, one may notice the substantial reduction in the number of parameters. Moreover, the similarity in the two linear combination or two factor linear subspace of the X_k between the two data sets was fairly high, suggesting that the factors identified have physical significance.

Figure 2.1 shows the results, in terms of residual mean square (RMS) error, of predicting the response measurements for gasolines in Set II from a knowledge of their composition estimates from both Set I and Set II. The upper curve (MRI) corresponds to predictions on Set II using coefficients of the linear combinations (factors) and corresponding regression coefficients estimated from Set I. The second curve (MRFI) corresponds to a fit for Set II obtained using coefficients of the linear combinations (factors) of the X_k estimated from Set I but re-estimating the regression coefficients of the Y_k on these factors using Set II. The lower curve (MRII) corresponds to the model estimated entirely using data from Set II.

EXAMPLE 2.2. The second example we discuss here is based on the work of Gudmundsson (1977), mentioned earlier in Section 2.1. Klein, Ball, Hazlewood and Vandome (1961) constructed a detailed econometric model of the United Kingdom. The data consist of quarterly observations from 1948 to 1956 of 37 time series of response variables and 32 time series of predictor variables, some of which may be lagged values of response variables. This could be conceptually handled through the reduced-rank regression model but the computational burden would be enormous. Therefore, some grouping of variables based on economic considerations was applied. But the focus of Gudmundsson's paper was to construct some linear combinations of variables, representing various aspects of the general economic situation. Because most series are likely to be related to the state of the economy in general, it is anticipated that some useful linear combinations are possible.

The following subset of time series variables was used in the reduced-rank regression calculations. The dependent variables are

y_{1t} Index of total industrial production
y_{2t} Consumption of food, drinks and tobacco at constant prices
y_{3t} Price index of total consumption
y_{4t} Total civilian labor force
y_{5t} Index of weekly wage rates
y_{6t} Index of volume of total imports
y_{7t} Index of volume of total exports

and the independent variables are $x_{1t} = y_{1t-1}$, $x_{2t} = y_{2t-1}$, $x_{3t} = y_{6t-1}$, $x_{4t} = y_{7t-1}$, x_{5t} = price index of total imports lagged by two time periods and $x_{6t} = t$, the time index.

The independent variables are normalized before fitting the reduced-rank regression model. Gudmundsson chose to estimate the linear combinations of predictor variables, $x^*_{it} = \beta'_i X_t$, $i = 1, \ldots, r$, as the coefficients β_i which maximize the sum of the squared sample correlation coefficients of x^*_i and the endogenous variables $y_1, \ldots, y_m$, subject to normalization and orthogonality conditions for the x^*_i. The first three linear combinations were judged by Gudmundsson to be sufficient for use in explaining all the endogenous variables, but no formal justification was provided for this choice. This implicitly is assuming that the rank of the regression coefficient matrix is three. The method used for estimating the β_i was mildly criticized by Theobald (1978). From the estimation results to be presented in Section 2.3, it follows that these estimates can be viewed as ML estimates under normality and the (possibly unrealistic) assumption that the covariance matrix of the error vector ϵ_t takes the form $\Sigma_{\epsilon\epsilon} = \sigma^2 I_m$.

Table 2.1 presents the information on the resulting (estimates of the) linear combinations ($X^*_t = BX_t$) and the sample correlations of the linear combinations with all the response variables. It can be seen from the first part of the table that the linear combinations are dominated by x_6, the time trend variable and by the variable x_1, the lagged index of industrial production. But from the correlations, it would appear that the first linear combination which accounts for most of the variation, is fairly highly correlated with all the response variables. Such a combination can also be taken as an index that represents the general economic situation. Gudmundsson also compared the performance of the three linear combinations in predicting all 37 variables used in Klein et al.'s original study. The comparisons were made using the root mean square error criterion. The reduced-rank model performed better than the econometric model for production variables, roughly comparable for consumption, imports and exports, but not as good as the econometric model for prices and employment series. This has to be expected because the three linear combinations were constructed from a subset of the independent variables that did not have representation by price or employment. But considering the magnitude of dimension reduction, this is somewhat a small price to pay.

Table 2.1 Coefficients in the linear combinations $X_t^* = BX_t$ and sample correlations with response variables Y_t for econometric data example.

Coefficients of Independent Variables in $X_t^* = BX_t$

	x_{1t}	x_{2t}	x_{3t}	x_{4t}	x_{5t}	x_{6t}
x_{1t}^*	.148	.095	.086	.013	−.058	.720
x_{2t}^*	−.797	−.193	−.013	−.226	.302	.936
x_{3t}^*	−.406	−.271	−.238	.319	−.271	.767

Correlation Coefficients of X_t^* with Endogenous Variables (Y_t)

	y_{1t}	y_{2t}	y_{3t}	y_{4t}	y_{5t}	y_{6t}	y_{7t}	sum of squared corr. coefs.
x_{1t}^*	.974	.941	.966	.984	.979	.897	.868	6.252
x_{2t}^*	−.083	−.008	.225	−.042	.138	−.110	−.142	0.110
x_{3t}^*	.034	.156	−.065	.015	.016	−.192	.028	0.068

EXAMPLE 2.3. Glasbey (1992) used the reduced-rank model in a recent application to demonstrate the relationship between measurements on solar radiation taken over various sites in Scotland and the physical characteristics of the sites. The aim of the analysis was to summarize the differences in solar radiation among sites, and if possible to relate them to other physical characteristics of the sites. The sites, radiometric stations ten in number, were located on the range of Pentland Hills to the south of Edinburgh (Figure 2.2) and monthly averages of solar radiation measurements on a log-scale for a period of 19 months form the 19×10 data matrix $\boldsymbol{Y}$. The log-scale was used to make variability over the seasons to be approximately the same. Because of the proximity of these stations one might expect these response variables to be interrelated over sites, but any possible spatial correlations over the sites will be ignored. The $\boldsymbol{X}$ matrix is formed with five site characteristics, latitude (x_1), longitude (x_2), altitude (x_3), and percentage sun loss in summer (x_4) and in winter (x_5). Thus, the data matrix

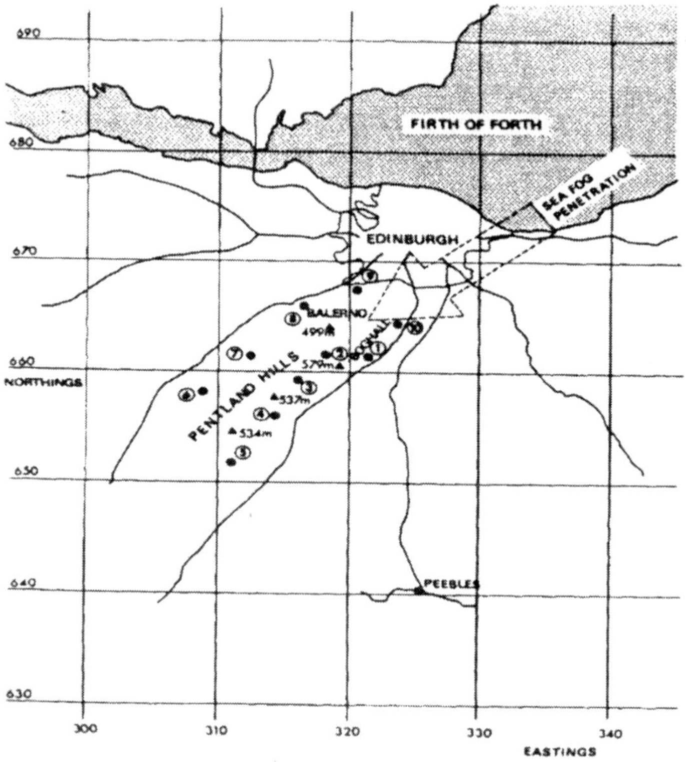

Figure 2.2. Microclimate survey area with locations of 10 radiometric stations (from Glasbey, 1992)

$\boldsymbol{X}$ is of dimension 5×10. This is a novel application of the reduced-rank model and does not conform to the dimensional details presented earlier (i.e., $m = 19$, $n = 5$, and $T = 10$ in this example, so that $m > T$). At the first stage after adjusting for monthly means, simple models for the response variables at each site that account for most of the variability in $\boldsymbol{Y}$ are constructed via the method of principal components and singular value decomposition. Two terms or components were considered to be sufficient in the singular value representation for $\boldsymbol{Y}$ on the basis that they account for a high percent (85%) of the overall variation of the site values about the monthly mean over all sites. Then these two summary vectors (1×10) of the responses were regressed on $\boldsymbol{X}$, the matrix of physical characteristics. The result can be shown to be in a form of reduced-rank regression, where the rank of the matrix is assumed to be equal to two. Over 80 percent of the variation in month-corrected $\boldsymbol{Y}$ is supposedly accounted by the two linear combinations of $\boldsymbol{X}$:

$$
\begin{aligned}
x_1^* &= -319.3 + \underset{(1.2)}{5.7x_1} + \underset{(0.6)}{0.2x_2} - \underset{(.0013)}{.0022x_3} + \underset{(.012)}{.004x_4} - \underset{(.0019)}{.0076x_5} \\
x_2^* &= 160.2 + \underset{(2.1)}{2.6x_1} + \underset{(1.1)}{4.8x_2} - \underset{(.0024)}{.0037x_3} + \underset{(.022)}{.017x_4} + \underset{(.0034)}{.0019x_5}
\end{aligned}
$$

where the numbers in parentheses are the estimated standard errors.

The first combination x_1^* is primarily a latitude effect, with some negative adjustments for altitude and percentage sun loss in winter. The second combination x_2^* is a longitude effect but adjusted for altitude. The more northerly sites (see Figure 2.2) have positive values and the southern sites have negative values for the regression coefficient on x_1^* and western sites have the positive values on x_2^*. In general, Glasbey (1992) concludes that the sites at higher altitudes received less radiation, winter sun loss reduced radiation levels but the summer loss is not known to be effective.

Based on the above examples alone, it would appear that the reduced-rank methods have strong potential for both theoretical study and applications. A few other recent applications of reduced-rank regression include the following. In a study by Ryan et al. (1992), reduced-rank methods were applied to the analysis of response data (DNA, RNA, and protein) from a joint toxicity bioassay experiment to determine the joint effects of toxic compounds that include copper and zinc on the growth of larval fathead minnows. In the field of financial economics, several asset pricing theories have been proposed for testing the efficiency of portfolios and, as described by Zhou (1991, 1995), these theories generally imply certain rank restrictions on the regression coefficient matrix that relates asset returns to market factors. Empirical verification using asset returns data on industry portfolios has been made through tests for reduced-rank regression.

2.3 Estimation of Parameters in the Reduced-Rank Model

The reduced-rank regression model described in Section 2.1 has the following parameters to be estimated from a sample of observations: the matrix A which is of dimension $m \times r$, the matrix B which is of dimension $r \times n$, and the $m \times m$ covariance matrix $\Sigma_{\epsilon\epsilon}$ of the error term. Estimation of the constraint vectors $\ell_1, \dots, \ell_{m-r}$ will be taken up following these. Before we proceed with the details, we note that in the reduced-rank model with $C = AB$ the component matrices A and B are determined only up to nonsingular linear transformations. To obtain a particular unique set of parameter values for A and B from the elements of C, certain normalization conditions are typically imposed. These conditions, as we have chosen here, will follow from the estimation criterion. To clearly exhibit the algebraic features of the problem, we will initially formulate the criterion for determination of A and B in terms of "population" quantities and later we will demonstrate how the maximum likelihood estimates from sample data are obtained through use of the initial population results.

As we shall see shortly, reduced-rank estimation will be obtained as a certain reduced-rank approximation of the full-rank least squares estimate of the regression coefficient matrix. Therefore, for presentation of reduced-rank estimation, we first need an important matrix result called the Householder–Young or Eckart–Young Theorem (Eckart and Young, 1936), which presents the tools for approximating a full-rank matrix by a matrix of lower rank. The solution is related to the singular value decomposition of the full-rank matrix. As a preliminary to this main theorem, we state a needed result given in Rao (1973, p. 63).

Lemma 2.1 Let A be an $m \times m$ symmetric matrix with eigenvalues $\lambda_1 \geq \cdots \geq \lambda_m$, and denote the corresponding normalized eigenvectors as $P_1, \ldots, P_m$. The supremum of $\sum_{i=1}^{r} X_i' A X_i = \text{tr}(X'AX)$, with $X = [X_1, \ldots, X_r]$, over all sets of $r \leq m$ mutually orthonormal vectors $X_1, \ldots, X_r$ is equal to $\sum_{i=1}^{r} \lambda_i$ and is attained when the $X_i = P_i$, $i = 1, \ldots, r$.

Proof: Let $P = [P_1, \ldots, P_m]$, so that P is an orthogonal matrix with $P'AP = \Lambda = \text{diag}(\lambda_1, \ldots, \lambda_m)$. Any set of r mutually orthonormal vectors $X_1, \ldots, X_r$ can be expressed as $X_i = \sum_{j=1}^{m} c_{ij} P_j \equiv P c_i$, $i = 1, \ldots, r$, where $c_i = (c_{i1}, \ldots, c_{im})'$, with $X_i' X_i = c_i' P' P c_i = c_i' c_i = 1$ and $X_i' X_l = c_i' P' P c_l = c_i' c_l = 0$, for $i \neq l$. Then

$$\sum_{i=1}^{r} X_i' A X_i = \sum_{i=1}^{r} c_i' P' A P c_i = \sum_{i=1}^{r} c_i' \Lambda c_i = \sum_{i=1}^{r} \sum_{j=1}^{m} \lambda_j c_{ij}^2 = \sum_{j=1}^{m} \lambda_j \Big(\sum_{i=1}^{r} c_{ij}^2 \Big),$$

where the coefficient of each λ_j, $\sum_{i=1}^{r} c_{ij}^2$, is ≤ 1 and the sum of these coefficients is r, since $\sum_{j=1}^{m} \sum_{i=1}^{r} c_{ij}^2 = \sum_{i=1}^{r} c_i' c_i = r$. To maximize the sum above, the optimum choice is thus to make the coefficients of the λ_j equal to 1 for $j = 1, \ldots, r$, and hence 0 for $j > r$. This is attained only by the choice $X_i = P_i$, $i = 1, \ldots, r$, that is, $c_{ij} = 1$ for $i = j$ and $c_{ij} = 0$ for $i \neq j$, where P_i is a normalized eigenvector of A corresponding to the eigenvalue λ_i. The eigenvectors P_i are (essentially) unique, of course, if the eigenvalues λ_i are distinct, but the choice of eigenvectors will not be unique if some eigenvalues among the first r are not distinct.

Theorem 2.1 Let S be a matrix of order $m \times n$ and of rank m. The Euclidean norm, $\text{tr}[(S-P)(S-P)']$, is minimum among matrices P of the same order as S but of rank r $(\leq m)$, when $P = MM'S$, where M is $m \times r$ and the columns of M are the first r (normalized) eigenvectors of SS', that is, the normalized eigenvectors corresponding to the r largest eigenvalues of SS'.

Proof: Let $P = MN$ where M is an $m \times r$ matrix and N is an $r \times n$ matrix. Without loss of generality, assume that M is orthonormal with $M'M = I_r$. Minimizing the criterion given in the theorem over N, for a given M, yields $N = (M'M)^{-1}M'S = M'S$, by similar argument that led to the least squares solution in (1.7). Substituting this in the criterion yields, after some simplification,

$$\text{tr}[(S-P)(S-P)'] = \text{tr}[SS'(I_m - MM')] = \text{tr}(SS') - \text{tr}(M'SS'M). \quad (2.9)$$

Minimizing the above quantity with respect to M is equivalent to maximizing

$$\text{tr}(M'SS'M) \quad \text{subject to} \quad M'M = I_r. \quad (2.10)$$

Based on the result in Lemma 2.1, it follows that the above goal is achieved by choosing the columns of M to be the (orthonormal) eigenvectors of SS' that correspond to the first (largest) r eigenvalues.

Remark–Singular Value Decomposition. The positive square roots of the eigenvalues of SS' are referred to as the singular values of the matrix S. In general, an $m \times n$ matrix S, of rank s, can be expressed in the *singular value decomposition* as $S = V\Lambda U'$, where $\Lambda = \text{diag}(\lambda_1, \ldots, \lambda_s)$ with $\lambda_1^2 \geq \cdots \geq \lambda_s^2 > 0$ being the nonzero eigenvalues of SS', $V = [V_1, \ldots, V_s]$ is an $m \times s$ matrix such that $V'V = I_s$, and $U = [U_1, \ldots, U_s]$ is $n \times s$ such that $U'U = I_s$. The columns V_i are the normalized eigenvectors of SS' corresponding to the λ_i^2, and $U_i = \frac{1}{\lambda_i} S'V_i$, $i = 1, \ldots, s$. For the situation in Theorem 2.1 above, from the singular value decomposition of S we see that the criterion to be minimized over rank r approximations $P = MN$ is $\text{tr}[(V\Lambda U' - MN)(V\Lambda U' - MN)'] = \text{tr}[(\Lambda - V'MNU)(\Lambda - V'MNU)']$. The result in Theorem 2.1 states that the minimizing matrix factors M and N are given by $M = [V_1, \ldots, V_r] \equiv V_{(r)}$ and $N = M'S = V_{(r)}'S = V_{(r)}'V\Lambda U' = [\lambda_1 U_1, \ldots, \lambda_r U_r]' \equiv \Lambda_{(r)} U_{(r)}'$, with $U_{(r)} = [U_1, \ldots, U_r]$ and $\Lambda_{(r)} = \text{diag}(\lambda_1, \ldots, \lambda_r)$. So $P = MN = V_{(r)}\Lambda_{(r)}U_{(r)}'$ yields the "optimum" rank r approximation to $S = V\Lambda U'$, and the resulting minimum value of the trace criterion is $\sum_{i=r+1}^{m} \lambda_i^2$.

We restate the multivariate regression model before we proceed with the estimation of the relevant parameters. From (2.5), the reduced-rank regression model is

$$Y_k = ABX_k + \epsilon_k, \qquad k = 1, \ldots, T, \quad (2.11)$$

where A is of dimension $m \times r$, B of dimension $r \times n$ and the vectors ϵ_k are assumed to be independent with mean zero and covariance matrix $\Sigma_{\epsilon\epsilon}$, where $\Sigma_{\epsilon\epsilon}$ is positive-definite. Estimation of A and B in (2.11) is based on the following result (Brillinger, 1981, Section 10.2) which essentially uses Theorem 2.1. For this "population" result, the vector of predictor variables X_k in (2.11) will be treated as a stochastic vector for ease of exposition.

Theorem 2.2 Suppose the $(m+n)$-dimensional random vector $(Y', X')'$ has mean vector 0 and covariance matrix with $\Sigma_{yx} = \Sigma'_{xy} = \text{Cov}(Y, X)$, and $\Sigma_{xx} = \text{Cov}(X)$ nonsingular. Then for any positive-definite matrix Γ, an $m \times r$ matrix A and $r \times n$ matrix B, for $r \leq \min(m, n)$, which minimize

$$\text{tr}\{E[\Gamma^{1/2}(Y - ABX)(Y - ABX)'\Gamma^{1/2}]\} \tag{2.12}$$

are given by

$$A^{(r)} = \Gamma^{-1/2}[V_1, \ldots, V_r] = \Gamma^{-1/2}V, \qquad B^{(r)} = V'\Gamma^{1/2}\Sigma_{yx}\Sigma_{xx}^{-1}, \tag{2.13}$$

where $V = [V_1, \ldots, V_r]$ and V_j is the (normalized) eigenvector that corresponds to the jth largest eigenvalue λ_j^2 of the matrix $\Gamma^{1/2}\Sigma_{yx}\Sigma_{xx}^{-1}\Sigma_{xy}\Gamma^{1/2}$ $(j = 1, 2, \ldots, r)$.

Proof: Setting $S = \Gamma^{1/2}\Sigma_{yx}\Sigma_{xx}^{-1/2}$ and $P = \Gamma^{1/2}AB\Sigma_{xx}^{1/2}$, the result follows easily from Theorem 2.1, because the criterion (2.12) can be expressed as

$$\begin{aligned}
&\text{tr}\{\Gamma^{1/2}(\Sigma_{yy} - AB\Sigma_{xy} - \Sigma_{yx}B'A' + AB\Sigma_{xx}B'A')\Gamma^{1/2}\} \\
&\quad = \text{tr}\{\Gamma^{1/2}(\Sigma_{yy} - \Sigma_{yx}\Sigma_{xx}^{-1}\Sigma_{xy})\Gamma^{1/2}\} \\
&\qquad + \text{tr}\{\Gamma^{1/2}(\Sigma_{yx}\Sigma_{xx}^{-1/2} - AB\Sigma_{xx}^{1/2})(\Sigma_{yx}\Sigma_{xx}^{-1/2} - AB\Sigma_{xx}^{1/2})'\Gamma^{1/2}\}
\end{aligned} \tag{2.12$'$}$$

where $\Sigma_{yy} = \text{Cov}(Y)$. Hence we see that minimization of this criterion, with respect to A and B, is equivalent to minimization of the criterion considered in Theorem 2.1. In terms of quantities used in Theorem 2.1, $M = \Gamma^{1/2}A^{(r)}$ and $N = B^{(r)}\Sigma_{xx}^{1/2}$. At the minimum the criterion (2.12) has the value $\text{tr}(\Sigma_{yy}\Gamma) - \sum_{j=1}^{r}\lambda_j^2$.

Several remarks follow regarding Theorem 2.2 [see Izenman (1975, 1980)]. As mentioned before, nothing appears to be gained by estimating the equations jointly in the multivariate regression model. But a true multivariate feature enters the model when we suspect that the coefficient matrix may not have full rank. Hence, the display of superscripts for A and B in Theorem 2.2 to denote the prescribed rank r. We also observed earlier that the decomposition $C = AB$ is not unique, because for any nonsingular $r \times r$ matrix P, $C = AP^{-1}PB$ and hence to determine A and B uniquely, we must impose some normalization conditions. For the expressions in (2.13), the eigenvectors V_j are normalized to satisfy $V_j'V_j = 1$, and this is equivalent to normalization for A and B as follows:

$$B\Sigma_{xx}B' = \Lambda^2, \qquad A'\Gamma A = I_r \tag{2.14}$$

where $\Lambda^2 = \text{diag}(\lambda_1^2, \ldots, \lambda_r^2)$ and I_r is an $r \times r$ identity matrix. Thus, the number of independent regression parameters in the reduced-rank model (2.11) is $r(m+n-r)$ compared to mn parameters in the full-rank model. Hence, a substantial reduction in the number of parameters is possible.

The elements of the reduced-rank approximation of the matrix C are given as

$$C^{(r)} = A^{(r)}B^{(r)} = \Gamma^{-1/2}\left(\sum_{j=1}^{r} V_j V_j'\right)\Gamma^{1/2}\Sigma_{yx}\Sigma_{xx}^{-1} = P_\Gamma \Sigma_{yx}\Sigma_{xx}^{-1} \qquad (2.15)$$

where P_Γ is an idempotent matrix for any Γ, but need not be symmetric. Observe that, analogous to (1.7), $\Sigma_{yx}\Sigma_{xx}^{-1}$ is the usual full-rank (population) regression coefficient matrix. When $r = m$, $\sum_{j=1}^{m} V_j V_j' = I_m$ and therefore, $C^{(r)}$ reduces to the full-rank coefficient matrix.

Rao (1979) has shown a stronger result that the solution for A and B in Theorem 2.2 also simultaneously minimizes all the eigenvalues of the matrix in (2.12) provided that $\Gamma = \Sigma_{yy}^{-1}$. This was proved using the Poincaire separation theorem. Robinson (1974) has proved a similar result, which indicates that the solutions of A and B in Theorem 2.2, when the corresponding sample quantities are substituted and with $\Gamma = \tilde{\Sigma}_{\epsilon\epsilon}^{-1}$, are Gaussian estimates under the reduced-rank model (2.11), that is, maximum likelihood estimates under the assumption of normality of the ϵ_k. In the remainder of this book, we may often refer to such estimates derived from the multivariate normal likelihood as maximum likelihood estimates, even when the distributional assumption of normality is not being made explicitly. We shall sketch the details below to establish the maximum likelihood estimates of the parameters in the reduced-rank regression model (2.11). Before we proceed with this, we first state a needed result from Rao (1979, Theorem 2.3) which provides, in some sense, a stronger form of the result in Theorem 2.1.

Lemma 2.2 Let S be an $m \times n$ matrix of rank m and P be an $m \times n$ matrix of rank $\leq r$ $(\leq m)$. Then

$$\lambda_i(S - P) \geq \lambda_{r+i}(S) \qquad \text{for any } i,$$

where $\lambda_i(S)$ denotes the ith largest singular value of the matrix S (as defined in the remark following the proof of Theorem 2.1), and $\lambda_{r+i}(S)$ is defined to be zero for $r + i > m$. The equality is attained for all i if and only if $P = V_{(r)}\Lambda_{(r)}U_{(r)}'$, where $S = V\Lambda U'$ represents the singular value decomposition of S as described in the remark.

For ML estimation of model (2.11), let $\boldsymbol{Y} = [Y_1, \ldots, Y_T]$ and $\boldsymbol{X} = [X_1, \ldots, X_T]$, where T denotes the number of vector observations. Assuming the ϵ_k are independent and identically distributed (iid), following a multivariate normal distribution with mean 0 and covariance matrix $\Sigma_{\epsilon\epsilon}$, the log-likelihood, apart from irrelevant constants, is

$$L(C, \Sigma_{\epsilon\epsilon}) = \left(\frac{T}{2}\right) \left[\log|\Sigma_{\epsilon\epsilon}^{-1}| - \mathrm{tr}(\Sigma_{\epsilon\epsilon}^{-1} W) \right], \tag{2.16}$$

where $|\Sigma_{\epsilon\epsilon}^{-1}|$ is the determinant of the matrix $\Sigma_{\epsilon\epsilon}^{-1}$ and $W = (1/T)(\boldsymbol{Y} - C\boldsymbol{X})(\boldsymbol{Y} - C\boldsymbol{X})'$. When $\Sigma_{\epsilon\epsilon}$ is unknown, the solution obtained by maximizing the above log-likelihood is $\hat{\Sigma}_{\epsilon\epsilon} = W$. Hence, the concentrated log-likelihood is $L(C, \hat{\Sigma}_{\epsilon\epsilon}) = -(T/2)(\log|W| + m)$. Maximizing this expression (with the structure $C = AB$, of course) is equivalent to minimizing $|W|$ and hence minimizing $|\tilde{\Sigma}_{\epsilon\epsilon}^{-1} W|$ since $|\tilde{\Sigma}_{\epsilon\epsilon}^{-1}|$ is a positive constant, where

$$\tilde{\Sigma}_{\epsilon\epsilon} = (1/T)(\boldsymbol{Y}\boldsymbol{Y}' - \boldsymbol{Y}\boldsymbol{X}'(\boldsymbol{X}\boldsymbol{X}')^{-1}\boldsymbol{X}\boldsymbol{Y}') \equiv (1/T)(\boldsymbol{Y} - \tilde{C}\boldsymbol{X})(\boldsymbol{Y} - \tilde{C}\boldsymbol{X})'$$

and $\tilde{C} = \boldsymbol{Y}\boldsymbol{X}'(\boldsymbol{X}\boldsymbol{X}')^{-1}$ is the (full-rank) LS estimate of C.

Robinson (1974) has shown that the solution in Theorem 2.2 when sample quantities are substituted, namely

$$\hat{A}^{(r)} = \Gamma^{-1/2}[\hat{V}_1, \ldots, \hat{V}_r], \qquad \hat{B}^{(r)} = [\hat{V}_1, \ldots, \hat{V}_r]'\Gamma^{1/2}\hat{\Sigma}_{yx}\hat{\Sigma}_{xx}^{-1}, \tag{2.17}$$

where $\hat{\Sigma}_{yx} = (1/T)\boldsymbol{Y}\boldsymbol{X}'$, $\hat{\Sigma}_{xx} = (1/T)\boldsymbol{X}\boldsymbol{X}'$, and $\hat{V}_j$ is the eigenvector that corresponds to the jth largest eigenvalue $\hat{\lambda}_j^2$ of $\Gamma^{1/2}\hat{\Sigma}_{yx}\hat{\Sigma}_{xx}^{-1}\hat{\Sigma}_{xy}\Gamma^{1/2}$, with the choice $\Gamma = \tilde{\Sigma}_{\epsilon\epsilon}^{-1}$, minimizes simultaneously all the eigenvalues of $\tilde{\Sigma}_{\epsilon\epsilon}^{-1} W$ and hence minimizes $|\tilde{\Sigma}_{\epsilon\epsilon}^{-1} W|$. To establish this, write

$$\begin{aligned} W &= \frac{1}{T}(\boldsymbol{Y} - \tilde{C}\boldsymbol{X} + (\tilde{C} - AB)\boldsymbol{X})(\boldsymbol{Y} - \tilde{C}\boldsymbol{X} + (\tilde{C} - AB)\boldsymbol{X})' \\ &= \frac{1}{T}(\boldsymbol{Y} - \tilde{C}\boldsymbol{X})(\boldsymbol{Y} - \tilde{C}\boldsymbol{X})' + \frac{1}{T}(\tilde{C} - AB)\boldsymbol{X}\boldsymbol{X}'(\tilde{C} - AB)' \\ &\equiv \tilde{\Sigma}_{\epsilon\epsilon} + (\tilde{C} - AB)\hat{\Sigma}_{xx}(\tilde{C} - AB)'. \end{aligned}$$

Thus $|\tilde{\Sigma}_{\epsilon\epsilon}^{-1} W| = |I_m + \tilde{\Sigma}_{\epsilon\epsilon}^{-1}(\tilde{C} - AB)\hat{\Sigma}_{xx}(\tilde{C} - AB)'| = \prod_{i=1}^m (1 + \delta_i^2)$, where δ_i^2 are the eigenvalues of the matrix $\tilde{\Sigma}_{\epsilon\epsilon}^{-1}(\tilde{C} - AB)\hat{\Sigma}_{xx}(\tilde{C} - AB)'$. So minimizing $|\tilde{\Sigma}_{\epsilon\epsilon}^{-1} W|$ is equivalent to simultaneously minimizing all the eigenvalues of $\tilde{\Sigma}_{\epsilon\epsilon}^{-1/2}(\tilde{C} - AB)\hat{\Sigma}_{xx}(\tilde{C} - AB)'\tilde{\Sigma}_{\epsilon\epsilon}^{-1/2} \equiv (S - P)(S - P)'$, with $S = \tilde{\Sigma}_{\epsilon\epsilon}^{-1/2}\tilde{C}\hat{\Sigma}_{xx}^{1/2}$ and $P = \tilde{\Sigma}_{\epsilon\epsilon}^{-1/2} AB\hat{\Sigma}_{xx}^{1/2}$. From the result of Lemma 2.2, we see that the simultaneous minimization is achieved with P chosen as the rank r approximation of S obtained through the singular value decomposition of S, that is, $P = \hat{V}_{(r)}\hat{V}_{(r)}' S = \hat{V}_{(r)}\hat{V}_{(r)}'\tilde{\Sigma}_{\epsilon\epsilon}^{-1/2}\tilde{C}\hat{\Sigma}_{xx}^{1/2} \equiv \tilde{\Sigma}_{\epsilon\epsilon}^{-1/2}\hat{A}^{(r)}\hat{B}^{(r)}\hat{\Sigma}_{xx}^{1/2}$. So $\hat{C}^{(r)} \equiv \hat{A}^{(r)}\hat{B}^{(r)} = \tilde{\Sigma}_{\epsilon\epsilon}^{1/2}\hat{V}_{(r)}\hat{V}_{(r)}'\tilde{\Sigma}_{\epsilon\epsilon}^{-1/2}\tilde{C}$, with $\hat{A}^{(r)}$ and $\hat{B}^{(r)}$ as given in (2.17), is the ML estimate.

Now the correspondence between the maximum likelihood estimators and the quantities $A^{(r)}$ and $B^{(r)}$ given in Theorem 2.2 is evident. It must be noted here that $\tilde{\Sigma}_{\epsilon\epsilon}$ is the ML estimate of $\Sigma_{\epsilon\epsilon}$ obtained in the full-rank regression model, that is, $\tilde{\Sigma}_{\epsilon\epsilon} = (1/T)(\boldsymbol{Y} - \tilde{C}\boldsymbol{X})(\boldsymbol{Y} - \tilde{C}\boldsymbol{X})'$ with

$\tilde{C} = \hat{\Sigma}_{yx}\hat{\Sigma}_{xx}^{-1}$ equal to the full-rank estimate of C as given in equation (1.7). The ML estimate of $\Sigma_{\epsilon\epsilon}$ under the reduced-rank structure is given by $\hat{\Sigma}_{\epsilon\epsilon} = (1/T)(\boldsymbol{Y} - \hat{C}^{(r)}\boldsymbol{X})(\boldsymbol{Y} - \hat{C}^{(r)}\boldsymbol{X})'$, with $\hat{C}^{(r)} = \hat{A}^{(r)}\hat{B}^{(r)}$ obtained from (2.17). From the developments above, we see that $\hat{\Sigma}_{\epsilon\epsilon}$ can be represented as

$$\begin{aligned} \hat{\Sigma}_{\epsilon\epsilon} &= \tilde{\Sigma}_{\epsilon\epsilon} + (\tilde{C} - \hat{C}^{(r)})\hat{\Sigma}_{xx}(\tilde{C} - \hat{C}^{(r)})' \\ &= \tilde{\Sigma}_{\epsilon\epsilon} + (I_m - P)\tilde{C}\hat{\Sigma}_{xx}\tilde{C}'(I_m - P)' \\ &= \tilde{\Sigma}_{\epsilon\epsilon} + \tilde{\Sigma}_{\epsilon\epsilon}^{1/2}(I_m - \hat{V}_{(r)}\hat{V}_{(r)}')\hat{R}(I_m - \hat{V}_{(r)}\hat{V}_{(r)}')\tilde{\Sigma}_{\epsilon\epsilon}^{1/2}, \end{aligned} \quad (2.18)$$

where $P = \tilde{\Sigma}_{\epsilon\epsilon}^{1/2}\hat{V}_{(r)}\hat{V}_{(r)}'\tilde{\Sigma}_{\epsilon\epsilon}^{-1/2}$, $\hat{R} = \tilde{\Sigma}_{\epsilon\epsilon}^{-1/2}\hat{\Sigma}_{yx}\hat{\Sigma}_{xx}^{-1}\hat{\Sigma}_{xy}\tilde{\Sigma}_{\epsilon\epsilon}^{-1/2}$, and $\hat{V}_{(r)} = [\hat{V}_1, \dots, \hat{V}_r]$. Note, also, that the sample version of the criterion in (2.12) and of the term on the right-hand side of the corresponding identity in (2.12′) (with $\Gamma = \tilde{\Sigma}_{\epsilon\epsilon}^{-1}$) is

$$\operatorname{tr}\{\tilde{\Sigma}_{\epsilon\epsilon}^{-1}\frac{1}{T}(\boldsymbol{Y} - AB\boldsymbol{X})(\boldsymbol{Y} - AB\boldsymbol{X})'\} = m + \operatorname{tr}\{\tilde{\Sigma}_{\epsilon\epsilon}^{-1}(\tilde{C} - AB)\hat{\Sigma}_{xx}(\tilde{C} - AB)'\}.$$

So the ML estimates $\hat{A}^{(r)}$ and $\hat{B}^{(r)}$ can be viewed as providing the "optimum" reduced-rank approximation of the full-rank least squares estimator $\tilde{C} = \hat{\Sigma}_{yx}\hat{\Sigma}_{xx}^{-1}$ corresponding to this criterion. The choice $\Gamma = \tilde{\Sigma}_{\epsilon\epsilon}^{-1}$ in the criterion, which leads to the ML estimates $\hat{A}^{(r)}$ and $\hat{B}^{(r)}$, provides an asymptotically efficient estimator under the assumptions that the errors ϵ_k are independent and normally distributed as $N(0, \Sigma_{\epsilon\epsilon})$.

Remark–Equivalence of Two Estimators. Let the solution (estimates) for the component matrices A and B given by Theorem 2.2 when sample quantities are substituted, with the choice $\Gamma = \hat{\Sigma}_{yy}^{-1}$ where $\hat{\Sigma}_{yy} = (1/T)\boldsymbol{Y}\boldsymbol{Y}'$, be denoted as $\hat{A}_*^{(r)}$ and $\hat{B}_*^{(r)}$, respectively. Then we mention that while this results in different component estimates than the estimates $\hat{A}^{(r)}$ and $\hat{B}^{(r)}$ provided in (2.17) for the choice $\Gamma = \tilde{\Sigma}_{\epsilon\epsilon}^{-1}$, it yields the *same* ML estimate for the overall coefficient matrix $C = AB$, that is, $\hat{C}^{(r)} = \hat{A}^{(r)}\hat{B}^{(r)} \equiv \hat{A}_*^{(r)}\hat{B}_*^{(r)}$. This connection in solutions between the choices $\Gamma = \hat{\Sigma}_{yy}^{-1}$ and $\Gamma = \tilde{\Sigma}_{\epsilon\epsilon}^{-1}$ will be discussed in more detail in Section 2.4.

It must be further observed that the $(m-r)$ linear restrictions on $C^{(r)} = A^{(r)}B^{(r)}$ are

$$\ell_j' = V_j'\Gamma^{1/2}, \qquad j = r+1, \dots, m, \quad (2.19)$$

because

$$V_j'\Gamma^{1/2}C^{(r)} = V_j'\Gamma^{1/2}\Gamma^{-1/2}[V_1, \dots, V_r]B^{(r)} = 0', \qquad j = r+1, \dots, m.$$

Thus, the last $(m - r)$ eigenvectors of the matrix given in Theorem 2.2 also provide the solution to the complementary part of the problem. Recall that the linear combination $\ell_j'Y_k = \ell_j'\epsilon_k$ in (2.3) does not depend on the predictor variables X_k. The necessary calculations for both aspects of the (population) problem can be achieved through the computation of all eigenvalues and eigenvectors of the matrix $\Gamma^{1/2}\Sigma_{yx}\Sigma_{xx}^{-1}\Sigma_{xy}\Gamma^{1/2}$, given in Theorem 2.2.

A comment worth noting here refers to the solution to the reduced-rank problem. In the construction of Theorem 2.1, observe that the optimal matrix N was derived in terms of M and then the optimal matrix was determined. This is equivalent to deriving the matrix B in terms of A and then determining the optimal matrix A. Conversely, one could fix B and solve for A in terms of B first. This is the regression step in the calculation procedure. Because

$$Y_k = A(BX_k) + \epsilon_k = AX_k^* + \epsilon_k,$$

given $X_k^* = BX_k$, $A = \Sigma_{yx}B'(B\Sigma_{xx}B')^{-1}$, the usual regression coefficient matrix obtained by regressing Y_k on X_k^*. If we substitute A in terms of B in the criterion (2.12) given in Theorem 2.2, it reduces to $\mathrm{tr}\{\Sigma_{yy}\Gamma - (B\Sigma_{xx}B')^{-1}B\Sigma_{xy}\Gamma\Sigma_{yx}B'\}$. From this it follows that the columns of $\Sigma_{xx}^{1/2}B'$ may be chosen as the eigenvectors corresponding to the r largest eigenvalues of $\Sigma_{xx}^{-1/2}\Sigma_{xy}\Gamma\Sigma_{yx}\Sigma_{xx}^{-1/2}$. To see all this in a convenient way, notice that, in terms of the notation of Theorem 2.1, the criterion in (2.12′) to be minimized, $\mathrm{tr}[(S-P)(S-P)']$, can also be expressed as $\mathrm{tr}[(S-P)'(S-P)] \equiv \mathrm{tr}[(S'-P')(S'-P')']$ with $P' = N'M'$. Thus, we may treat this as simply interchanging S with S' in the reduced-rank approximation problem. Hence, following the proof of Theorem 2.1, the minimizing value of M', for given N', is obtained as $M' = (NN')^{-1}NS'$. With the correspondences $S = \Gamma^{1/2}\Sigma_{yx}\Sigma_{xx}^{-1/2}$, $M = \Gamma^{1/2}A$, and $N = B\Sigma_{xx}^{1/2}$ used in the proof of Theorem 2.2, this leads to the optimal value $M' \equiv A'\Gamma^{1/2} = (B\Sigma_{xx}B')^{-1}B\Sigma_{xy}\Gamma^{1/2}$, and hence $A = \Sigma_{yx}B'(B\Sigma_{xx}B')^{-1}$ as indicated above. Then as in Theorem 2.1, the optimal value of $N' \equiv \Sigma_{xx}^{1/2}B'$ is determined as the matrix whose columns are the eigenvalues of $S'S = \Sigma_{xx}^{-1/2}\Sigma_{xy}\Gamma\Sigma_{yx}\Sigma_{xx}^{-1/2}$ corresponding to its r largest eigenvalues. Hence we see that this approach calls for eigenvectors of $S'S$ instead of SS' as in the result of Theorem 2.1. The approach presented here, of fixing B first, is appealing as various hypothesized values can be set for B and also it is easier to calculate quantities needed for making inferences, such as standard errors of the estimator of A, and so on, as in the full-rank set up.

The result presented in Theorem 2.2 and the above remark describe how an explicit solution for the component matrices A and B in (2.12) can be obtained through computation of eigenvalues and eigenvectors of either SS' or $S'S$. For the extended reduced-rank models to be considered in later chapters such an explicit solution is not possible, and iterative procedures need to be used. Although the reduced-rank model (2.11) is nonlinear in the parameter matrices A and B, the structure can be regarded as bilinear, and thus certain simplifications in the computations are possible. For the population problem, consider the first order equations resulting from minimization of the criterion (2.12), that is, consider the first partial derivatives of the criterion (2.12) with respect to A and B, set equal to zero. These first order equations are readily found to be

$$\Sigma_{yx}B' - AB\Sigma_{xx}B' = 0 \qquad \text{and} \qquad A'\Gamma\Sigma_{yx} - (A'\Gamma A)B\Sigma_{xx} = 0.$$

As observed previously, from the first equations above we see that for given B, the solution for A can be calculated as $A = \Sigma_{yx}B'(B\Sigma_{xx}B')^{-1}$. In addition, similarly, from the second equations above we find that for given A, the solution for B can be obtained as $B = (A'\Gamma A)^{-1}A'\Gamma\Sigma_{yx}\Sigma_{xx}^{-1}$, or $B = A'\Gamma\Sigma_{yx}\Sigma_{xx}^{-1}$ assuming that the normalization conditions in (2.14) for A are imposed. The iterative procedure, similar to calculations involved in a method known as *partial least squares*, calls for iterating between the two solutions (i.e., the solution for A in terms of B and the solution for B in terms of A) and at each step of the iteration imposing the normalization conditions (2.14). A similar iterative procedure was suggested by Lyttkens (1972) for the computation of canonical variates and canonical correlations.

It might also be noted at this point that alternate normalizations could be considered. One alternate normalization, with B partitioned as $B = [B_1, B_2]$ where B_1 is $r \times r$, is obtained by assuming that, since B is of rank r, its leading $r \times r$ submatrix B_1 is nonsingular (the components of X_k can always be arranged so that this holds). Then we can write $C = AB = (AB_1)[I_r, B_1^{-1}B_2] \equiv A_0\,[I_r, B_0]$ and the component parameter matrices A_0 and B_0 are uniquely identified. (A similar alternate normalization could be imposed by specifying the upper $r \times r$ submatrix of A equal to I_r). The appeal of using partial least squares increases under either of these alternate normalizations, since either normalization can be directly imposed on the corresponding first order equations above. These notions also carry over to the sample situation of constructing estimates $\hat{A}^{(r)}$ and $\hat{B}^{(r)}$, iteratively, by using sample versions of the above partial least squares estimating equations. An additional appeal in the sample setting is that it is very convenient to accommodate extra zero constraints on the parameter elements of A and B, if desired, in the partial least squares estimation whereas this is not possible in the eigenvalue-eigenvector solutions as in (2.17).

Remark–Case of Special Covariance Structure. The preceding results on ML estimation apply to the typical case where the error covariance matrix $\Sigma_{\epsilon\epsilon}$ is unknown and is a completely general positive-definite matrix. We may also briefly consider the special case where the error covariance matrix in model (2.11) is assumed to have the specified form $\Sigma_{\epsilon\epsilon} = \sigma^2\,\Psi_0$ where σ^2 is an unknown scalar parameter but Ψ_0 is a known (specified) $m\times m$ matrix. Then from (2.16), the ML estimates of A and B are obtained by minimizing

$$\frac{1}{\sigma^2}\,\mathrm{tr}[\Psi_0^{-1}(\boldsymbol{Y}-A B\boldsymbol{X})(\boldsymbol{Y}-A B\boldsymbol{X})']$$
$$=\frac{1}{\sigma^2}\mathrm{tr}[\Psi_0^{-1}(\boldsymbol{Y}-\tilde{C}\boldsymbol{X})(\boldsymbol{Y}-\tilde{C}\boldsymbol{X})']+\frac{T}{\sigma^2}\mathrm{tr}[\Psi_0^{-1}(\tilde{C}-AB)\hat{\Sigma}_{xx}(\tilde{C}-AB)'].$$

So from Theorem 2.1 it follows immediately that the ML solution is $\hat{A}=\Psi_0^{1/2}\hat{V}$, $\hat{B}=\hat{V}'\Psi_0^{-1/2}\tilde{C}$, and so $\hat{C}=\hat{A}\hat{B}=\Psi_0^{1/2}\hat{V}\hat{V}'\Psi_0^{-1/2}\tilde{C}$, where $\hat{V}=[\hat{V}_1,\ldots,\hat{V}_r]$ and the $\hat{V}_j$ are the normalized eigenvectors of the matrix

$$\Psi_0^{-1/2}\tilde{C}\hat{\Sigma}_{xx}\tilde{C}'\Psi_0^{-1/2}\equiv\Psi_0^{-1/2}\hat{\Sigma}_{yx}\hat{\Sigma}_{xx}^{-1}\hat{\Sigma}_{xy}\Psi_0^{-1/2}$$

corresponding to the r largest eigenvalues $\hat{\lambda}_1^2>\cdots>\hat{\lambda}_r^2$. The corresponding ML estimate of σ^2 is given by

$$\begin{aligned}\hat{\sigma}^2 &= \frac{1}{Tm}\mathrm{tr}[(\boldsymbol{Y}-\hat{A}\hat{B}\boldsymbol{X})'\Psi_0^{-1}(\boldsymbol{Y}-\hat{A}\hat{B}\boldsymbol{X})]\\ &\equiv \frac{1}{m}\{\mathrm{tr}[\Psi_0^{-1}\tilde{\Sigma}_{\epsilon\epsilon}]+\sum_{i=r+1}^{m}\hat{\lambda}_i^2\}.\end{aligned}\tag{2.20}$$

In particular, assuming $\Sigma_{\epsilon\epsilon}=\sigma^2 I_m$, so $\Psi_0=I_m$, corresponds to minimizing the sum of squares criterion $\mathrm{tr}[(\boldsymbol{Y}-AB\boldsymbol{X})(\boldsymbol{Y}-AB\boldsymbol{X})']$. This gives the reduced-rank ML or least squares estimators as $\hat{A}=\hat{V}$, $\hat{B}=\hat{V}'\tilde{C}$, and $\hat{C}=\hat{V}\hat{V}'\tilde{C}$, with $\hat{\sigma}^2=\frac{1}{Tm}\mathrm{tr}[(\boldsymbol{Y}-\hat{A}\hat{B}\boldsymbol{X})'(\boldsymbol{Y}-\hat{A}\hat{B}\boldsymbol{X})]\equiv\frac{1}{m}\{\mathrm{tr}[\tilde{\Sigma}_{\epsilon\epsilon}]+\sum_{i=r+1}^{m}\hat{\lambda}_i^2\}$, where the $\hat{V}_j$ are normalized eigenvectors of the matrix $\tilde{C}\hat{\Sigma}_{xx}\tilde{C}'\equiv\hat{\Sigma}_{yx}\hat{\Sigma}_{xx}^{-1}\hat{\Sigma}_{xy}$. This reduced-rank LS estimator was used by Davies and Tso (1982) and Gudmundsson (1977), in particular. The corresponding population reduced-rank LS result for random vectors Y and X amounts to setting $\Gamma=I_m$ in the criterion (2.12) of Theorem 2.2, with the problem reduced to finding A and B which minimize $\mathrm{tr}\{E[(Y-ABX)(Y-ABX)']\}\equiv E[(Y-ABX)'(Y-ABX)]$. The optimal (LS) solution is $A=V_{(r)}$, $B=V_{(r)}'\Sigma_{yx}\Sigma_{xx}^{-1}$ where the V_j are normalized eigenvectors of $\Sigma_{yx}\Sigma_{xx}^{-1}\Sigma_{xy}$. In Section 2.1, we referenced the related works by Rao (1964), Fortier (1966), and van den Wollenberg (1977) that used different terminologies, such as redundancy analysis, regarding the reduced-rank problem, but the procedures discussed in these papers can be recognized basically as being equivalent to the use of this reduced-rank LS method. It was also noted by Fortier (1966) that the linear predictive factors, that is, the transformed predictor variables $BX=V'\Sigma_{yx}\Sigma_{xx}^{-1}X$, resulting from the LS approach can be viewed simply as the principal components (to be discussed in the next section) of the unrestricted LS prediction vector $\hat{Y}=\Sigma_{yx}\Sigma_{xx}^{-1}X$, noting that $\mathrm{Cov}(\hat{Y})=\Sigma_{yx}\Sigma_{xx}^{-1}\Sigma_{xy}$ and the V_j in $V'\Sigma_{yx}\Sigma_{xx}^{-1}X\equiv V'\hat{Y}$ are the eigenvectors of this covariance matrix.

It should be mentioned, perhaps, that estimation results under the LS criterion are sensitive to the choice of scaling of the response variables Y_k, and in applications of the LS method it is often suggested that the component variables y_{ik} be standardized to have unit variances before applying the LS procedure. Of course, if Y_k is "standardized" in the more general way as $\Sigma_{yy}^{-1/2} Y_k$ so that the standardized vector has the identity covariance matrix, then the LS procedure applied to $\Sigma_{yy}^{-1/2} Y_k$ is equivalent to the canonical correlation analysis results for the original variables Y_k. Even when the form $\Sigma_{\epsilon\epsilon} = \sigma^2 I_m$ does not hold, the reduced-rank LS estimator may be useful relative to the reduced-rank ML estimator in situations where the number of vector observations T is not large relative to the dimensions m and n of the response and predictor vectors Y_k and X_k, since then the estimator $\tilde{\Sigma}_{\epsilon\epsilon}$ may not be very accurate or may not even be nonsingular (e.g., as in the case of Example 2.3 where $T = 10 < m = 19$).

2.4 Relation to Principal Components and Canonical Correlation Analysis

2.4.1 Principal Components Analysis

Principal component analysis, originally developed by Hotelling (1933), is usually concerned with explaining the covariance structure of a large number of interdependent variables through a small number of linear combinations of the original variables. The two objectives of the analysis are data reduction and interpretation which coincide with the objectives of the reduced-rank regression. A major difference between the two approaches is that the principal components are descriptive tools that do not usually rest on any modeling or distributional assumptions.

Let the $m \times 1$ random vector Y have the covariance matrix Σ_{yy} with eigenvalues $\lambda_1^2 \geq \lambda_2^2 \geq \cdots \geq \lambda_m^2 \geq 0$. Consider the linear combinations $y_1^* = \ell_1' Y, \ldots, y_m^* = \ell_m' Y$, normalized so that $\ell_i' \ell_i = 1$, $i = 1, \ldots, m$. These linear combinations have

$$\begin{aligned} \operatorname{Var}(y_i^*) &= \ell_i' \Sigma_{yy} \ell_i, \quad i = 1, 2, \ldots, m, \\ \operatorname{Cov}(y_i^*, y_j^*) &= \ell_i' \Sigma_{yy} \ell_j, \quad i, j = 1, 2, \ldots, m. \end{aligned} \tag{2.21}$$

In a principal components analysis, the linear combinations y_i^* are chosen so that they are not correlated but their variances are as large as possible. That is, $y_1^* = \ell_1' Y$ is chosen to have the largest variance among linear combinations with $\ell_1' \ell_1 = 1$, $y_2^* = \ell_2' Y$ is chosen to have the largest variance subject to $\ell_2' \ell_2 = 1$ and the condition that $\operatorname{Cov}(y_2^*, y_1^*) = \ell_2' \Sigma_{yy} \ell_1 = 0$, and so on. It can be easily shown (e.g., from Lemma 2.1) that the ℓ_i's are chosen to be the (normalized) eigenvectors that correspond to the eigenvalues λ_i^2 of the matrix Σ_{yy}, so that the normalized eigenvectors satisfy $\ell_i' \Sigma_{yy} \ell_i =$

$\lambda_i^2 \equiv \mathrm{Var}(y_i^*)$. Often, the first r ($r < m$) components y_i^*, corresponding to the r largest eigenvalues, will contain nearly all the variation that arises from all m variables, in that $\sum_{i=1}^r \mathrm{Var}(y_i^*) \equiv \sum_{i=1}^r \ell_i' \Sigma_{yy} \ell_i$ represents a very large proportion of the "total variance" $\sum_{i=1}^m \mathrm{Var}(y_i^*) \equiv \mathrm{tr}(\Sigma_{yy})$. For such cases, in studying variations of Y, attention can be directed on these r linear combinations resulting in a reduction in dimension of the variables involved.

The principal component problem can be represented as a reduced-rank problem, in the sense that a reduced-rank model can be written for this situation as

$$Y_k = ABY_k + \epsilon_k \tag{2.22}$$

(with $X_k \equiv Y_k$). From the solutions of Theorem 2.2, by setting $\Gamma = I_m$, we have $A^{(r)} = [V_1, \ldots, V_r]$ and $B^{(r)} = V'$ because $\Sigma_{yx} = \Sigma_{yy}$. The V_js refer to the eigenvectors of Σ_{yy}. Thus, a correspondence is easily established. The ℓ_i in (2.21) are the same as the V_i that result from the solution for the matrices A and B in (2.22). If both approaches provide the same solutions, one might ask, then, what is the advantage of formulating the principal components analysis in terms of reduced-rank regression model. The inclusion of the error term ϵ_k appears to be superfluous but the model (2.22) could be a useful framework to perform the residual analysis to detect outliers, influential observations, and so on, for the principal components method.

Application to Functional and Structural Relationships Models

Consider further the linear functional and structural relationships model discussed briefly in Section 2.1,

$$Y_k - \mu = W_k + \epsilon_k, \qquad k = 1, \ldots, T, \tag{2.23}$$

where the $\{W_k\}$ are assumed to satisfy the linear relations $LW_k = 0$ for all k, with L an unknown $(m{-}r) \times m$ matrix of full rank and the normalization $LL' = I_{m-r}$ is assumed. In the functional case where the W_k are treated as fixed, with $E(Y_k - \mu) = W_k$, it is assumed that $\sum_{k=1}^T W_k = 0$, whereas in the structural case the W_k are assumed to be random vectors with $E(W_k) = 0$ and $\mathrm{Cov}(W_k) = \Theta = \Lambda\Lambda'$ where Λ is an $m \times r$ full-rank matrix ($r < m$). In both cases, it is also assumed that the error covariance matrix has the structure $\Sigma_{\epsilon\epsilon} = \sigma^2 I_m$ where σ^2 is unknown. Thus, in the structural case the Y_k have covariance matrix of the form $\mathrm{Cov}(Y_k) = \Theta + \Sigma_{\epsilon\epsilon} = \Lambda\Lambda' + \sigma^2 I_m$.

Let $\bar{Y} = (1/T)\sum_{k=1}^T Y_k$ and $\tilde{\Sigma}_{yy} = (1/T)\sum_{k=1}^T (Y_k - \bar{Y})(Y_k - \bar{Y})'$ denote the sample mean vector and sample covariance matrix of the Y_k. Also let $\hat{\lambda}_1^2 > \hat{\lambda}_2^2 > \cdots > \hat{\lambda}_m^2 > 0$ denote the eigenvalues of $\tilde{\Sigma}_{yy}$ with corresponding normalized eigenvectors $\hat{\ell}_i$, $i = 1, \ldots, m$. Then $\hat{y}_{ik}^* = \hat{\ell}_i' Y_k$ are the sample principal components of the Y_k. In the functional case, it has been shown originally by Tintner (1945), also see Anderson (1984b), that a ML estimate of the unknown constraints matrix L is given by the matrix $\hat{L} = \hat{L}_2' \equiv$

$[\hat{\ell}_{r+1}, \ldots, \hat{\ell}_m]'$ whose rows correspond to the last (smallest) $m-r$ principal components in the sample. We also set $\hat{L}_1 = [\hat{\ell}_1, \ldots, \hat{\ell}_r]$, corresponding to the first (largest) r sample principal components. Then, the ML estimates of the unknown but fixed values W_k are given by

$$\hat{W}_k = \hat{L}_1 \hat{L}_1'(Y_k - \bar{Y}), \qquad k = 1, \ldots, T, \tag{2.24}$$

with $\hat{\mu} = \bar{Y}$, and $\hat{\sigma}^2 = \frac{1}{m} \sum_{i=r+1}^{m} \hat{\lambda}_i^2$. Thus, in effect, under ML estimation the variability described by the first r sample principal components (as measured by $\hat{\lambda}_1^2, \ldots, \hat{\lambda}_r^2$) is assigned to the variation in the mean values W_k and the remaining variability, $\sum_{i=r+1}^{m} \hat{\lambda}_i^2$, is assigned to the variance σ^2 of the errors ϵ_k. The above estimator $\hat{\sigma}^2$ of σ^2 is biased downward, however.

We verify these results by using the framework of the ML estimation results established in Section 2.3 for the reduced-rank regression model. As discussed in Section 2.1, the functional model $Y_k - \mu = W_k + \epsilon_k$, $LW_k = 0$, can be written equivalently as

$$Y_k - \mu = AF_k + \epsilon_k, \qquad k = 1, \ldots, T, \tag{2.25}$$

that is, $W_k = AF_k$, where A is $m \times r$ such that $LA = 0$ and the F_k are $r \times 1$ vectors of unknown constants. Hence, in matrix form we have

$$\boldsymbol{Y} = \mu \mathbf{1}' + \boldsymbol{W} + \boldsymbol{\epsilon} = \mu \mathbf{1}' + AB\boldsymbol{X} + \boldsymbol{\epsilon} \equiv \mu \mathbf{1}' + C\boldsymbol{X} + \boldsymbol{\epsilon}, \tag{2.26}$$

where $\boldsymbol{X} = I_T$, $\boldsymbol{W} = [W_1, \ldots, W_T] \equiv C = AB$ with $B = [F_1, \ldots, F_T]$, and $\mathbf{1} = (1, \ldots, 1)'$ denotes a $T \times 1$ vector of ones. For estimation, the condition $\boldsymbol{W}\mathbf{1} = \sum_{k=1}^{T} W_k = 0$, equivalently, $\sum_{k=1}^{T} F_k = 0$, is imposed. The unrestricted LS estimators of the model are $\tilde{\mu} = \bar{Y}$, $\tilde{W}_k = Y_k - \bar{Y}$, $k = 1, \ldots, T$, or $\tilde{C} = \boldsymbol{Y} - \bar{Y}\mathbf{1}'$ (with $\tilde{\Sigma}_{\epsilon\epsilon} \equiv 0$ since $Y_k - \tilde{\mu} - \tilde{W}_k \equiv 0$). Then notice that $\tilde{C}\hat{\Sigma}_{xx}\tilde{C}' = \frac{1}{T}(\boldsymbol{Y} - \bar{Y}\mathbf{1}')(\boldsymbol{Y} - \bar{Y}\mathbf{1}')' = \tilde{\Sigma}_{yy}$, with eigenvalues $\hat{\lambda}_1^2 > \hat{\lambda}_2^2 > \cdots > \hat{\lambda}_m^2 > 0$ and corresponding normalized eigenvectors $\hat{\ell}_i$, $i = 1, \ldots, m$. Then with $\hat{L}_1 = [\hat{\ell}_1, \ldots, \hat{\ell}_r]$, according to the ML estimation results for the special case $\Sigma_{\epsilon\epsilon} = \sigma^2 I_m$ presented at the end of Section 2.3, we have that the ML estimates are $\hat{A} = \hat{L}_1$, $\hat{B} = \hat{L}_1'\tilde{C} \equiv \hat{L}_1'(\boldsymbol{Y} - \bar{Y}\mathbf{1}')$, so that $\hat{\boldsymbol{W}} = \hat{C} = \hat{A}\hat{B} = \hat{L}_1\hat{L}_1'(\boldsymbol{Y} - \bar{Y}\mathbf{1}')$. That is,

$$\hat{W}_k = \hat{L}_1 \hat{L}_1'(Y_k - \bar{Y}), \qquad k = 1, \ldots, T,$$

as in (2.24), with $\hat{\mu} = \tilde{\mu} = \bar{Y}$. In addition, the ML estimate of σ^2, from (2.20), is given by $\hat{\sigma}^2 = \frac{1}{m} \sum_{i=r+1}^{m} \hat{\lambda}_i^2$ as stated above. (Actually, the above obtained results follow from a slight extension of results for the basic model considered in this chapter, which explicitly allows for a constant term. The general case of this extension will be considered in Section 3.1.)

In the structural case of the model (2.23), $\hat{L} = \hat{L}_2'$ is also a ML estimate of L whereas the ML estimate of σ^2 is $\hat{\sigma}^2 = \frac{1}{m-r} \sum_{i=r+1}^{m} \hat{\lambda}_i^2$. The ML estimate of $\mathrm{Cov}(W_k) = \Theta = \Lambda\Lambda'$ is given by

$$\hat{\Theta} = \hat{L}_1 \hat{D}_1 \hat{L}_1' - \hat{\sigma}^2 \hat{L}_1 \hat{L}_1' = \hat{L}_1(\hat{D}_1 - \hat{\sigma}^2 I_m)\hat{L}_1',$$

where $\hat{D}_1 = \text{diag}(\hat{\lambda}_1^2, \ldots, \hat{\lambda}_r^2)$. So the ML estimate of $\text{Cov}(Y_k) = \Sigma_{yy}$ is

$$\hat{\Sigma}_{yy} = \hat{\Theta} + \hat{\sigma}^2 I_m \equiv \hat{L}_1 \hat{D}_1 \hat{L}_1' + \hat{\sigma}^2 \hat{L}_2 \hat{L}_2',$$

noting that $\hat{L}_1 \hat{L}_1' + \hat{L}_2 \hat{L}_2' = I_m$. These estimators were obtained by Lawley (1953) and by Theobald (1975), also see Anderson (1984b).

Thus, in the above linear relationships models we see that the sample principal components arise fundamentally in the ML solution. The first r principal components are the basis for estimation of the systematic part W_k of Y_k, e.g., $\hat{W}_k = \hat{L}_1 \hat{L}_1'(Y_k - \bar{Y})$ in the functional case, and $(Y_k - \bar{Y}) - \hat{L}_1 \hat{L}_1'(Y_k - \bar{Y}) = \hat{L}_2 \hat{L}_2'(Y_k - \bar{Y})$ can be considered as the estimate of the error. However, a distinction in this application of principal components analysis is that the error in approximation of Y_k through its first r principal components $\hat{L}_1' Y_k$ is not necessarily small, but should have the attributes of the error term ϵ_k, namely mean 0 and covariance matrix with equal variances σ^2. So in this analysis the remaining (last $m-r$) values $\hat{\lambda}_{r+1}^2, \ldots, \hat{\lambda}_m^2$ are not necessarily "negligible" relative to $\hat{\lambda}_1^2, \ldots, \hat{\lambda}_r^2$, as would be hoped in some typical uses of principal components analysis. In the linear functional or structural analysis, the representation of Y_k by such a model form might be considered to be good provided only that $\hat{\lambda}_{r+1}^2, \ldots, \hat{\lambda}_m^2$ are roughly equal (but not necessarily small).

2.4.2 Canonical Correlation Analysis

Canonical correlation analysis was introduced by Hotelling (1935, 1936) as a method of summarizing relationships between two sets of variables. The objective is to find that linear combination of one set of variables which is most highly correlated with any linear combination of a second set of variables. The technique has received attention from theoretical statisticians as early as Bartlett (1938), but ter Braak (1990) identifies that because of difficulty in interpretation of the canonical variates, its use is not widespread in some disciplines. By exploring the connection between the reduced-rank regression and canonical correlations, we provide an additional method of interpreting the canonical variates. Because the role of any descriptive tool such as canonical correlations is in better model building, the connection we display below provides an avenue for interpreting and further using the canonical correlations.

More formally, in canonical correlation analysis, for random vectors X and Y of dimensions n and m, respectively, the purpose is to obtain an $r \times n$ matrix G and an $r \times m$ matrix H, so that the r-vector variates $\xi = GX$ and $\omega = HY$ will capture as much of the covariances between X and Y as possible. The following results from Izenman (1975) summarize how this is done.

Theorem 2.3 We assume the same conditions as given in Theorem 2.2 and further we assume Σ_{yy} is nonsingular. Then the $r \times n$ matrix G and the $r \times m$ matrix H, with $H\Sigma_{yy}H' = I_r$, that minimize simultaneously all the eigenvalues of

$$E[(HY - GX)(HY - GX)'] \tag{2.27}$$

are given by

$$G = V_*'\Sigma_{yy}^{-1/2}\Sigma_{yx}\Sigma_{xx}^{-1}, \qquad H = V_*'\Sigma_{yy}^{-1/2},$$

where $V_* = [V_1^*, \ldots, V_r^*]$ and V_j^* is the (normalized) eigenvector corresponding to the jth largest eigenvalue ρ_j^2 of the matrix

$$R_* = \Sigma_{yy}^{-1/2}\Sigma_{yx}\Sigma_{xx}^{-1}\Sigma_{xy}\Sigma_{yy}^{-1/2}. \tag{2.28}$$

Denoting by g_j' and h_j' the jth rows of the matrices G and H, respectively, the jth pair of canonical variates are defined to be $\xi_j = g_j'X$ and $\omega_j = h_j'Y$, $j = 1, 2, \ldots, m$ $(m \leq n)$. The correlation between ξ_j and ω_j is the jth canonical correlation coefficient and hence, since $\text{Cov}(\xi_j, \omega_j) = V_j^{*\prime}R_*V_j^* = \rho_j^2$, $\text{Var}(\xi_j) = V_j^{*\prime}R_*V_j^* = \rho_j^2$, and $\text{Var}(\omega_j) = V_j^{*\prime}V_j^* = 1$, we have $\text{Corr}(\xi_j, \omega_j) = \rho_j$ and the ρ_js are the canonical correlations between X and Y. The canonical variates possess the property that ξ_1 and ω_1 have the largest correlation among all possible linear combinations of X and Y, ξ_2 and ω_2 have the largest possible correlation among all linear combinations of X and Y that are uncorrelated with ξ_1 and ω_1, and so on.

We shall compare the results of Theorem 2.3 with those of Theorem 2.2. When $\Gamma = \Sigma_{yy}^{-1}$, the matrix $B_*^{(r)} = V_*'\Gamma^{1/2}\Sigma_{yx}\Sigma_{xx}^{-1}$ from (2.13) is identical to G and $H = V_*'\Sigma_{yy}^{-1/2}$ is equal to $A_*^{(r)-}$, a reflexive generalized inverse of $A_*^{(r)} = \Sigma_{yy}^{1/2}V_*$ [see Rao (1973, p. 26)]. The transformed variates $\xi^{(r)} = B_*^{(r)}X$ and $\omega^{(r)} = A_*^{(r)-}Y$ have $V_*'R_*V_* = \Lambda_*^2 \equiv \text{diag}(\rho_1^2, \ldots, \rho_r^2)$ as their cross-covariance matrix. The matrix R_* is a multivariate version of the simple squared correlation coefficient between two variables $(m = n = 1)$ and also of the squared multiple correlation coefficient between a single dependent variable $(m = 1)$ and a number of predictor variables.

The model, with the choice of $\Gamma = \Sigma_{yy}^{-1}$, is referred to as the reduced-rank model corresponding to the canonical variates situation by Izenman (1975). Although a number of choices for Γ are possible, we want to comment on the choice of $\Gamma = \Sigma_{\epsilon\epsilon}^{-1}$, that leads to maximum likelihood estimates [see Robinson (1974) and Section 2.3]. Because $\Sigma_{\epsilon\epsilon} = \Sigma_{yy} - \Sigma_{yx}\Sigma_{xx}^{-1}\Sigma_{xy}$, the correspondence with results for the choice $\Gamma = \Sigma_{yy}^{-1}$ is fairly simple. In the first choice of $\Gamma = \Sigma_{yy}^{-1}$, the squared canonical correlations are indeed the eigenvalues of $\Sigma_{yx}\Sigma_{xx}^{-1}\Sigma_{xy}$ with respect to Σ_{yy}. For the choice of $\Gamma = \Sigma_{\epsilon\epsilon}^{-1}$, we compute eigenvalues, say λ_j^2, of the same matrix $\Sigma_{yx}\Sigma_{xx}^{-1}\Sigma_{xy}$ but with respect to $\Sigma_{yy} - \Sigma_{yx}\Sigma_{xx}^{-1}\Sigma_{xy}$, that is, roots of the determinantal equation $|\lambda_j^2(\Sigma_{yy} - \Sigma_{yx}\Sigma_{xx}^{-1}\Sigma_{xy}) - \Sigma_{yx}\Sigma_{xx}^{-1}\Sigma_{xy}| = 0$. Since the ρ_j^2 for the first choice $\Gamma = \Sigma_{yy}^{-1}$ satisfy $|\rho_j^2\Sigma_{yy} - \Sigma_{yx}\Sigma_{xx}^{-1}\Sigma_{xy}| = 0$, algebraically it follows that

$$\rho_j^2 = \lambda_j^2/(1 + \lambda_j^2). \tag{2.29}$$

The canonical variates can be recovered from the calculation of eigenvalues and eigenvectors for the choice $\Gamma = \Sigma_{\epsilon\epsilon}^{-1}$ through

$$\xi^{(r)} = B^{(r)}X = V'\Sigma_{\epsilon\epsilon}^{-1/2}\Sigma_{yx}\Sigma_{xx}^{-1}X \quad \text{and} \quad \omega^{(r)} = A^{(r)-}Y = V'\Sigma_{\epsilon\epsilon}^{-1/2}Y$$

where $V = [V_1, \ldots, V_r]$, λ_j^2 and V_j are the eigenvalues and (normalized) eigenvectors, respectively, of the matrix $\Sigma_{\epsilon\epsilon}^{-1/2}\Sigma_{yx}\Sigma_{xx}^{-1}\Sigma_{xy}\Sigma_{\epsilon\epsilon}^{-1/2}$, and $\lambda_j = \rho_j/(1-\rho_j^2)^{1/2}$ where ρ_j is the jth largest canonical correlation. It must be noted that the corresponding solutions from (2.13) for the choice $\Gamma = \Sigma_{\epsilon\epsilon}^{-1}$ satisfy $A^{(r)} = \Sigma_{\epsilon\epsilon}^{1/2}V = \Sigma_{\epsilon\epsilon}^{1/2}\Sigma_{\epsilon\epsilon}^{1/2}\Sigma_{yy}^{-1/2}V_*(I_r - \Lambda_*^2)^{-1/2} = \Sigma_{\epsilon\epsilon}\Sigma_{yy}^{-1/2}V_*(I_r - \Lambda_*^2)^{-1/2}$ and $B^{(r)} = V'\Sigma_{\epsilon\epsilon}^{-1/2}\Sigma_{yx}\Sigma_{xx}^{-1} = (I_r - \Lambda_*^2)^{-1/2}V_*'\Sigma_{yy}^{-1/2}\Sigma_{yx}\Sigma_{xx}^{-1}$, because $V_* = \Sigma_{yy}^{1/2}\Sigma_{\epsilon\epsilon}^{-1/2}V(I_r - \Lambda_*^2)^{1/2}$ and so $V = \Sigma_{\epsilon\epsilon}^{1/2}\Sigma_{yy}^{-1/2}V_*(I_r - \Lambda_*^2)^{-1/2}$, where $V_* = [V_1^*, \ldots, V_r^*]$ and the V_j^* are normalized eigenvectors of the matrix R_* in (2.28). Also notice then that

$$\begin{aligned} A^{(r)} &= \Sigma_{\epsilon\epsilon}\Sigma_{yy}^{-1/2}V_*(I_r - \Lambda_*^2)^{-1/2} \\ &= (\Sigma_{yy} - \Sigma_{yx}\Sigma_{xx}^{-1}\Sigma_{xy})\Sigma_{yy}^{-1/2}V_*(I_r - \Lambda_*^2)^{-1/2} \\ &= \Sigma_{yy}^{1/2}(I_m - \Sigma_{yy}^{-1/2}\Sigma_{yx}\Sigma_{xx}^{-1}\Sigma_{xy}\Sigma_{yy}^{-1/2})V_*(I_r - \Lambda_*^2)^{-1/2} \\ &= \Sigma_{yy}^{1/2}V_*(I_r - \Lambda_*^2)^{1/2} \end{aligned}$$

and $B^{(r)} = (I_r - \Lambda_*^2)^{-1/2}V_*'\Sigma_{yy}^{-1/2}\Sigma_{yx}\Sigma_{xx}^{-1}$, whereas the solutions of A and B for the choice $\Gamma = \Sigma_{yy}^{-1}$ are given by $A_*^{(r)} = \Sigma_{yy}^{1/2}V_*$ and $B_*^{(r)} = V_*'\Sigma_{yy}^{-1/2}\Sigma_{yx}\Sigma_{xx}^{-1}$. Thus, compared to these latter solutions, only the columns and rows of the solution matrices A and B, respectively, are scaled differently for the choice $\Gamma = \Sigma_{\epsilon\epsilon}^{-1}$, but the solution for the overall coefficient matrix remains the same with $A^{(r)}B^{(r)} = A_*^{(r)}B_*^{(r)}$.

The connection described above between reduced-rank regression quantities in Theorem 2.2 and canonical correlation analysis in the population also carries over to the sample situation of ML estimation of the reduced-rank model. In particular, let $\hat{\rho}_j^2$ denote the sample squared canonical correlations between the X_k and Y_k, that is, the eigenvalues of $\hat{R}_* = \hat{\Sigma}_{yy}^{-1/2}\hat{\Sigma}_{yx}\hat{\Sigma}_{xx}^{-1}\hat{\Sigma}_{xy}\hat{\Sigma}_{yy}^{-1/2}$ where $\hat{\Sigma}_{yy} = \frac{1}{T}\boldsymbol{Y}\boldsymbol{Y}'$, $\hat{\Sigma}_{yx} = \frac{1}{T}\boldsymbol{Y}\boldsymbol{X}'$, and $\hat{\Sigma}_{xx} = \frac{1}{T}\boldsymbol{X}\boldsymbol{X}'$, and let $\hat{V}_j^*$ denote the corresponding (normalized) eigenvectors, $j = 1, \ldots, m$. (Recall from the remark in Section 1.2 that, in practice, in the regression analysis the predictor and response variables $\boldsymbol{X}$ and $\boldsymbol{Y}$ will typically be centered by subtraction of appropriate sample mean vectors before the sample canonical correlation calculations above are performed.) Then the ML estimators of A and B in the reduced-rank model can be expressed in an equivalent form to (2.17) as

$$\hat{A}^{(r)} = \tilde{\Sigma}_{\epsilon\epsilon}\hat{\Sigma}_{yy}^{-1/2}\hat{V}_*(I_r - \hat{\Lambda}_*^2)^{-1/2} = \hat{\Sigma}_{yy}^{1/2}\hat{V}_*(I_r - \hat{\Lambda}_*^2)^{1/2},$$

$$\hat{B}^{(r)} = (I_r - \hat{\Lambda}_*^2)^{-1/2}\hat{V}_*'\hat{\Sigma}_{yy}^{-1/2}\hat{\Sigma}_{yx}\hat{\Sigma}_{xx}^{-1},$$

with $\hat{V}_* = [\hat{V}_1^*, \ldots, \hat{V}_r^*]$, $\hat{\Lambda}_*^2 = \text{diag}(\hat{\rho}_1^2, \ldots, \hat{\rho}_r^2)$, and $\tilde{\Sigma}_{\epsilon\epsilon} = \hat{\Sigma}_{yy} - \hat{\Sigma}_{yx}\hat{\Sigma}_{xx}^{-1}\hat{\Sigma}_{xy}$.

Through the linkage between the two procedures, we have demonstrated how the descriptive tool of canonical correlation can also be used for modeling and prediction purposes.

2.5 Asymptotic Distribution of Estimators in Reduced-Rank Model

The small sample distribution of the reduced-rank estimators associated with the canonical correlations, although available, is somewhat difficult to use. Therefore, we focus on the large sample behavior of the estimators and we derive in this section the asymptotic theory for the estimators of the parameters A and B in model (2.11). The asymptotic results follow from Robinson (1973). From Theorem 2.2, for any positive-definite Γ, the population quantities can be expressed as

$$A = \Gamma^{-1/2}[V_1, \ldots, V_r] = \Gamma^{-1/2}V = [\alpha_1, \ldots, \alpha_r], \tag{2.30a}$$

$$B = V'\Gamma^{1/2}\Sigma_{yx}\Sigma_{xx}^{-1} = [\beta_1, \ldots, \beta_r]'. \tag{2.30b}$$

The estimators of A and B, from (2.17), corresponding to these populations quantities are

$$\hat{A} = \Gamma^{-1/2}[\hat{V}_1, \ldots, \hat{V}_r] = \Gamma^{-1/2}\hat{V} = [\hat{\alpha}_1, \ldots, \hat{\alpha}_r], \tag{2.31a}$$

$$\hat{B} = \hat{V}'\Gamma^{1/2}\hat{\Sigma}_{yx}\hat{\Sigma}_{xx}^{-1} = [\hat{\beta}_1, \ldots, \hat{\beta}_r]', \tag{2.31b}$$

where $\hat{V}_j$ is the normalized eigenvector corresponding to the jth largest eigenvalue $\hat{\lambda}_j^2$ of $\Gamma^{1/2}\hat{\Sigma}_{yx}\hat{\Sigma}_{xx}^{-1}\hat{\Sigma}_{xy}\Gamma^{1/2}$, with $\hat{\Sigma}_{yx} = (1/T)\boldsymbol{YX}'$ and $\hat{\Sigma}_{xx} = (1/T)\boldsymbol{XX}'$. We assume that $\Sigma_{xx} = \lim_{T\to\infty} \frac{1}{T}\boldsymbol{XX}'$ exists and is positive-definite, and $\Sigma_{yx} = C\Sigma_{xx}$. In the following result, we assume $\Gamma = \Sigma_{\epsilon\epsilon}^{-1}$ is known, for convenience. The argument and asymptotic results for estimators $\hat{A}$ and $\hat{B}$ would be somewhat affected by the use of $\Gamma = \tilde{\Sigma}_{\epsilon\epsilon}^{-1}$ where $\tilde{\Sigma}_{\epsilon\epsilon} = (1/T)(\boldsymbol{YY}' - \boldsymbol{YX}'(\boldsymbol{XX}')^{-1}\boldsymbol{XY}')$, mainly because the normalization restrictions on $\hat{A}$ then depend on $\tilde{\Sigma}_{\epsilon\epsilon}^{-1}$. These effects will be discussed subsequently. Recall that for the choice of $\Gamma = \tilde{\Sigma}_{\epsilon\epsilon}^{-1}$ the estimators $\hat{A}$ and $\hat{B}$ in (2.31) are the maximum likelihood estimators of A and B under the assumption of normality of the error terms ϵ_k [Robinson (1974, Theorem 1), and Section 2.3].

Our main result on the asymptotic distribution of $\hat{A}$ and $\hat{B}$ when $\Gamma = \Sigma_{\epsilon\epsilon}^{-1}$ is assumed to be known is contained in the next theorem.

Theorem 2.4 For the model (2.11) where the ϵ_k satisfy the usual assumptions, let $(\text{vec}(A)', \text{vec}(B)')' \in \Theta$, a compact set defined by the normalization conditions (2.14). Then, with $\hat{A}$, $\hat{B}$ and A, B given by (2.31) and

(2.30), respectively, $\text{vec}(\hat{A})$ and $\text{vec}(\hat{B})$ converge in probability to $\text{vec}(A)$ and $\text{vec}(B)$, respectively, as $T \to \infty$. Also the vector variates $T^{1/2}(\hat{\alpha}_j - \alpha_j)$ and $T^{1/2}(\hat{\beta}_j - \beta_j)$ $(j = 1, \ldots, r)$ have a joint limiting distribution as $T \to \infty$ which is singular multivariate normal with null mean vectors and the following asymptotic covariance matrices:

$$E[T(\hat{\alpha}_j - \alpha_j)(\hat{\alpha}_\ell - \alpha_\ell)'] \to \begin{cases} \displaystyle\sum_{i \neq j=1}^{m} \frac{\lambda_j^2 + \lambda_i^2}{(\lambda_j^2 - \lambda_i^2)^2} \alpha_i \alpha_i' & (j = \ell), \\ \displaystyle -\frac{\lambda_j^2 + \lambda_\ell^2}{(\lambda_j^2 - \lambda_\ell^2)^2} \alpha_\ell \alpha_j' & (j \neq \ell), \end{cases}$$

$$E[T(\hat{\beta}_j - \beta_j)(\hat{\beta}_\ell - \beta_\ell)'] \to \begin{cases} \displaystyle\sum_{i \neq j=1}^{r} \frac{3\lambda_j^2 - \lambda_i^2}{(\lambda_j^2 - \lambda_i^2)^2} \beta_i \beta_i' + \Sigma_{xx}^{-1} & (j = \ell), \\ \displaystyle -\frac{\lambda_j^2 + \lambda_\ell^2}{(\lambda_j^2 - \lambda_\ell^2)^2} \beta_\ell \beta_j' & (j \neq \ell), \end{cases}$$

$$E[T(\hat{\alpha}_j - \alpha_j)(\hat{\beta}_\ell - \beta_\ell)'] \to \begin{cases} \displaystyle 2\lambda_j^2 \sum_{i \neq j=1}^{r} \frac{1}{(\lambda_j^2 - \lambda_i^2)^2} \alpha_i \beta_i' & (j = \ell), \\ \displaystyle -\frac{2\lambda_\ell^2}{(\lambda_j^2 - \lambda_\ell^2)^2} \alpha_\ell \beta_j' & (j \neq \ell), \end{cases}$$

where the λ_j^2, $j = 1, \ldots, r$, are the nonzero eigenvalues, assumed to be distinct, of the matrix $\Gamma^{1/2} \Sigma_{yx} \Sigma_{xx}^{-1} \Sigma_{xy} \Gamma^{1/2}$, with $\Gamma = \Sigma_{\epsilon\epsilon}^{-1}$.

Proof: With the use of perturbation expansion of matrices (Izenman, 1975), the eigenvectors $\hat{V}_j$ of the matrix $M = \Sigma_{\epsilon\epsilon}^{-1/2} \hat{\Sigma}_{yx} \hat{\Sigma}_{xx}^{-1} \hat{\Sigma}_{xy} \Sigma_{\epsilon\epsilon}^{-1/2}$ can be expanded around the eigenvectors V_j of $N = \Sigma_{\epsilon\epsilon}^{-1/2} \Sigma_{yx} \Sigma_{xx}^{-1} \Sigma_{xy} \Sigma_{\epsilon\epsilon}^{-1/2}$ to give

$$\hat{V}_j \sim V_j + \sum_{i \neq j}^{m} \frac{1}{(\lambda_j^2 - \lambda_i^2)} V_i [V_j'(M - N) V_i] \qquad (j = 1, \ldots, r). \tag{2.32}$$

Then since $\hat{\Sigma}_{xx}$ and $\hat{\Sigma}_{yx}$ converge in probability to Σ_{xx} and Σ_{yx} (Anderson, 1971, p. 195), M converges to N and hence $\hat{V}_j$ converges in probability to V_j as $T \to \infty$. Consistency of $\hat{A}$ and $\hat{B}$ then follows directly from the relations (2.30) and (2.31).

To establish asymptotic normality, using $\hat{\Sigma}_{xy} = \hat{\Sigma}_{xx} C' + T^{-1} \sum_{k=1}^{T} X_k \epsilon_k'$, we can write

$$\hat{\Sigma}_{yx} \hat{\Sigma}_{xx}^{-1} \hat{\Sigma}_{xy} - \Sigma_{yx} \Sigma_{xx}^{-1} \Sigma_{xy} = C U_T + U_T' C' + U_T' \hat{\Sigma}_{xx}^{-1} U_T + C(\hat{\Sigma}_{xx} - \Sigma_{xx}) C',$$

where $U_T = T^{-1}\sum_{k=1}^{T} X_k \epsilon_k'$. Hence, following Robinson (1973), from the perturbation expansion in (2.32), we have

$$\begin{aligned} T^{1/2}(\hat{V}_j - V_j) &= T^{1/2}\sum_{i\neq j}^{m} \frac{1}{(\lambda_j^2-\lambda_i^2)} V_i[(V_i'\Sigma_{\epsilon\epsilon}^{-1/2}\otimes V_j'VB) \\ &\quad + (V_j'\Sigma_{\epsilon\epsilon}^{-1/2}\otimes V_i'VB)]\operatorname{vec}(U_T) + O_p(T^{-1/2}) \\ &\equiv T^{1/2}D_j'\operatorname{vec}(U_T) + O_p(T^{-1/2}) \qquad (j=1,\ldots,r), \qquad (2.33) \end{aligned}$$

where we have used the relation $\operatorname{vec}(ABC) = (C'\otimes A)\operatorname{vec}(B)$ and the fact that $\Sigma_{\epsilon\epsilon}^{-1/2}C = \Sigma_{\epsilon\epsilon}^{-1/2}AB = VB$.

Define $D = [D_1,\ldots,D_r]$, so that $T^{1/2}\operatorname{vec}(\hat{V}-V) \sim T^{1/2}D'\operatorname{vec}(U_T)$. Then since $T^{1/2}\operatorname{vec}(U_T) = \operatorname{vec}(T^{-1/2}\sum_{k=1}^{T} X_k\epsilon_k')$ converges in distribution to $N(0,\Sigma_{\epsilon\epsilon}\otimes\Sigma_{xx})$ as $T\to\infty$ (Anderson, 1971, p. 200), the asymptotic distribution of $T^{1/2}\operatorname{vec}(\hat{V}-V)$ is $N(0,G)$, where $G = D'(\Sigma_{\epsilon\epsilon}\otimes\Sigma_{xx})D$ has blocks $G_{j\ell} = D_j'(\Sigma_{\epsilon\epsilon}\otimes\Sigma_{xx})D_\ell$, with

$$G_{jj} = \sum_{i\neq j=1}^{m} \frac{\lambda_j^2+\lambda_i^2}{(\lambda_j^2-\lambda_i^2)^2} V_iV_i' \qquad (j=1,\ldots,r),$$

$$G_{j\ell} = -\frac{\lambda_j^2+\lambda_\ell^2}{(\lambda_j^2-\lambda_\ell^2)^2} V_\ell V_j' \qquad (j\neq\ell),$$

if we note that $V_i'V_j = 0$ for $i\neq j$, $V_i'V_i = 1$, and $B\Sigma_{xx}B' = \Lambda^2$. The asymptotic distribution of $T^{1/2}\operatorname{vec}(\hat{A}-A)$ follows easily from the above and relation (2.31) with $T^{1/2}\operatorname{vec}(\hat{A}-A) \simeq T^{1/2}(I_r\otimes\Sigma_{\epsilon\epsilon}^{1/2})\operatorname{vec}(\hat{V}-V)$ converging in distribution to $N\{0,(I_r\otimes\Sigma_{\epsilon\epsilon}^{1/2})G(I_r\otimes\Sigma_{\epsilon\epsilon}^{1/2})\}$. Hence, if we note the relation $\Sigma_{\epsilon\epsilon}^{1/2}V_i = \alpha_i$, the individual covariance terms that correspond to the above distribution are as given in the theorem.

To evaluate the asymptotic distribution of $T^{1/2}\operatorname{vec}(\hat{B}'-B')$ observe that

$$\begin{aligned} T^{1/2}\operatorname{vec}(\hat{B}'-B') &= T^{1/2}\operatorname{vec}(\hat{\Sigma}_{xx}^{-1}\hat{\Sigma}_{xy}\Sigma_{\epsilon\epsilon}^{-1/2}\hat{V} - \Sigma_{xx}^{-1}\Sigma_{xy}\Sigma_{\epsilon\epsilon}^{-1/2}V) \\ &\simeq T^{1/2}\operatorname{vec}\{\Sigma_{xx}^{-1}\Sigma_{xy}\Sigma_{\epsilon\epsilon}^{-1/2}(\hat{V}-V)\} \\ &\quad + T^{1/2}\operatorname{vec}\{(\hat{\Sigma}_{xx}^{-1}\hat{\Sigma}_{xy}\Sigma_{\epsilon\epsilon}^{-1/2} - \Sigma_{xx}^{-1}\Sigma_{xy}\Sigma_{\epsilon\epsilon}^{-1/2})V\} \\ &= T^{1/2}(I_r\otimes\Sigma_{xx}^{-1}\Sigma_{xy}\Sigma_{\epsilon\epsilon}^{-1/2})\operatorname{vec}(\hat{V}-V) \\ &\quad + T^{1/2}(V'\Sigma_{\epsilon\epsilon}^{-1/2}\otimes\hat{\Sigma}_{xx}^{-1})\operatorname{vec}(T^{-1}\textstyle\sum_{k=1}^{T} X_k\epsilon_k') \\ &\simeq \{(I_r\otimes B'V')D' + (V'\Sigma_{\epsilon\epsilon}^{-1/2}\otimes\Sigma_{xx}^{-1})\}\,T^{1/2}\operatorname{vec}(U_T) \qquad (2.34) \end{aligned}$$

where we again have used the fact that $\Sigma_{\epsilon\epsilon}^{-1/2}\Sigma_{yx}\Sigma_{xx}^{-1} = \Sigma_{\epsilon\epsilon}^{-1/2}C = VB$. Hence $T^{1/2}\operatorname{vec}(\hat{B}'-B')$ converges in distribution to $N(0,H)$, where

$$\begin{aligned} H &= (I_r \otimes B'V')D'(\Sigma_{\epsilon\epsilon} \otimes \Sigma_{xx})D(I_r \otimes VB) + (I_r \otimes \Sigma_{xx}^{-1}) \\ &\quad + (I_r \otimes B'V')D'(\Sigma_{\epsilon\epsilon} \otimes \Sigma_{xx})(\Sigma_{\epsilon\epsilon}^{-1/2}V \otimes \Sigma_{xx}^{-1}) \\ &\quad + (V'\Sigma_{\epsilon\epsilon}^{-1/2} \otimes \Sigma_{xx}^{-1})(\Sigma_{\epsilon\epsilon} \otimes \Sigma_{xx})D(I_r \otimes VB), \end{aligned} \tag{2.35}$$

and we used the fact that $V'V = I_r$.

The elements of the matrix H can be evaluated more explicitly as follows: the (j,ℓ)th block of $(I_r \otimes B'V')D'(\Sigma_{\epsilon\epsilon} \otimes \Sigma_{xx})D(I_r \otimes VB)$ is

$$B'V'G_{j\ell}VB = \begin{cases} \sum(\lambda_j^2 + \lambda_i^2)(\lambda_j^2 - \lambda_i^2)^{-2}\beta_i\beta_i' & (j = \ell), \\ -(\lambda_j^2 + \lambda_\ell^2)(\lambda_j^2 - \lambda_\ell^2)^{-2}\beta_\ell\beta_j' & (j \neq \ell), \end{cases}$$

while the (j,ℓ)th block of $(I_r \otimes B'V')D'(\Sigma_{\epsilon\epsilon} \otimes \Sigma_{xx})(\Sigma_{\epsilon\epsilon}^{-1/2}V \otimes \Sigma_{xx}^{-1})$ is

$$B'V'D_j'(\Sigma_{\epsilon\epsilon}^{1/2}V_\ell \otimes I_n) = \begin{cases} \sum(\lambda_j^2 - \lambda_i^2)^{-1}\beta_i\beta_i' & (j = \ell), \\ (\lambda_j^2 - \lambda_\ell^2)^{-1}\beta_\ell\beta_j' & (j \neq \ell), \end{cases}$$

where the sums are over $i = 1, \ldots, r$, $i \neq j$. Combining terms and simplifying, we obtain the results stated in the theorem.

Finally, from (2.33) and (2.34), asymptotic covariances between the elements of $T^{1/2}\,\mathrm{vec}(\hat{A} - A)$ and $T^{1/2}\,\mathrm{vec}(\hat{B}' - B')$ are given by the elements of the matrix

$$(I_r \otimes \Sigma_{\epsilon\epsilon}^{1/2})D'(\Sigma_{\epsilon\epsilon} \otimes \Sigma_{xx})[\, D(I_r \otimes VB) + (\Sigma_{\epsilon\epsilon}^{-1/2}V \otimes \Sigma_{xx}^{-1})\,].$$

The (j,ℓ)th block of this matrix can be determined explicitly in a manner similar to the previous derivation for the blocks of the matrix H, and can be shown to have elements as given in the statement of the theorem.

Robinson (1974) develops the asymptotic theory for estimates of the coefficients in a general multivariate regression model which includes as a special case the reduced-rank regression model (2.5). The results presented here do not agree with the distribution results of Robinson. The difference occurs in the distribution of $T^{1/2}\,\mathrm{vec}(\hat{B}' - B')$. Robinson claims $T^{1/2}\,\mathrm{vec}(\hat{B}' - B')$ has asymptotic covariance matrix $(I_r \otimes \Sigma_{xx}^{-1})$. However, $(I_r \otimes \Sigma_{xx}^{-1})$ is the asymptotic covariance matrix for the estimator $\tilde{B} = V'\Sigma_{\epsilon\epsilon}^{-1/2}\hat{\Sigma}_{yx}\hat{\Sigma}_{xx}^{-1}$ of B which would be appropriate if V, or equivalently A, were known. As can be seen from (2.34) and (2.35), the actual asymptotic covariance matrix of the estimator $\hat{B}$ includes additional contributions besides $(I_r \otimes \Sigma_{xx}^{-1})$, which correspond to the presence of the first term on the right-hand side of (2.34).

When $\mathrm{rank}(C) = 1$, that is, $C = \alpha\beta'$, the asymptotic distributions of the estimators $\hat{\alpha}$, $\hat{\beta}$ implied by Theorem 2.4 agree with those of Robinson

(1974). The results agree in the rank one situation because of the asymptotic independence of $\hat{\alpha}$, $\hat{\beta}$ for this case; the first term on the right-hand side of (2.34) vanishes in this case since $(I_r \otimes B'V')D' = \beta V_1' D_1' = 0$. This asymptotic independence does not hold in general.

The asymptotic theory for the estimators $\hat{A}$ and $\hat{B}$ is useful for inferences concerning whether certain components of X_k contribute in the formation of explanatory indexes BX_k and also whether certain components of Y_k are influenced by certain of these indices. Because of the connection between the component matrices and the quantities involved in the canonical correlation analysis, inferences regarding the latter quantities can also be made via Theorem 2.4.

An additional quantity of interest is the overall regression coefficient matrix $C = AB$ and the asymptotic distribution of the estimator $\hat{C} = \hat{A}\hat{B}$ follows directly from the results of Theorem 2.4 [see Lütkepohl (1993, p. 199)]. It is easily observed, under conditions of Theorem 2.4, that $\hat{C}$ is a consistent estimator of C. The asymptotic distribution of $\hat{C}$ can be derived in the following way. Based on the relation

$$T^{1/2}(\hat{C}-C) = T^{1/2}(\hat{A}\hat{B} - AB) = T^{1/2}(\hat{A}-A)B + T^{1/2}A(\hat{B}-B) + o_p(1),$$

it follows that

$$\begin{aligned} T^{1/2}\operatorname{vec}(\hat{C}' - C') &= T^{1/2}[(I_m \otimes B'), (A \otimes I_n)] \begin{bmatrix} \operatorname{vec}(\hat{A}' - A') \\ \operatorname{vec}(\hat{B}' - B') \end{bmatrix} + o_p(1) \\ &= T^{1/2} M \begin{bmatrix} \operatorname{vec}(\hat{A}' - A') \\ \operatorname{vec}(\hat{B}' - B') \end{bmatrix} + o_p(1) \end{aligned}$$

where $M = [(I_m \otimes B'), (A \otimes I_n)]$. Collecting the terms in Theorem 2.4 for the variances and covariances in the limiting distribution of the components in the vectors $T^{1/2}(\hat{\alpha}_j - \alpha_j)$ and $T^{1/2}(\hat{\beta}_j - \beta_j)$, $j = 1, \ldots, r$, we denote this covariance matrix as

$$W = \lim_{T\to\infty} \operatorname{Cov}\begin{bmatrix} T^{1/2}\operatorname{vec}(\hat{A}' - A') \\ T^{1/2}\operatorname{vec}(\hat{B}' - B') \end{bmatrix} = \begin{bmatrix} W_{11} & W_{12} \\ W_{21} & W_{22} \end{bmatrix},$$

where W_{11} denotes the asymptotic covariance matrix of $T^{1/2}\operatorname{vec}(\hat{A}' - A')$, W_{12} is the asymptotic covariance matrix between $T^{1/2}\operatorname{vec}(\hat{A}' - A')$ and $T^{1/2}\operatorname{vec}(\hat{B}' - B')$, and W_{22} is the asymptotic covariance matrix of $T^{1/2}\operatorname{vec}(\hat{B}' - B')$; for example, $W_{22} = H$, the expression in (2.35) from the proof of Theorem 2.4. It must be noted that care is needed in identifying the elements of W_{11} and W_{12}, because the results in the proof of Theorem 2.4 refer to $\operatorname{vec}(\hat{A} - A)$ not $\operatorname{vec}(\hat{A}' - A')$. The asymptotic distribution of $\hat{C}$ now follows directly as

$$T^{1/2}\operatorname{vec}(\hat{C}' - C') \xrightarrow{d} N(0, MWM'). \tag{2.36}$$

When the rank of the matrix C is assumed to be full, the distribution in (2.36) is equivalent to that in (1.15).

The result in (2.36) for the reduced-rank estimator $\hat{C}$ can be compared with the asymptotic covariance matrix of $T^{1/2}\,\mathrm{vec}(\tilde{C}'-C')$ for the full-rank estimator $\tilde{C}$ (which ignores the reduced-rank structure). This covariance matrix is given by $\Sigma_{\epsilon\epsilon}\otimes\Sigma_{xx}^{-1}$ from Result 1.1 of Section 1.3. For the general rank r case, the covariance matrix in (2.36) does not take a particularly convenient form for comparison with that of the full-rank estimator. However, we consider the special case of $r=1$ for which explicit results are easily obtained. For this case we have already noted that $(I_r\otimes B'V')D'=0$, and so we see from (2.35) that W_{22} reduces to $W_{22}=H=\Sigma_{xx}^{-1}$. It also follows that $W_{12}=0$ and we can find that $W_{11}=\Sigma_{\epsilon\epsilon}^{1/2}(I_m-V_1V_1')\Sigma_{\epsilon\epsilon}^{1/2}(\beta'\Sigma_{xx}\beta)^{-1}$, with $C=\alpha\beta'$. Then using $\alpha=\Sigma_{\epsilon\epsilon}^{1/2}V_1$, it follows that the asymptotic covariance matrix in (2.36) for the rank one case simplifies to

$$\begin{aligned}
MWM' &= \{(\Sigma_{\epsilon\epsilon}-\Sigma_{\epsilon\epsilon}^{1/2}V_1V_1'\Sigma_{\epsilon\epsilon}^{1/2})\otimes\beta(\beta'\Sigma_{xx}\beta)^{-1}\beta'\} \\
&\quad + (\Sigma_{\epsilon\epsilon}^{1/2}V_1V_1'\Sigma_{\epsilon\epsilon}^{1/2}\otimes\Sigma_{xx}^{-1}) \\
&= (\Sigma_{\epsilon\epsilon}\otimes\Sigma_{xx}^{-1}) \\
&\quad - \{(\Sigma_{\epsilon\epsilon}-\Sigma_{\epsilon\epsilon}^{1/2}V_1V_1'\Sigma_{\epsilon\epsilon}^{1/2})\otimes(\Sigma_{xx}^{-1}-\beta(\beta'\Sigma_{xx}\beta)^{-1}\beta')\}.
\end{aligned}$$

This expression can be directly compared with the corresponding covariance matrix expression $\Sigma_{\epsilon\epsilon}\otimes\Sigma_{xx}^{-1}$ of the full-rank estimator, and the reduction in covariance matrix of the reduced-rank estimator is readily revealed for this special case.

One can also look at the effect of reduced-rank estimation on prediction. The prediction error of the predictor $\hat{Y}_0=\tilde{C}X_0$ at X_0 for full-rank LS estimation is $Y_0-\tilde{C}X_0=\epsilon_0-(\tilde{C}-C)X_0$, so that the prediction error covariance matrix is $(1+X_0'(\boldsymbol{X}\boldsymbol{X}')^{-1}X_0)\Sigma_{\epsilon\epsilon}$. In the rank one case, using (2.36) the covariance matrix of the prediction error $Y_0-\hat{C}X_0=\epsilon_0-(\hat{C}-C)X_0\equiv\epsilon_0-(I_m\otimes X_0')\mathrm{vec}(\hat{C}'-C')$ using the reduced-rank estimator $\hat{C}=\hat{\alpha}\hat{\beta}'$ is (approximately)

$$\begin{aligned}
&\Sigma_{\epsilon\epsilon}+\frac{1}{T}(I_m\otimes X_0')\,MWM'\,(I_m\otimes X_0) \\
&\quad = (1+X_0'(\boldsymbol{X}\boldsymbol{X}')^{-1}X_0)\Sigma_{\epsilon\epsilon} \\
&\qquad - \left\{X_0'(\boldsymbol{X}\boldsymbol{X}')^{-1}X_0-\frac{(\beta'X_0)^2}{(\beta'(\boldsymbol{X}\boldsymbol{X}')^{-1}\beta)}\right\}(\Sigma_{\epsilon\epsilon}-\Sigma_{\epsilon\epsilon}^{1/2}V_1V_1'\Sigma_{\epsilon\epsilon}^{1/2}).
\end{aligned}$$

So the prediction error covariance matrix is decreased (by the amount equal to the second term on the right side above) by use of the reduced-rank estimator relative to the full-rank LS estimator.

We now discuss the effect on asymptotic distribution results from the use of the estimator $\Gamma=\tilde{\Sigma}_{\epsilon\epsilon}^{-1}$ in construction of the estimators $\hat{A}$ and $\hat{B}$

in (2.31). Restricting consideration to the rank one case, the perturbation methods applied in the proof of Theorem 2.4 can also be used to obtain the asymptotic covariance matrix of estimators $\hat{\alpha}$ and $\hat{\beta}$ based on $\tilde{\Sigma}_{\epsilon\epsilon}^{-1}$, subject to $\hat{\alpha}'\tilde{\Sigma}_{\epsilon\epsilon}^{-1}\hat{\alpha} = 1$. First, define $\tilde{M} = \tilde{\Sigma}_{\epsilon\epsilon}^{-1/2}\hat{\Sigma}_{yx}\hat{\Sigma}_{xx}^{-1}\hat{\Sigma}_{xy}\tilde{\Sigma}_{\epsilon\epsilon}^{-1/2}$ and let $\tilde{V}_1$ be the normalized eigenvector of $\tilde{M}$ corresponding to its largest eigenvalue $\tilde{\lambda}_1^2$. Similar to (2.32), we have the expansion of $\tilde{V}_1$ around V_1 as

$$\tilde{V}_1 \sim V_1 + \sum_{i=2}^{m} \frac{1}{\lambda_1^2} V_i[V_i'(\tilde{M} - N)V_i] \equiv V_1 + \frac{1}{\lambda_1^2}(I_m - V_1V_1')(\tilde{M} - N)V_1 .$$

Then write $\tilde{M} - N = (\tilde{M} - M) + (M - N)$, where $M - N$ is treated as in the proof of Theorem 2.4, and

$$\begin{aligned}\tilde{M} - M &= \tilde{\Sigma}_{\epsilon\epsilon}^{-1/2}\hat{\Sigma}_{yx}\hat{\Sigma}_{xx}^{-1}\hat{\Sigma}_{xy}\tilde{\Sigma}_{\epsilon\epsilon}^{-1/2} - \Sigma_{\epsilon\epsilon}^{-1/2}\hat{\Sigma}_{yx}\hat{\Sigma}_{xx}^{-1}\hat{\Sigma}_{xy}\Sigma_{\epsilon\epsilon}^{-1/2} \\ &= (\tilde{\Sigma}_{\epsilon\epsilon}^{-1/2} - \Sigma_{\epsilon\epsilon}^{-1/2})\Sigma_{yx}\Sigma_{xx}^{-1}\Sigma_{xy}\Sigma_{\epsilon\epsilon}^{-1/2} \\ &\quad + \Sigma_{\epsilon\epsilon}^{-1/2}\Sigma_{yx}\Sigma_{xx}^{-1}\Sigma_{xy}(\tilde{\Sigma}_{\epsilon\epsilon}^{-1/2} - \Sigma_{\epsilon\epsilon}^{-1/2}) + O_p(T^{-1}).\end{aligned}$$

Also notice that $\Sigma_{yx}\Sigma_{xx}^{-1}\Sigma_{xy} = \alpha\beta'\Sigma_{xx}\beta\alpha' = \lambda_1^2\alpha\alpha'$ in the rank one case.

Because $\alpha = \Sigma_{\epsilon\epsilon}^{1/2}V_1$ and $C = \alpha\beta'$, we have $(I_m - V_1V_1')\Sigma_{\epsilon\epsilon}^{-1/2}C = 0$. Using this and the approximation $\tilde{\Sigma}_{\epsilon\epsilon}^{-1/2} - \Sigma_{\epsilon\epsilon}^{-1/2} = -\Sigma_{\epsilon\epsilon}^{-1/2}(\tilde{\Sigma}_{\epsilon\epsilon}^{1/2} - \Sigma_{\epsilon\epsilon}^{1/2})\Sigma_{\epsilon\epsilon}^{-1/2} + O_p(T^{-1})$, we obtain a representation analogous to (2.33) as

$$\begin{aligned}T^{1/2}(\tilde{V}_1 - V_1) &= \frac{1}{\lambda_1^2}(I_m - V_1V_1')\{\Sigma_{\epsilon\epsilon}^{-1/2}T^{1/2}U_T'\beta + \lambda_1^2T^{1/2}(\tilde{\Sigma}_{\epsilon\epsilon}^{-1/2} - \Sigma_{\epsilon\epsilon}^{-1/2})\alpha\} \\ &= \frac{1}{\lambda_1^2}[(I_m - V_1V_1')\Sigma_{\epsilon\epsilon}^{-1/2} \otimes \beta']\, T^{1/2}\text{vec}(U_T) \\ &\quad - (I_m - V_1V_1')\Sigma_{\epsilon\epsilon}^{-1/2}T^{1/2}(\tilde{\Sigma}_{\epsilon\epsilon}^{1/2} - \Sigma_{\epsilon\epsilon}^{1/2})\Sigma_{\epsilon\epsilon}^{-1/2}\alpha + O_p(T^{-1/2}).\end{aligned}$$

The estimator of $\alpha = \Sigma_{\epsilon\epsilon}^{1/2}V_1$ is $\hat{\alpha} = \tilde{\Sigma}_{\epsilon\epsilon}^{1/2}\tilde{V}_1$, so that $\hat{\alpha} - \alpha = \Sigma_{\epsilon\epsilon}^{1/2}(\tilde{V}_1 - V_1) + (\tilde{\Sigma}_{\epsilon\epsilon}^{1/2} - \Sigma_{\epsilon\epsilon}^{1/2})V_1 + O_p(T^{-1})$. Therefore, from above we have

$$\begin{aligned}T^{1/2}(\hat{\alpha} - \alpha) &= \frac{1}{\lambda_1^2}[\Sigma_{\epsilon\epsilon}^{1/2}(I_m - V_1V_1')\Sigma_{\epsilon\epsilon}^{-1/2} \otimes \beta']\, T^{1/2}\text{vec}(U_T) \\ &\quad + \alpha\alpha'\Sigma_{\epsilon\epsilon}^{-1}T^{1/2}(\tilde{\Sigma}_{\epsilon\epsilon}^{1/2} - \Sigma_{\epsilon\epsilon}^{1/2})\Sigma_{\epsilon\epsilon}^{-1/2}\alpha + O_p(T^{-1/2}) \\ &= \frac{1}{\lambda_1^2}[\Sigma_{\epsilon\epsilon}^{1/2}(I_m - V_1V_1')\Sigma_{\epsilon\epsilon}^{-1/2} \otimes \beta']\, T^{1/2}\text{vec}(U_T) \\ &\quad + \frac{1}{2}(\alpha'\Sigma_{\epsilon\epsilon}^{-1} \otimes \alpha\alpha'\Sigma_{\epsilon\epsilon}^{-1})\, T^{1/2}\text{vec}(\tilde{\Sigma}_{\epsilon\epsilon} - \Sigma_{\epsilon\epsilon}),\end{aligned} \qquad (2.37)$$

where we have used the relation $(\tilde{\Sigma}_{\epsilon\epsilon}^{1/2} - \Sigma_{\epsilon\epsilon}^{1/2})\Sigma_{\epsilon\epsilon}^{1/2} + \Sigma_{\epsilon\epsilon}^{1/2}(\tilde{\Sigma}_{\epsilon\epsilon}^{1/2} - \Sigma_{\epsilon\epsilon}^{1/2}) = \tilde{\Sigma}_{\epsilon\epsilon} - \Sigma_{\epsilon\epsilon} + O_p(T^{-1})$.

From Muirhead (1982, Chap. 3) and Schott (1997, Chap. 9), $T^{1/2}\text{vec}(\tilde{\Sigma}_{\epsilon\epsilon} - \Sigma_{\epsilon\epsilon})$ has asymptotic covariance matrix equal to $2D_m^*(\Sigma_{\epsilon\epsilon} \otimes \Sigma_{\epsilon\epsilon})D_m^*$, where $D_m^* = D_m(D_m D_m')^{-1}D_m'$ and D_m is the $m^2 \times \frac{1}{2}m(m+1)$ duplication matrix such that $D_m \text{vech}(\Sigma_{\epsilon\epsilon}) = \text{vec}(\Sigma_{\epsilon\epsilon})$. Here, $\text{vech}(\Sigma_{\epsilon\epsilon})$ represents the $\frac{1}{2}m(m+1) \times 1$ vector consisting of the distinct elements of $\Sigma_{\epsilon\epsilon}$ (i.e., those on and below the diagonal). In addition, we know that $T^{1/2}\text{vec}(U_T)$ converges in distribution to $N(0, \Sigma_{\epsilon\epsilon} \otimes \Sigma_{xx})$, and U_T and $\tilde{\Sigma}_{\epsilon\epsilon}$ are asymptotically independent. These results can be used in (2.37) to obtain the asymptotic covariance matrix of $\hat{\alpha}$ as

$$\begin{aligned} T\text{Cov}(\hat{\alpha} - \alpha) &= \left(\frac{1}{\lambda_1^2}\right)^2 [\Sigma_{\epsilon\epsilon}^{1/2}(I_m - V_1V_1')\Sigma_{\epsilon\epsilon}^{-1/2} \otimes \beta'](\Sigma_{\epsilon\epsilon} \otimes \Sigma_{xx}) \\ &\quad \times [\Sigma_{\epsilon\epsilon}^{-1/2}(I_m - V_1V_1')\Sigma_{\epsilon\epsilon}^{1/2} \otimes \beta] + \frac{1}{4}(\alpha'\Sigma_{\epsilon\epsilon}^{-1} \otimes \alpha\alpha'\Sigma_{\epsilon\epsilon}^{-1}) \\ &\quad \times 2D_m^*(\Sigma_{\epsilon\epsilon} \otimes \Sigma_{\epsilon\epsilon})D_m^*(\Sigma_{\epsilon\epsilon}^{-1}\alpha \otimes \Sigma_{\epsilon\epsilon}^{-1}\alpha\alpha') \\ &= \frac{1}{\lambda_1^2}\Sigma_{\epsilon\epsilon}^{1/2}(I_m - V_1V_1')\Sigma_{\epsilon\epsilon}^{1/2} + \frac{1}{2}\alpha\alpha', \end{aligned}$$

since $D_m^*(\Sigma_{\epsilon\epsilon}^{-1}\alpha \otimes \Sigma_{\epsilon\epsilon}^{-1}\alpha) = \Sigma_{\epsilon\epsilon}^{-1}\alpha \otimes \Sigma_{\epsilon\epsilon}^{-1}\alpha$.

Next we consider the estimator $\hat{\beta}' = \tilde{V}_1'\tilde{\Sigma}_{\epsilon\epsilon}^{-1/2}\hat{\Sigma}_{yx}\hat{\Sigma}_{xx}^{-1} = \hat{\alpha}'\tilde{\Sigma}_{\epsilon\epsilon}^{-1}\hat{\Sigma}_{yx}\hat{\Sigma}_{xx}^{-1}$. From the approximation $\tilde{\Sigma}_{\epsilon\epsilon}^{-1}\hat{\alpha} - \Sigma_{\epsilon\epsilon}^{-1}\alpha = \Sigma_{\epsilon\epsilon}^{-1}(\hat{\alpha} - \alpha) + (\tilde{\Sigma}_{\epsilon\epsilon}^{-1} - \Sigma_{\epsilon\epsilon}^{-1})\alpha + O_p(T^{-1})$, we obtain the representation

$$\begin{aligned} T^{1/2}(\hat{\beta} - \beta) &= \Sigma_{xx}^{-1}\Sigma_{xy}\{\Sigma_{\epsilon\epsilon}^{-1}T^{1/2}(\hat{\alpha} - \alpha) + T^{1/2}(\tilde{\Sigma}_{\epsilon\epsilon}^{-1} - \Sigma_{\epsilon\epsilon}^{-1})\alpha\} \\ &\quad + T^{1/2}(\hat{\Sigma}_{xx}^{-1}\hat{\Sigma}_{xy} - \Sigma_{xx}^{-1}\Sigma_{xy})\Sigma_{\epsilon\epsilon}^{-1}\alpha + O_p(T^{-1/2}) \\ &= \Sigma_{xx}^{-1}\Sigma_{xy}\{\Sigma_{\epsilon\epsilon}^{-1}T^{1/2}(\hat{\alpha} - \alpha) - \Sigma_{\epsilon\epsilon}^{-1}T^{1/2}(\tilde{\Sigma}_{\epsilon\epsilon} - \Sigma_{\epsilon\epsilon})\Sigma_{\epsilon\epsilon}^{-1}\alpha\} \\ &\quad + T^{1/2}(\hat{\Sigma}_{xx}^{-1}\hat{\Sigma}_{xy} - \Sigma_{xx}^{-1}\Sigma_{xy})\Sigma_{\epsilon\epsilon}^{-1}\alpha, \end{aligned}$$

using the approximation $\tilde{\Sigma}_{\epsilon\epsilon}^{-1} - \Sigma_{\epsilon\epsilon}^{-1} = -\Sigma_{\epsilon\epsilon}^{-1}(\tilde{\Sigma}_{\epsilon\epsilon} - \Sigma_{\epsilon\epsilon})\Sigma_{\epsilon\epsilon}^{-1} + O_p(T^{-1})$. Thus, from the approximation for $T^{1/2}(\hat{\alpha} - \alpha)$ given in (2.37) we have

$$\begin{aligned} T^{1/2}(\hat{\beta} - \beta) &= -\frac{1}{2}\beta\alpha'\Sigma_{\epsilon\epsilon}^{-1}T^{1/2}(\tilde{\Sigma}_{\epsilon\epsilon} - \Sigma_{\epsilon\epsilon})\Sigma_{\epsilon\epsilon}^{-1}\alpha \\ &\quad + T^{1/2}(\hat{\Sigma}_{xx}^{-1}\hat{\Sigma}_{xy} - \Sigma_{xx}^{-1}\Sigma_{xy})\Sigma_{\epsilon\epsilon}^{-1}\alpha + O_p(T^{-1/2}) \\ &= -\frac{1}{2}(\alpha'\Sigma_{\epsilon\epsilon}^{-1} \otimes \beta\alpha'\Sigma_{\epsilon\epsilon}^{-1})\, T^{1/2}\text{vec}(\tilde{\Sigma}_{\epsilon\epsilon} - \Sigma_{\epsilon\epsilon}) \\ &\quad + (\alpha'\Sigma_{\epsilon\epsilon}^{-1} \otimes \Sigma_{xx}^{-1})\, T^{1/2}\text{vec}(U_T). \end{aligned} \tag{2.38}$$

This representation yields the asymptotic covariance matrix for $\hat{\beta}$ as

$$\begin{aligned} T\mathrm{Cov}(\hat{\beta}-\beta) &= (\alpha'\Sigma_{\epsilon\epsilon}^{-1} \otimes \Sigma_{xx}^{-1})(\Sigma_{\epsilon\epsilon} \otimes \Sigma_{xx})(\Sigma_{\epsilon\epsilon}^{-1}\alpha \otimes \Sigma_{xx}^{-1}) \\ &\quad + \frac{1}{4}(\alpha'\Sigma_{\epsilon\epsilon}^{-1} \otimes \beta\alpha'\Sigma_{\epsilon\epsilon}^{-1}) 2D_m^*(\Sigma_{\epsilon\epsilon} \otimes \Sigma_{\epsilon\epsilon}) D_m^*(\Sigma_{\epsilon\epsilon}^{-1}\alpha \otimes \Sigma_{\epsilon\epsilon}^{-1}\alpha\beta') \\ &= \Sigma_{xx}^{-1} + \frac{1}{2}\beta\beta'. \end{aligned}$$

Furthermore, the asymptotic cross-covariance matrix between $\hat{\alpha}$ and $\hat{\beta}$ is also obtained from the representations (2.37) and (2.38) as

$$\begin{aligned} T\mathrm{Cov}(\hat{\alpha}-\alpha, \hat{\beta}-\beta) &= \frac{1}{\lambda_1^2}[\Sigma_{\epsilon\epsilon}^{1/2}(I_m - V_1V_1')\Sigma_{\epsilon\epsilon}^{-1/2} \otimes \beta'](\Sigma_{\epsilon\epsilon} \otimes \Sigma_{xx}) \\ &\quad \times (\Sigma_{\epsilon\epsilon}^{-1}\alpha \otimes \Sigma_{xx}^{-1}) - \frac{1}{4}(\alpha'\Sigma_{\epsilon\epsilon}^{-1} \otimes \alpha\alpha'\Sigma_{\epsilon\epsilon}^{-1}) \\ &\quad \times 2D_m^*(\Sigma_{\epsilon\epsilon} \otimes \Sigma_{\epsilon\epsilon}) D_m^*(\Sigma_{\epsilon\epsilon}^{-1}\alpha \otimes \Sigma_{\epsilon\epsilon}^{-1}\alpha\beta') \\ &= \frac{1}{\lambda_1^2}[\Sigma_{\epsilon\epsilon}^{1/2}(I_m - V_1V_1')\Sigma_{\epsilon\epsilon}^{-1/2}\alpha \otimes \beta'] - \frac{1}{2}\alpha\beta' = -\frac{1}{2}\alpha\beta'. \end{aligned}$$

Therefore, we find that $T^{1/2}\{(\hat{\alpha}-\alpha)', (\hat{\beta}-\beta)'\}'$ has asymptotic covariance matrix given by

$$W = \begin{bmatrix} W_{11} & W_{12} \\ W_{21} & W_{22} \end{bmatrix} = \begin{bmatrix} (\Sigma_{\epsilon\epsilon} - \alpha\alpha')(\beta'\Sigma_{xx}\beta)^{-1} + \frac{1}{2}\alpha\alpha' & -\frac{1}{2}\alpha\beta' \\ -\frac{1}{2}\beta\alpha' & \Sigma_{xx}^{-1} + \frac{1}{2}\beta\beta' \end{bmatrix}.$$

Compared to the previous case where $\Gamma = \Sigma_{\epsilon\epsilon}^{-1}$ is taken as known, the above includes the additional term $\frac{1}{2}\delta\delta'$ with $\delta = (\alpha', -\beta')'$. However, it is also readily seen that the asymptotic covariance matrix, MWM' where $M = [I_m \otimes \beta, \alpha \otimes I_n]$, of the overall regression coefficient matrix estimator $\hat{C} = \hat{\alpha}\hat{\beta}'$ is not affected by use of $\tilde{\Sigma}_{\epsilon\epsilon}^{-1}$ instead of $\Sigma_{\epsilon\epsilon}^{-1}$, since $M\delta = (\alpha \otimes \beta) - (\alpha \otimes \beta) = 0$. Therefore, whereas the asymptotic distribution of the individual component estimators $\hat{\alpha}$ and $\hat{\beta}$ is affected by use of $\tilde{\Sigma}_{\epsilon\epsilon}^{-1}$ in place of $\Sigma_{\epsilon\epsilon}^{-1}$, that of the overall estimator $\hat{C} = \hat{\alpha}\hat{\beta}'$ is not changed.

2.6 Identification of Rank of the Regression Coefficient Matrix

It is obviously important to be able to determine the rank of the coefficient matrix C for it is a key element in the structure of the reduced-rank regression problem. The asymptotic results are derived assuming that the rank condition is true. However, the relationship between canonical correlation analysis and reduced-rank regression allows one to check on the rank by testing if certain correlations are zero.

Bartlett (1947) suggested $T\sum_{j=r+1}^{m}\log(1+\hat{\lambda}_j^2)$ as the appropriate statistic for testing the significance of the last $(m-r)$ canonical correlations, where $\hat{\lambda}_j = \hat{\rho}_j/(1-\hat{\rho}_j^2)^{1/2}$ and $\hat{\rho}_j$ is the jth largest sample canonical correlation between the Y_k and X_k. From the discussion presented in Section 2.4, it follows that if the hypothesis that the last $(m-r)$ population canonical correlations are zero is true, then it is equivalent to assuming that rank$(C) = r$ (strictly, that rank$(C) \leq r$). Bartlett's statistic follows from the likelihood ratio method of test construction [see Anderson (1951)], as we now indicate.

Under the likelihood ratio (LR) testing approach, it is easy to see, similar to the results presented at the end of Section 1.3, that the LR test statistic for testing rank$(C) = r$ is $\lambda = U^{T/2}$, where $U = |S|/|S_1|$,

$$S = (\boldsymbol{Y} - \tilde{C}\boldsymbol{X})(\boldsymbol{Y} - \tilde{C}\boldsymbol{X})', \qquad S_1 = (\boldsymbol{Y} - \hat{C}^{(r)}\boldsymbol{X})(\boldsymbol{Y} - \hat{C}^{(r)}\boldsymbol{X})',$$

S is the residual sum of squares matrix from fitting the full-rank regression model, while S_1 is the residual sum of squares matrix from fitting the model under the hypothesis of the rank condition on C. It must be noted that $\tilde{C}$ and $\hat{C}^{(r)}$ are related as $\hat{C}^{(r)} = P\tilde{C}$ where P is an idempotent matrix of rank r. Specifically, recall from Sections 2.3 and 2.4 that the ML estimates under the null hypothesis of rank$(C) = r$ are $\hat{A}^{(r)} = \tilde{\Sigma}_{\epsilon\epsilon}^{1/2}\hat{V}$, $\hat{B}^{(r)} = \hat{V}'\tilde{\Sigma}_{\epsilon\epsilon}^{-1/2}\hat{\Sigma}_{yx}\hat{\Sigma}_{xx}^{-1}$, with $\hat{C}^{(r)} = \hat{A}^{(r)}\hat{B}^{(r)}$, where $\hat{V} = [\hat{V}_1, \ldots, \hat{V}_r]$ and the $\hat{V}_j$ are (normalized) eigenvectors of the matrix $\hat{R} = \tilde{\Sigma}_{\epsilon\epsilon}^{-1/2}\hat{\Sigma}_{yx}\hat{\Sigma}_{xx}^{-1}\hat{\Sigma}_{xy}\tilde{\Sigma}_{\epsilon\epsilon}^{-1/2}$ associated with the r largest eigenvalues $\hat{\lambda}_j^2$, and $\hat{\Sigma}_{\epsilon\epsilon} = (1/T)S_1 = (1/T)(\boldsymbol{Y} - \hat{C}^{(r)}\boldsymbol{X})(\boldsymbol{Y} - \hat{C}^{(r)}\boldsymbol{X})'$. The unrestricted ML estimates are, of course, $\tilde{C} = \hat{\Sigma}_{yx}\hat{\Sigma}_{xx}^{-1}$ and $\tilde{\Sigma}_{\epsilon\epsilon} = (1/T)(\boldsymbol{Y} - \tilde{C}\boldsymbol{X})(\boldsymbol{Y} - \tilde{C}\boldsymbol{X})'$ as derived in Section 1.2. Notice that, since $\sum_{j=1}^{m}\hat{V}_j\hat{V}_j' = I_m$, we have

$$\begin{aligned}
\tilde{C} - \hat{C}^{(r)} &= (I_m - \tilde{\Sigma}_{\epsilon\epsilon}^{1/2}\hat{V}\hat{V}'\tilde{\Sigma}_{\epsilon\epsilon}^{-1/2})\hat{\Sigma}_{yx}\hat{\Sigma}_{xx}^{-1} \\
&= \tilde{\Sigma}_{\epsilon\epsilon}^{1/2}\left[\sum_{j=r+1}^{m}\hat{V}_j\hat{V}_j'\right]\tilde{\Sigma}_{\epsilon\epsilon}^{-1/2}\hat{\Sigma}_{yx}\hat{\Sigma}_{xx}^{-1}.
\end{aligned}$$

Therefore, we find that

$$\begin{aligned}
\frac{1}{T}S_1 &= \frac{1}{T}(\boldsymbol{Y} - \tilde{C}\boldsymbol{X} + (\tilde{C} - \hat{C}^{(r)})\boldsymbol{X})(\boldsymbol{Y} - \tilde{C}\boldsymbol{X} + (\tilde{C} - \hat{C}^{(r)})\boldsymbol{X})' \\
&= \tilde{\Sigma}_{\epsilon\epsilon} + \tilde{\Sigma}_{\epsilon\epsilon}^{1/2}\left[\sum_{j=r+1}^{m}\hat{V}_j\hat{V}_j'\right]\hat{R}\left[\sum_{j=r+1}^{m}\hat{V}_j\hat{V}_j'\right]\tilde{\Sigma}_{\epsilon\epsilon}^{1/2} \\
&= \tilde{\Sigma}_{\epsilon\epsilon} + \tilde{\Sigma}_{\epsilon\epsilon}^{1/2}\left[\sum_{j=r+1}^{m}\hat{\lambda}_j^2\hat{V}_j\hat{V}_j'\right]\tilde{\Sigma}_{\epsilon\epsilon}^{1/2},
\end{aligned} \tag{2.39}$$

where the $\hat{\lambda}_j^2$, $j = r+1, \ldots, m$, are the $(m-r)$ smallest eigenvalues of $\hat{R}$. Hence, using (2.39) the LR statistic is

$$\lambda = \left(\frac{|S_1|}{|S|}\right)^{-T/2} = \left(\frac{|(1/T)S_1|}{|\tilde{\Sigma}_{\epsilon\epsilon}|}\right)^{-T/2}$$

$$= \left(\frac{|\tilde{\Sigma}_{\epsilon\epsilon}|\,|I_m + \sum_{j=r+1}^{m} \hat{\lambda}_j^2 \hat{V}_j \hat{V}_j'|}{|\tilde{\Sigma}_{\epsilon\epsilon}|}\right)^{-T/2} = [\ \prod_{j=r+1}^{m}(1+\hat{\lambda}_j^2)\ ]^{-T/2}.$$

Therefore, the criterion $\lambda = U^{T/2}$ is such that

$$-2\log(\lambda) = T \sum_{j=r+1}^{m} \log(1+\hat{\lambda}_j^2) = -T \sum_{j=r+1}^{m} \log(1-\hat{\rho}_j^2), \qquad (2.40)$$

since $1+\hat{\lambda}_j^2 = 1/(1-\hat{\rho}_j^2)$. This statistic, $-2\log(\lambda)$, follows asymptotically the $\chi^2_{(m-r)(n-r)}$ distribution under the null hypothesis [see Anderson (1951, Theorem 3)]. The test statistic is asymptotically equivalent to $T\sum_{j=r+1}^{m} \hat{\lambda}_j^2 \equiv T\mathrm{tr}\{\tilde{\Sigma}_{\epsilon\epsilon}^{-1}(\tilde{C}-\hat{C}^{(r)})\hat{\Sigma}_{xx}(\tilde{C}-\hat{C}^{(r)})'\}$, which follows the above χ^2 distribution and this result was obtained independently by Hsu (1941). A simple correction factor for the LR statistic in (2.40), to improve the approximation to the $\chi^2_{(m-r)(n-r)}$ distribution, is given by $-2\{[T-n+(n-m-1)/2]/T\}\log(\lambda) = -[T-n+(n-m-1)/2]\sum_{j=r+1}^{m}\log(1-\hat{\rho}_j^2)$. This is similar to the corrected form used in (1.16) for the LR test in the classical full-rank model.

Thus, to specify the appropriate rank of the matrix C we use the test statistic $\mathcal{M} = -[T-(m+n+1)/2]\sum_{j=r+1}^{m}\log(1-\hat{\rho}_j^2)$, for $r = 0, 1, \ldots, m-1$, and reject H_0 : rank$(C) = r$ when $\mathcal{M}$ is greater than an upper critical value determined by the $\chi^2_{(m-r)(n-r)}$ distribution. The smallest value of r for which H_0 is not rejected provides a reasonable choice for the rank. Additional guidance in selecting the appropriate rank could be obtained by considering the LR test of H_0 : rank$(C) = r$ versus the more refined alternative hypothesis H_1 : rank$(C) = r+1$. This test statistic is given by $-T\log(1-\hat{\rho}_{r+1}^2)$. An alternate procedure for testing the rank of a matrix in more general settings has recently been proposed by Cragg and Donald (1996), based on the LDU-decomposition of the unrestricted estimate of the matrix. Other useful tools for the specification of the rank include the use of information-theoretic model selection criteria, such as the AIC criterion of Akaike (1974) and the BIC criterion of Schwarz (1978), or cross-validation methods (Stone, 1974) based on measures of the predictive performance of models of various ranks.

Remark–Relation of LR Test Between Known and Unknown Constraints. We briefly compare the results relating to the LR test for specified constraints on the regression coefficient matrix C (see Section 1.3) with the above LR test results for the rank of C, that is, for a specified number of constraints, but where these constraints are not known a priori. From Section 1.3, the LR test of $H_0 : F_1 C = 0$, where F_1 is a specified (known) $(m-r) \times m$ full-rank matrix, is obtained as $\mathcal{M} = [(T-n) + \frac{1}{2}(n - m_1 - 1)] \sum_{i=1}^{m_1} \log(1 + \hat{\lambda}_i^2)$, with $m_1 = m-r$, where $\hat{\lambda}_i^2$ are the $m-r$ eigenvalues of $S_*^{-1} H_* = (F_1 S F_1')^{-1} F_1 \tilde{C} \boldsymbol{X}\boldsymbol{X}' \tilde{C}' F_1' \equiv (F_1 \tilde{\Sigma}_{\epsilon\epsilon} F_1')^{-1} F_1 \tilde{C} \hat{\Sigma}_{xx} \tilde{C}' F_1'$. The asymptotic distribution of $\mathcal{M}$ under H_0 is $\chi^2_{(m-r)n}$. In the case where F_1 is unknown, $F_1 C = 0$ (for some F_1) is equivalent to the hypothesis that the rank of C is r. As has been shown in this section, the LR test for this case is given by $\mathcal{M} = [(T-n) + \frac{1}{2}(n - m - 1)] \sum_{j=r+1}^{m} \log(1 + \hat{\lambda}_j^2)$, where $\hat{\lambda}_j^2$ are the $m-r$ smallest eigenvalues of $\hat{R} = \tilde{\Sigma}_{\epsilon\epsilon}^{-1/2} \tilde{C} \hat{\Sigma}_{xx} \tilde{C}' \tilde{\Sigma}_{\epsilon\epsilon}^{-1/2}$, and the asymptotic distribution of $\mathcal{M}$ is $\chi^2_{(m-r)(n-r)}$. In addition, the corresponding ML estimate of the unknown constraints matrix $F_1 = [\ell_{r+1}, \ldots, \ell_m]'$ in $F_1 C = 0$ is given by $\hat{F}_1 = [\hat{V}_{r+1}, \ldots, \hat{V}_m]' \tilde{\Sigma}_{\epsilon\epsilon}^{-1/2}$, where $\hat{V}_j$ are the normalized eigenvectors of $\hat{R}$ associated with the $\hat{\lambda}_j^2$. Hence we see a strong similarity in the form of the LR test statistic results, where in the case of unknown constraints the LR statistic can be viewed as taking the same form as in the known constraints case, but with the constraints estimated by ML to have a particular form. (Note that if the ML estimate $\hat{F}_1$ is substituted for F_1 in the expression $S_*^{-1} H_*$ for the case of known F_1, this expression reduces to $[\hat{V}_{r+1}, \ldots, \hat{V}_m]' \tilde{\Sigma}_{\epsilon\epsilon}^{-1/2} \tilde{C} \hat{\Sigma}_{xx} \tilde{C}' \tilde{\Sigma}_{\epsilon\epsilon}^{-1/2} [\hat{V}_{r+1}, \ldots, \hat{V}_m]$.) Because the constraints are estimated, the degrees of freedom of $(m-r)n$ in the known constraints case are adjusted to $(m-r)(n-r)$ in the case of unknown constraints, a decrease of $(m-r)r$. This reflects the increase in the number of unknown (unconstrained) parameters from rn in the known constraints case to $r(m+n-r)$ in the unknown constraints case.

The correspondence between correlation and regression is known to be fully exploited in the modeling of data in the case of multiple regression. The results above demonstrate the usefulness of the canonical correlations as descriptive statistics in aiding the multivariate regression modeling of large data sets. However, we shall demonstrate in future chapters that such a neat connection does not always appear to hold in such an explicit way for more complex models, but nevertheless canonical correlation methods can still be used as descriptive tools and in model specification for these more complicated situations.

2.7 Numerical Example Using Biochemical Data

We illustrate the reduced-rank regression methods developed in this chapter using the example on biochemical data that was examined in Section 1.5. As indicated in the analysis presented in Section 1.5, these data suggest that there may be a reduced-rank structure for the regression coefficient matrix (excluding the constant term) of the predictor variables

$X_k = (x_{1k}, x_{2k}, x_{3k})'$. To examine this feature, the rank of the coefficient matrix first needs to be determined. The LR test statistic $\mathcal{M} = -[(T-1) - (m+n+1)/2] \sum_{j=r+1}^{m} \log(1 - \hat{\rho}_j^2)$ with $n = 3$, discussed in relation to (2.40), is used for $r = 0, 1, 2$, to test the rank, that is, to test H_0 : rank $\leq r$, and the results are given in Table 2.2. It would appear from these results that the possibility that the rank is either one or two could be entertained. Only estimation results for the rank 2 situation will be presented in detail here.

Table 2.2 Summary of results for LR tests on rank of coefficient matrix for biochemical data.

Eigenvalues $\hat{\lambda}_j^2$	Canonical Correlations $\hat{\rho}_j$	Rank r	$\mathcal{M}$ = LR Statistic	d.f.	Critical Value (5%)
4.121	0.897		56.89	15	24.99
0.519	0.584	1	11.98	8	15.51
0.018	0.132	2	0.49	3	7.81

To obtain the maximum likelihood estimates, we find the normalized eigenvectors of the matrix $\hat{R} = \tilde{\Sigma}_{\epsilon\epsilon}^{-1/2} \hat{\Sigma}_{yx} \hat{\Sigma}_{xx}^{-1} \hat{\Sigma}_{xy} \tilde{\Sigma}_{\epsilon\epsilon}^{-1/2}$ associated with the two largest eigenvalues, $\hat{\lambda}_1^2$ and $\hat{\lambda}_2^2$, in Table 2.2. (Note that in constructing the sample covariance matrices $\hat{\Sigma}_{xx}$ and $\hat{\Sigma}_{xy}$, the variables Y_k and X_k are adjusted for sample means.) The normalized eigenvectors are

$$\hat{V}_1' = (-0.290,\ 0.234,\ 0.025,\ 0.906,\ 0.199)$$

and

$$\hat{V}_2' = (0.425,\ 0.269,\ -0.841,\ 0.047,\ 0.194),$$

respectively. We then compute the ML estimates from $\hat{A}^{(2)} = \tilde{\Sigma}_{\epsilon\epsilon}^{1/2} \hat{V}_{(2)}$ and $\hat{B}^{(2)} = \hat{V}_{(2)}' \tilde{\Sigma}_{\epsilon\epsilon}^{-1/2} \hat{\Sigma}_{yx} \hat{\Sigma}_{xx}^{-1}$, with $\hat{V}_{(2)} = [\hat{V}_1, \hat{V}_2]$. Using the results of Theorem 2.4, the asymptotic standard errors of the elements of these matrix estimates are also obtained. These ML estimates and associated standard errors are given below as

$$\hat{A}' = \begin{bmatrix} -1.085 & 0.333 & 0.231 & 0.397 & 1.109 \\ (0.269) & (0.043) & (0.041) & (0.013) & (0.439) \\ 1.335 & -0.126 & -0.266 & -0.092 & 0.875 \\ (0.655) & (0.120) & (0.086) & (0.050) & (1.216) \end{bmatrix}$$

$$\hat{B} = \begin{bmatrix} 1.361 & -1.326 & 1.190 \\ (0.348) & (0.218) & (0.344) \\ -0.908 & 0.494 & 1.240 \\ (0.297) & (0.176) & (0.332) \end{bmatrix}$$

To appreciate the recovery of information by using only two linear combinations $\hat{B}X_k$ of $X_k = (x_{1k}, x_{2k}, x_{3k})'$ to predict Y_k, the reduced-rank matrix $\hat{C} = \hat{A}\hat{B}$ can be compared with the full-rank LS estimate $\tilde{C}$ given in Section 1.5. Notice again that we are excluding the first column of intercepts from this comparison. The reduced-rank estimate of the regression coefficient matrix is

$$\hat{C} = \begin{bmatrix} -2.6893 & 2.0981 & 0.3649 \\ 0.5679 & -0.5040 & 0.2400 \\ 0.5558 & -0.4374 & -0.0555 \\ 0.6248 & -0.5726 & 0.3583 \\ 0.7142 & -1.0379 & 2.4044 \end{bmatrix}$$

Most of the coefficients of $\tilde{C}$ that were found to be statistically significant are recovered in the reduced-rank estimate $\hat{C}$ fairly well. The (approximate) unbiased estimate of the error covariance matrix from the reduced-rank analysis (with correlations shown above the diagonal) is given as

$$\overline{\Sigma}_{\epsilon\epsilon} = \frac{1}{33-3}\hat{\epsilon}\hat{\epsilon}' = \begin{bmatrix} 10.8813 & -0.4243 & -0.3090 & -0.3243 & -0.0841 \\ -0.8638 & 0.3808 & 0.8952 & 0.6743 & 0.3452 \\ -0.5204 & 0.2820 & 0.2606 & 0.6592 & 0.2912 \\ -0.4743 & 0.1845 & 0.1492 & 0.1966 & 0.0320 \\ -1.5012 & 1.1527 & 0.8045 & 0.0768 & 29.2890 \end{bmatrix}$$

where the residuals are obtained from the LS regression of Y_k on $\hat{B}X_k$ with intercept included. The diagonal elements of the above matrix, when compared to the entries of the unbiased estimate of the error covariance matrix under the full-rank LS regression (given in Section 1.5), indicate that there is little or no loss in fit after discarding one linear combination of the predictor variables.

For comparison purposes, corresponding ML results when the rank of the coefficient matrix is assumed to be one can be obtained by using as regressor only the (single) linear combination of predictor variables formed from the first row of $\hat{B}$. The diagonal elements of the resulting (approximate) unbiased estimate of the error covariance matrix from the rank one fit are given by 11.5149, 0.3773, 0.2913, 0.1949, and 28.7673. These values are very close to the residual variance estimates of both the rank 2 and full-rank models, suggesting that the rank one model could also provide an acceptable fit. Note that the single index or predictive factor in the estimated rank one model is roughly proportional to $x_1^* \approx (x_1 + x_3) - x_2$. Although a physical or scientific interpretation for this index is unclear, it does have statistical predictive importance. For illustration, Figure 2.3 displays scatter plots of each of the response variables y_i versus the first predictive index $x_1^* = \hat{\beta}_1' X$, with estimated regression line. It is seen that variables y_2, y_3, and y_4 have quite strong relationship with x_1^*, whereas the relationship with x_1^* is much weaker for y_5 and is of opposite sign for y_1.

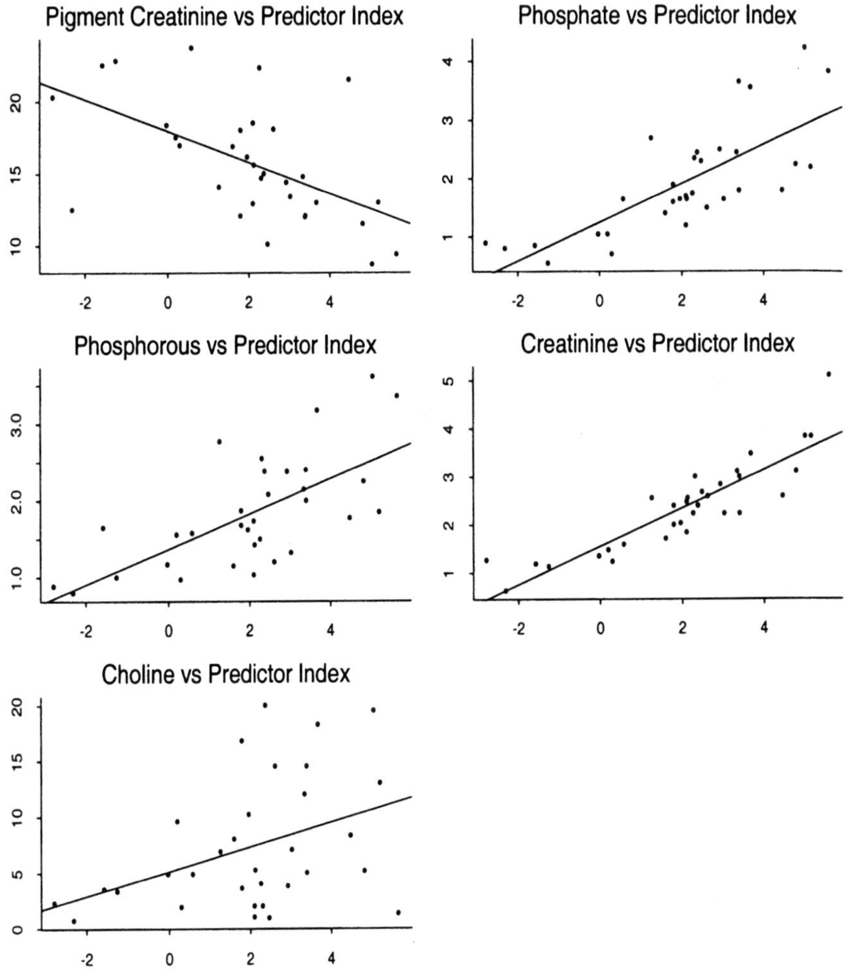

Figure 2.3. Scatter plots of response variables $y_1, \ldots, y_5$ for biochemical data versus first predictive variable index x_1^*

3

Reduced-Rank Regression Models With Two Sets of Regressors

3.1 Reduced-Rank Model of Anderson

In Chapter 2 we have demonstrated the utility of the reduced-rank model for analyzing data on a large number of variables. The interrelationship between the dependent and independent variables can be explained parsimoniously through the assumption of a lower rank for the regression coefficient matrix. The basic model that was described in Chapter 2 assumes that the predictor variables are all grouped into one set and therefore, they are all subject to the same canonical calculations. In this chapter we broaden the scope of the reduced-rank model by entertaining the possibility that the predictor variables can be divided into two distinct sets with separate reduced-rank structures. Such an extension will be shown to have some interesting applications.

An extension of the basic model to include more than one set of regressors was already suggested in the seminal work of Anderson (1951). Anderson considered the following model where the set of dependent variables is influenced by two sets of regressor variables, one having a reduced-rank coefficient matrix and the other having a full-rank coefficient matrix. Formally let

$$Y_k = CX_k + DZ_k + \epsilon_k, \qquad k = 1, \ldots, T, \tag{3.1}$$

where X_k is a vector of regressor variables, and C is of reduced rank, Z_k is a vector of additional variables that influence Y_k but the matrix D is of full rank.

This model corresponds to a latent structure as mentioned by Robinson (1974). The set of regressor variables X_k supposedly influences the dependent variables Y_k through a few unobservable latent variables, while the set of regressor variables Z_k directly influences Y_k. That is, one might suppose the model $Y_k = AY_k^* + DZ_k + U_k$ where Y_k^* is an $r_1 \times 1$ vector of unobservable latent variables which influence Y_k, with Y_k^* determined in part through a set of causal variables X_k as $Y_k^* = BX_k + V_k$. Then we obtain the reduced-form model equation as $Y_k = ABX_k + DZ_k + (AV_k + U_k) = CX_k + DZ_k + \epsilon_k$ with $C = AB$ of reduced rank. An alternative interpretation of the model is that Z_k may contain important variables, and hence the rank of its coefficient matrix is full, but X_k contains a large number of potential explanatory variables and possibly the dimension of X_k can be reduced. One elementary example of model (3.1) is its use to explicitly allow for a constant term in the reduced-rank model of Chapter 2 for the "true" predictor variables X_k, which can be done formally by taking Z_k to be the scalar 1 with D being the $m \times 1$ vector of constants.

The formulation of the model by Anderson was in terms of constraints as in (2.2) acting upon the reduced-rank coefficient matrix C. He considered estimation of these constraints and the inferential aspects of these estimated constraints. Because our focus here is on the component matrices, we shall outline only their associated inferential results. In model (3.1), the vector of response variables Y_k is of dimension $m \times 1$, X_k is of dimension $n_1 \times 1$, the vector of additional predictor variables Z_k is of dimension $n_2 \times 1$ and the ϵ_k are $m \times 1$ independent random vectors of errors with mean 0 and positive-definite covariance matrix $\Sigma_{\epsilon\epsilon}$, for $k = 1, 2, \ldots, T$. Assume that

$$\text{rank}(C) = r_1 \leq \min(m, n_1) \tag{3.2}$$

and hence, as in (2.4), $C = AB$ where A is an $m \times r_1$ matrix and B is an $r_1 \times n_1$ matrix. Hence the model (3.1) can be written as

$$Y_k = ABX_k + DZ_k + \epsilon_k, \qquad k = 1, \ldots, T. \tag{3.3}$$

Define the data matrices as $\boldsymbol{Y} = [Y_1, \ldots, Y_T]$, $\boldsymbol{X} = [X_1, \ldots, X_T]$, and $\boldsymbol{Z} = [Z_1, \ldots, Z_T]$.

Estimation of A and B in (3.3) and estimation of D follow from the population result given below. The result is similar to Theorem 2.2 in Chapter 2.

Theorem 3.1 Suppose the $(m+n_1+n_2)$-dimensional random vector $(Y', X', Z')'$ has mean 0 and covariance matrix with $\Sigma_{yx} = \Sigma_{xy}' = \text{Cov}(Y, X)$, $\Sigma_{yz} = \Sigma_{zy}' = \text{Cov}(Y, Z)$, and $\begin{bmatrix} \Sigma_{xx} & \Sigma_{xz} \\ \Sigma_{zx} & \Sigma_{zz} \end{bmatrix} = \text{Cov}\begin{bmatrix} X \\ Z \end{bmatrix}$ is nonsingular. Then for any positive-definite matrix Γ, an $m \times r_1$ matrix A and $r_1 \times n_1$ matrix B, for $r_1 \leq \min(m, n_1)$, which minimize

$$\text{tr}\{E[\Gamma^{1/2}(Y - ABX - DZ)(Y - ABX - DZ)'\Gamma^{1/2}]\} \tag{3.4}$$

are given by

$$\begin{aligned} A^{(r_1)} &= \Gamma^{-1/2}[V_1, \ldots, V_{r_1}] = \Gamma^{-1/2}V, \\ B^{(r_1)} &= V'\Gamma^{1/2}\Sigma_{yx.z}\Sigma_{xx.z}^{-1} \end{aligned} \tag{3.5}$$

and $D = \Sigma_{yz}\Sigma_{zz}^{-1} - C^{(r_1)}\Sigma_{xz}\Sigma_{zz}^{-1}$ with $C^{(r_1)} = A^{(r_1)}B^{(r_1)}$, where $V = [V_1, \ldots, V_{r_1}]$ and V_j is the (normalized) eigenvector that corresponds to the jth largest eigenvalue λ_{1j}^2 of the matrix $\Gamma^{1/2}\Sigma_{yx.z}\Sigma_{xx.z}^{-1}\Sigma_{xy.z}\Gamma^{1/2}$ ($j = 1, 2, \ldots, r_1$). It must be observed that $\Sigma_{yx.z} = \Sigma_{yx} - \Sigma_{yz}\Sigma_{zz}^{-1}\Sigma_{zx}$, $\Sigma_{xy.z} = \Sigma'_{yx.z}$, $\Sigma_{xx.z} = \Sigma_{xx} - \Sigma_{xz}\Sigma_{zz}^{-1}\Sigma_{zx}$ are the partial covariance matrices obtained by eliminating the linear effects of Z from both Y and X.

Proof: With the following transformation,

$$X^* = X - \Sigma_{xz}\Sigma_{zz}^{-1}Z, \qquad D^* = D + C\Sigma_{xz}\Sigma_{zz}^{-1}, \tag{3.6}$$

we can write

$$Y - CX - DZ = Y - CX^* - D^*Z \tag{3.7}$$

where X^* and Z are orthogonal, that is, $\text{Cov}(X^*, Z) = 0$. The criterion in (3.4), it must then be observed, is the same as

$$\text{tr}\{E[\Gamma^{1/2}(Y - ABX^* - D^*Z)(Y - ABX^* - D^*Z)'\Gamma^{1/2}]\}.$$

Because X^* and Z are orthogonal, the above criterion can be decomposed into two parts:

$$\begin{aligned} &\text{tr}\{E[\Gamma^{1/2}(Y - ABX^*)(Y - ABX^*)'\Gamma^{1/2}]\} \\ &-(\text{tr}\{E[\Gamma^{1/2}YY'\Gamma^{1/2}]\} - \text{tr}\{E[\Gamma^{1/2}(Y - D^*Z)(Y - D^*Z)'\Gamma^{1/2}]\}) \end{aligned}$$

The first part is similar to (2.12) in Theorem 2.2 and therefore the solutions for A and B follow directly. To see this clearly, observe that $\Sigma_{yx^*} = \Sigma_{yx.z}$ and $\Sigma_{x^*x^*} = \Sigma_{xx.z}$. As far as the determination of D^* is concerned, observe that the part where D^* appears in the above criterion is simply the least squares criterion and therefore $D^* = \Sigma_{yz}\Sigma_{zz}^{-1}$. The optimal value of the solution for D is recovered from (3.6).

We shall now indicate several useful remarks in relation to the result of Theorem 3.1. As in the reduced-rank regression model (2.11) without the Z_k variables, there are some implied normalization conditions on the component matrices B and A in the solution of Theorem 3.1 as follows:

$$B\Sigma_{xx.z}B' = \Lambda_1^2, \qquad A'\Gamma A = I_{r_1}. \tag{3.8}$$

The solutions A and B in (3.5) are related to partial canonical vectors and λ_{1j}^2 are related to partial canonical correlations. For the most direct relation, the appropriate choice of Γ is the inverse of the partial covariance matrix of Y, adjusted for Z, $\Gamma = \Sigma_{yy.z}^{-1}$, where $\Sigma_{yy.z} = \Sigma_{yy} - \Sigma_{yz}\Sigma_{zz}^{-1}\Sigma_{zy}$.

Then, $\lambda_{1j} = \rho_{1j}$ represent the partial canonical correlations between Y and X after eliminating Z. Another choice for Γ is $\Gamma = \Sigma_{yy.(x,z)}^{-1}$, and for this choice it can be demonstrated, similar to the derivation in Section 2.4, that the associated eigenvalues λ_{1j}^2 in Theorem 3.1 are related to the partial canonical correlations ρ_{1j} as

$$\lambda_{1j}^2 = \rho_{1j}^2/(1 - \rho_{1j}^2). \tag{3.9}$$

The maximum likelihood estimates are obtained by substituting the sample quantities in the solutions of Theorem 3.1 and by the choice of $\Gamma = \tilde{\Sigma}_{\epsilon\epsilon}^{-1}$, where

$$\tilde{\Sigma}_{\epsilon\epsilon} \equiv \hat{\Sigma}_{yy.(x,z)} = \hat{\Sigma}_{yy} - [\hat{\Sigma}_{yx} \;\; \hat{\Sigma}_{yz}] \begin{bmatrix} \hat{\Sigma}_{xx} & \hat{\Sigma}_{xz} \\ \hat{\Sigma}_{zx} & \hat{\Sigma}_{zz} \end{bmatrix}^{-1} \begin{bmatrix} \hat{\Sigma}_{xy} \\ \hat{\Sigma}_{zy} \end{bmatrix} \tag{3.10}$$

is the estimate of the error covariance matrix $\Sigma_{\epsilon\epsilon}$ based on full-rank regression. In (3.10), the estimates are defined to be $\hat{\Sigma}_{yy} = \frac{1}{T}\boldsymbol{YY}'$, $\hat{\Sigma}_{yx} = \frac{1}{T}\boldsymbol{YX}'$, $\hat{\Sigma}_{yz} = \frac{1}{T}\boldsymbol{YZ}'$, $\hat{\Sigma}_{xx} = \frac{1}{T}\boldsymbol{XX}'$, $\hat{\Sigma}_{xz} = \frac{1}{T}\boldsymbol{XZ}'$, and $\hat{\Sigma}_{zz} = \frac{1}{T}\boldsymbol{ZZ}'$, where $\boldsymbol{Y}$, $\boldsymbol{X}$ and $\boldsymbol{Z}$ are data matrices. This yields the ML estimates

$$\hat{A}^{(r_1)} = \tilde{\Sigma}_{\epsilon\epsilon}^{1/2}\hat{V}, \qquad \hat{B}^{(r_1)} = \hat{V}'\tilde{\Sigma}_{\epsilon\epsilon}^{-1/2}\hat{\Sigma}_{yx.z}\hat{\Sigma}_{xx.z}^{-1} \tag{3.11}$$

and $\hat{D} = \hat{\Sigma}_{yz}\hat{\Sigma}_{zz}^{-1} - \hat{C}^{(r_1)}\hat{\Sigma}_{xz}\hat{\Sigma}_{zz}^{-1}$ with $\hat{C}^{(r_1)} = \hat{A}^{(r_1)}\hat{B}^{(r_1)}$, where $\hat{V} = [\hat{V}_1, \ldots, \hat{V}_{r_1}]$ and the $\hat{V}_j$ are the (normalized) eigenvectors of the matrix

$$\tilde{\Sigma}_{\epsilon\epsilon}^{-1/2}\hat{\Sigma}_{yx.z}\hat{\Sigma}_{xx.z}^{-1}\hat{\Sigma}_{xy.z}\tilde{\Sigma}_{\epsilon\epsilon}^{-1/2} \equiv \tilde{\Sigma}_{\epsilon\epsilon}^{-1/2}\tilde{C}\hat{\Sigma}_{xx.z}\tilde{C}'\tilde{\Sigma}_{\epsilon\epsilon}^{-1/2}$$

associated with the r_1 largest eigenvalues. In the above, $\hat{\Sigma}_{yx.z} = \hat{\Sigma}_{yx} - \hat{\Sigma}_{yz}\hat{\Sigma}_{zz}^{-1}\hat{\Sigma}_{zx}$, $\hat{\Sigma}_{xx.z} = \hat{\Sigma}_{xx} - \hat{\Sigma}_{xz}\hat{\Sigma}_{zz}^{-1}\hat{\Sigma}_{zx}$, and $\tilde{C} = \hat{\Sigma}_{yx.z}\hat{\Sigma}_{xx.z}^{-1}$ is the full-rank LS estimate of C.

The above results on the form of ML estimates can be verified by similar arguments as in Section 2.3 for the single regressor set model (2.11). Similar to Section 2.3, the ML estimates of A, B, and D are obtained by minimizing $|W|$, or equivalently, $|\tilde{\Sigma}_{\epsilon\epsilon}^{-1} W|$, where $W = (1/T)(\boldsymbol{Y} - AB\boldsymbol{X} - D\boldsymbol{Z})(\boldsymbol{Y} - AB\boldsymbol{X} - D\boldsymbol{Z})'$. Now with $C = AB$, we can write

$$\begin{aligned} \boldsymbol{Y} - C\boldsymbol{X} - D\boldsymbol{Z} &= (\boldsymbol{Y} - \tilde{C}\boldsymbol{X} - \tilde{D}\boldsymbol{Z}) + (\tilde{C} - C)\boldsymbol{X} + (\tilde{D} - D)\boldsymbol{Z} \\ &= (\boldsymbol{Y} - \tilde{C}\boldsymbol{X} - \tilde{D}\boldsymbol{Z}) + (\tilde{C} - C)(\boldsymbol{X} - \hat{\Sigma}_{xz}\hat{\Sigma}_{zz}^{-1}\boldsymbol{Z}) \\ &\quad + \{(\tilde{D} - D) + (\tilde{C} - C)\hat{\Sigma}_{xz}\hat{\Sigma}_{zz}^{-1}\}\boldsymbol{Z} \end{aligned}$$

where $\tilde{C}$ and $\tilde{D}$ are the unrestricted full-rank LS estimators, and the three terms on the right-hand side of the last expression are orthogonal. So we obtain the corresponding decomposition

$$\begin{aligned} W &= \tilde{\Sigma}_{\epsilon\epsilon} + (\tilde{C} - AB)\hat{\Sigma}_{xx.z}(\tilde{C} - AB)' \\ &\quad + (\tilde{D}^* - D - C\hat{\Sigma}_{xz}\hat{\Sigma}_{zz}^{-1})\hat{\Sigma}_{zz}(\tilde{D}^* - D - C\hat{\Sigma}_{xz}\hat{\Sigma}_{zz}^{-1})' \end{aligned}$$

where $\tilde{D}^* = \tilde{D} + \tilde{C}\hat{\Sigma}_{xz}\hat{\Sigma}_{zz}^{-1} \equiv \hat{\Sigma}_{yz}\hat{\Sigma}_{zz}^{-1}$, with the last equality holding from the normal equations for the unrestricted LS estimators $\tilde{C}$ and $\tilde{D}$. For any C, the third component in the above decomposition can be uniformly minimized, as the zero matrix, by the choice of $\hat{D} = \tilde{D}^* - C\hat{\Sigma}_{xz}\hat{\Sigma}_{zz}^{-1} = \hat{\Sigma}_{yz}\hat{\Sigma}_{zz}^{-1} - C\hat{\Sigma}_{xz}\hat{\Sigma}_{zz}^{-1}$. Thus, the ML estimation of minimizing $|\tilde{\Sigma}_{\epsilon\epsilon}^{-1} W|$ is reduced to minimizing

$$|I_m + \tilde{\Sigma}_{\epsilon\epsilon}^{-1/2}(\tilde{C} - AB)\hat{\Sigma}_{xx.z}(\tilde{C} - AB)'\tilde{\Sigma}_{\epsilon\epsilon}^{-1/2}|$$

with respect to A and B. This is the same minimization problem as for ML estimation in Section 2.3, and it again follows similarly from Lemma 2.2 that all eigenvalues of the matrix $\tilde{\Sigma}_{\epsilon\epsilon}^{-1/2}(\tilde{C} - AB)\hat{\Sigma}_{xx.z}(\tilde{C} - AB)'\tilde{\Sigma}_{\epsilon\epsilon}^{-1/2}$ will be simultaneously minimized by the choice of estimators given in (3.11), and hence the above determinant also will be minimized. Thus, these are ML estimators of A and B, and the corresponding ML estimator of D is as noted previously, $\hat{D} = \hat{\Sigma}_{yz}\hat{\Sigma}_{zz}^{-1} - \hat{C}^{(r_1)}\hat{\Sigma}_{xz}\hat{\Sigma}_{zz}^{-1}$ with $\hat{C}^{(r_1)} = \hat{A}^{(r_1)}\hat{B}^{(r_1)}$. The ML estimate of $\Sigma_{\epsilon\epsilon}$ under the reduced-rank structure is given by $\hat{\Sigma}_{\epsilon\epsilon} = (1/T)(\boldsymbol{Y} - \hat{C}^{(r_1)}\boldsymbol{X} - \hat{D}\boldsymbol{Z})(\boldsymbol{Y} - \hat{C}^{(r_1)}\boldsymbol{X} - \hat{D}\boldsymbol{Z})'$. Similar to the result in (2.18) of Section 2.3, $\hat{\Sigma}_{\epsilon\epsilon}$ can be represented as

$$\begin{aligned}\hat{\Sigma}_{\epsilon\epsilon} &= \tilde{\Sigma}_{\epsilon\epsilon} + (\tilde{C} - \hat{C}^{(r_1)})\hat{\Sigma}_{xx.z}(\tilde{C} - \hat{C}^{(r_1)})' \\ &= \tilde{\Sigma}_{\epsilon\epsilon} + (I_m - P)\tilde{C}\hat{\Sigma}_{xx.z}\tilde{C}'(I_m - P)'\end{aligned}$$

where $P = \tilde{\Sigma}_{\epsilon\epsilon}^{1/2}\hat{V}\hat{V}'\tilde{\Sigma}_{\epsilon\epsilon}^{-1/2}$, and $\tilde{C} = \hat{\Sigma}_{yx.z}\hat{\Sigma}_{xx.z}^{-1}$ is the full-rank LS estimate of C.

Because of the connection displayed in (3.9), for example, the rank of the matrix C can be identified by the number of nonzero partial canonical correlations between Y_k and X_k, eliminating Z_k. More precisely, based on a sample of T observations, the likelihood ratio test procedure for testing $H_0\colon \text{rank}(C) \leq r_1$ yields the test statistic

$$-2\log(\lambda) = -T \sum_{j=r_1+1}^{m} \log(1 - \hat{\rho}_{1j}^2) \tag{3.12}$$

where $\hat{\rho}_{1j}$ are the sample partial canonical correlations between Y_k and X_k, eliminating Z_k, that is, $\hat{\rho}_{1j}^2$ are the eigenvalues of the matrix

$$\hat{\Sigma}_{yy.z}^{-1/2}\hat{\Sigma}_{yx.z}\hat{\Sigma}_{xx.z}^{-1}\hat{\Sigma}_{xy.z}\hat{\Sigma}_{yy.z}^{-1/2}.$$

The derivation of the form of the LR statistic in (3.12) is similar to the derivation given in Section 2.6 for the reduced-rank model with one set of regressor variables. The restricted ML estimators are $\hat{C}^{(r_1)}$ and $\hat{D}$ as given above, and we let $S_1 = (\boldsymbol{Y} - \hat{C}^{(r_1)}\boldsymbol{X} - \hat{D}\boldsymbol{Z})(\boldsymbol{Y} - \hat{C}^{(r_1)}\boldsymbol{X} - \hat{D}\boldsymbol{Z})'$ denote the corresponding residual sum of squares matrix under H_0. Since

$\tilde{D} = \hat{\Sigma}_{yz}\hat{\Sigma}_{zz}^{-1} - \tilde{C}\hat{\Sigma}_{xz}\hat{\Sigma}_{zz}^{-1}$, where $\tilde{C} = \hat{\Sigma}_{yx.z}\hat{\Sigma}_{xx.z}^{-1}$ and $\tilde{D}$ denote the full-rank LS estimators, it follows that

$$(\tilde{C} - \hat{C}^{(r_1)})\boldsymbol{X} + (\tilde{D} - \hat{D})\boldsymbol{Z} = (\tilde{C} - \hat{C}^{(r_1)})(\boldsymbol{X} - \hat{\Sigma}_{xz}\hat{\Sigma}_{zz}^{-1}\boldsymbol{Z}),$$

and we also have that

$$\begin{aligned} \tilde{C} - \hat{C}^{(r_1)} &= (I_m - \tilde{\Sigma}_{\epsilon\epsilon}^{1/2}\hat{V}\hat{V}'\tilde{\Sigma}_{\epsilon\epsilon}^{-1/2})\tilde{C} \\ &= \tilde{\Sigma}_{\epsilon\epsilon}^{1/2}\left[\sum_{j=r_1+1}^{m}\hat{V}_j\hat{V}_j'\right]\tilde{\Sigma}_{\epsilon\epsilon}^{-1/2}\hat{\Sigma}_{yx.z}\hat{\Sigma}_{xx.z}^{-1}. \end{aligned}$$

Therefore, similar to the result in (2.39) of Section 2.6, we find that

$$\begin{aligned} \tfrac{1}{T}S_1 &= \tfrac{1}{T}S + \tfrac{1}{T}[(\tilde{C} - \hat{C}^{(r_1)})\boldsymbol{X} + (\tilde{D}-\hat{D})\boldsymbol{Z}][(\tilde{C} - \hat{C}^{(r_1)})\boldsymbol{X} + (\tilde{D}-\hat{D})\boldsymbol{Z}]' \\ &= \tilde{\Sigma}_{\epsilon\epsilon} + \tilde{\Sigma}_{\epsilon\epsilon}^{1/2}\left[\sum_{j=r_1+1}^{m}\hat{V}_j\hat{V}_j'\right]\hat{R}_1\left[\sum_{j=r_1+1}^{m}\hat{V}_j\hat{V}_j'\right]\tilde{\Sigma}_{\epsilon\epsilon}^{1/2} \\ &= \tilde{\Sigma}_{\epsilon\epsilon} + \tilde{\Sigma}_{\epsilon\epsilon}^{1/2}\left[\sum_{j=r_1+1}^{m}\hat{\lambda}_{1j}^2\hat{V}_j\hat{V}_j'\right]\tilde{\Sigma}_{\epsilon\epsilon}^{1/2}, \end{aligned}$$

where the $\hat{\lambda}_{1j}^2$, $j = r_1+1, \ldots, m$, are the $(m-r_1)$ smallest eigenvalues of the matrix $\hat{R}_1 = \tilde{\Sigma}_{\epsilon\epsilon}^{-1/2}\hat{\Sigma}_{yx.z}\hat{\Sigma}_{xx.z}^{-1}\hat{\Sigma}_{xy.z}\tilde{\Sigma}_{\epsilon\epsilon}^{-1/2}$. From results similar to those presented in the discussion at the end of Section 2.4 [e.g., see (2.29)], we know that $\hat{\lambda}_{1j}^2 = \hat{\rho}_{1j}^2/(1-\hat{\rho}_{1j}^2)$ and so $1+\hat{\lambda}_{1j}^2 = 1/(1-\hat{\rho}_{1j}^2)$. Therefore,

$$U^{-1} = |S_1|/|S| = |\tfrac{1}{T}S_1|/|\tilde{\Sigma}_{\epsilon\epsilon}| = \prod_{j=r_1+1}^{m}(1+\hat{\lambda}_{1j}^2) = \prod_{j=r_1+1}^{m}(1-\hat{\rho}_{1j}^2)^{-1}.$$

Thus, it follows that the LR criterion $-2\log(\lambda)$, where $\lambda = U^{T/2}$, is given as in (3.12). Anderson (1951) has shown that under the null hypothesis that $\operatorname{rank}(C) \le r_1$, the asymptotic distribution of the LR statistic is such that $-2\log(\lambda) \sim \chi^2_{(m-r_1)(n_1-r_1)}$. The LR statistic in (3.12) can be used to test the hypothesis $H_0\colon \operatorname{rank}(C) \le r_1$, and can be considered for different values of r_1 to obtain an appropriate choice of the rank of C. As in Section 2.6, a correction factor is commonly used for the LR statistic in (3.12), yielding the test statistic $\mathcal{M} = -[T-n+(n_1-m-1)/2]\sum_{j=r_1+1}^{m}\log(1-\hat{\rho}_{1j}^2)$, whose null distribution is more closely approximated by the asymptotic $\chi^2_{(m-r_1)(n_1-r_1)}$ distribution than is that of $-2\log(\lambda)$ in (3.12).

The model (3.1) written as

$$Y_k = ABX_k + DZ_k + \epsilon_k, \qquad k = 1, \ldots, T, \tag{3.13}$$

provides an alternative way of interpreting the estimation procedure. Given B, therefore BX_k, estimates of A and D are obtained by the usual multivariate least squares regression of Y_k on BX_k and Z_k. When these estimates are substituted in the likelihood criterion as functions of B, the resulting quantity will directly lead to the estimation of B, via the canonical vectors associated with the sample partial canonical correlations between Y_k and X_k after eliminating Z_k. This interpretation will be shown to be useful for a more general model that we consider in this chapter.

As in Section 2.3 for the single regressor reduced-rank model, the above discussion leads to an alternate form of the population solution to the reduced-rank problem for model (3.1) besides that given in Theorem 3.1. For given BX, we can substitute the least squares values of A and D (in particular, $A = \Sigma_{yx.z}B'(B\Sigma_{xx.z}B')^{-1}$) from the regression of Y on BX and Z into the criterion (3.4) or alternatively into its decomposition given in the proof of Theorem 3.1. This criterion then simplifies to $\mathrm{tr}\{\Sigma_{yy.z}\Gamma - (B\Sigma_{xx.z}B')^{-1}B\Sigma_{xy.z}\Gamma\Sigma_{yx.z}B'\}$, and similar to the discussion at the end of Section 2.3 it follows that the solution for B is such that the columns of $\Sigma_{xx.z}^{1/2}B'$ can be chosen as the eigenvectors corresponding to the r_1 largest eigenvalues of $\Sigma_{xx.z}^{-1/2}\Sigma_{xy.z}\Gamma\Sigma_{yx.z}\Sigma_{xx.z}^{-1/2}$.

Although the results presented in Theorem 3.1 and above provide an explicit solution for the component matrices A, B, and for D, using the eigenvalues and eigenvectors of appropriate matrices, an iterative partial least squares approach similar to that discussed for the single regressor model in Section 2.3 can also be useful and is computationally convenient. For the population problem, the first order equations that result from minimization of the criterion (3.4) are

$$\Sigma_{yx}B' - AB\Sigma_{xx}B' - D\Sigma_{zx}B' = 0$$

$$A'\Gamma\Sigma_{yx} - (A'\Gamma A)B\Sigma_{xx} - A'\Gamma D\Sigma_{zx} = 0$$

$$\Sigma_{yz} - AB\Sigma_{xz} - D\Sigma_{zz} = 0$$

For given A and B, the first step in the iterative process is to obtain D from the third equation as $D = (\Sigma_{yz} - AB\Sigma_{xz})\Sigma_{zz}^{-1}$. For given B (and D), the solution for A can be calculated from the first equation as $A = (\Sigma_{yx} - D\Sigma_{zx})B'(B\Sigma_{xx}B')^{-1}$, while for given A (and D), from the second equation the solution for B can be calculated as $B = (A'\Gamma A)^{-1}A'\Gamma(\Sigma_{yx} - D\Sigma_{zx})$. Alternatively, the optimal expression for D (from the third equation) can be substituted into the first equation, and the solution for A is then obtained for given B as $A = \Sigma_{yx.z}B'(B\Sigma_{xx.z}B')^{-1}$, and similarly, substituting the optimal expression for D into the second equation yields the solution of B for given A as $B = (A'\Gamma A)^{-1}A'\Gamma\Sigma_{yx.z}\Sigma_{xx.z}^{-1}$. Thus, in effect, D can be eliminated from the first two equations and the partial least squares iterations can be performed in terms of A and B only, completely analogous to the procedure in Section 2.3 but using partial covariance matrices. The solution for D can finally be recovered as given above. At each step of the iterations, one can consider using the normalization conditions (3.8) for A and B. This partial least squares procedure can also be carried over directly to the sample situation with the appropriate sample quantities being used to iteratively construct estimates $\hat{A}^{(r_1)}$, $\hat{B}^{(r_1)}$, and $\hat{D}$, which are given more explicitly in (3.11).

Recall the original motivation behind Anderson's article. While A and B, the components of C, are formed out of the largest partial canonical correlations and the associated vectors, the constraints on C which were of Anderson's focus can be obtained via the $(m - r_1)$ smallest (possibly zero) partial canonical correlations and the associated vectors. It may be clear by now that the reduced-rank model in (3.13) of this chapter can be accommodated through essentially the same set of quantities as defined for the reduced-rank model in Chapter 2, with the difference being that the covariance matrices used in Chapter 2 are replaced by partial covariance matrices. For this reason, we do not repeat the asymptotic theory for the estimators that was presented in Section 2.5. The results of Theorem 2.4 in that section could be restated here for the case $\Gamma = \Sigma_{\epsilon\epsilon}^{-1}$ known by substituting the covariance matrices with corresponding partial covariance matrices.

3.2 Application to One-Way ANOVA and Linear Discriminant Analysis

We consider the reduced-rank model (3.1) applied to the multivariate one-way ANOVA situation, and exhibit its relation to linear discriminant analysis. Suppose we have independent random samples from n multivariate normal distributions $N(\mu_i, \Sigma)$, denoted by $Y_{i1}, \ldots, Y_{iT_i}$, $i = 1, \ldots, n$, and set $T = T_1 + \cdots + T_n$. It may be hypothesized that the n mean vectors μ_i lie in some (unknown) r-dimensional linear subspace, with $r \leq \min(n-1, m)$. (We will assume $n < m$ for convenience of notation.) This is equivalent to the hypothesis that the $m \times (n-1)$ matrix $(\mu_1 - \mu, \ldots, \mu_{n-1} - \mu)$, or $(\mu_1 - \mu_n, \ldots, \mu_{n-1} - \mu_n)$, is of reduced rank r, where $\mu = \sum_{i=1}^{n} T_i \mu_i / T$.

Since $Y_{ij} = (\mu_i - \mu_n) + \mu_n + \epsilon_{ij}$, $j = 1, \ldots, T_i$, $i = 1, \ldots, n$, the model can be easily written in the reduced-rank model form (3.1), $Y_{ij} = CX_{ij} + DZ_{ij} + \epsilon_{ij}$. The correspondences are $C = (\mu_1 - \mu_n, \ldots, \mu_{n-1} - \mu_n)$, $D = \mu_n$, $Z_{ij} \equiv 1$, and X_{ij} is an $(n-1) \times 1$ vector with 1 in the ith position and zeros elsewhere for $i < n$ and $X_{nj} \equiv 0$. The reduced-rank restrictions imply $C = AB$ with $B = (\xi_1, \ldots, \xi_{n-1})$ of dimension $r \times (n-1)$, or $\mu_i - \mu_n = A\xi_i$, $i = 1, \ldots, n-1$; equivalently, $L(\mu_i - \mu_n) = 0$, $i = 1, \ldots, n-1$, where $L = (\ell_1', \ldots, \ell_{m-r}')'$ is $(m-r) \times m$ with row vectors ℓ_i' as in (2.1). In terms of data matrices, we have $\boldsymbol{Y} = C\boldsymbol{X} + D\boldsymbol{Z} + \boldsymbol{\epsilon}$, where $\boldsymbol{Y} = [Y_{11}, \ldots, Y_{1T_1}, \ldots, Y_{n1}, \ldots, Y_{nT_n}]$, $\boldsymbol{Z} = (1, \ldots, 1) \equiv \mathbf{1}_T'$, and $\boldsymbol{X} = [\boldsymbol{X}_{(1)}, \ldots, \boldsymbol{X}_{(n-1)}, 0]$ with $\boldsymbol{X}_{(i)} = [0, \ldots, 0, \mathbf{1}_{T_i}, 0, \ldots, 0]'$ where $\mathbf{1}_N$ denotes an N-dimensional vector of ones, so that $\boldsymbol{X}_{(i)}$ is an $(n-1) \times T_i$ matrix with ones in the ith row and zeros elsewhere. From the LS estimation results of Sections 1.2, 1.3, and 3.1, the full-rank or unrestricted LS estimates of C

and D are given as

$$\tilde{C} = \hat{\Sigma}_{yx.z}\hat{\Sigma}_{xx.z}^{-1} = [\bar{Y}_1 - \bar{Y}_n, \ldots, \bar{Y}_{n-1} - \bar{Y}_n], \tag{3.14a}$$

$$\tilde{D} = (\hat{\Sigma}_{yz} - \tilde{C}\hat{\Sigma}_{xz})\hat{\Sigma}_{zz}^{-1} = \bar{Y}_n, \tag{3.14b}$$

where $\bar{Y}_i = \sum_{j=1}^{T_i} Y_{ij}/T_i$ is the ith sample mean vector, $i = 1, \ldots, n$. The corresponding ML estimate of the error covariance matrix $\Sigma_{\epsilon\epsilon} \equiv \Sigma$ is given by $\tilde{\Sigma}_{\epsilon\epsilon} = \frac{1}{T}S$ with $S = \sum_{i=1}^{n}\sum_{j=1}^{T_i}(Y_{ij} - \bar{Y}_i)(Y_{ij} - \bar{Y}_i)'$ being the usual "within-group" sum of squares and cross-products matrix.

For the reduced-rank or restricted estimates of C and D, we need to consider the eigenvectors associated with the r largest eigenvalues of the matrix $\hat{\Sigma}_{yx.z}\hat{\Sigma}_{xx.z}^{-1}\hat{\Sigma}_{xy.z} \equiv \tilde{C}\hat{\Sigma}_{xx.z}\tilde{C}'$ with respect to the (unrestricted) residual covariance matrix estimate $\tilde{\Sigma}_{\epsilon\epsilon}$. Observe that $\hat{\Sigma}_{zz} = 1$ and $\hat{\Sigma}_{xz} = \frac{1}{T}(T_1, \ldots, T_{n-1})'$, and hence

$$\hat{\Sigma}_{xx.z} = \hat{\Sigma}_{xx} - \hat{\Sigma}_{xz}\hat{\Sigma}_{zz}^{-1}\hat{\Sigma}_{zx} = \frac{1}{T}[\text{diag}(T_1, \ldots, T_{n-1}) - \frac{1}{T}NN'],$$

where $N = (T_1, \ldots, T_{n-1})'$. Therefore, it follows that

$$\begin{aligned}\tilde{C}\hat{\Sigma}_{xx.z}\tilde{C}' &= \frac{1}{T}\left[\sum_{i=1}^{n-1} T_i(\bar{Y}_i - \bar{Y}_n)(\bar{Y}_i - \bar{Y}_n)' - T(\bar{Y} - \bar{Y}_n)(\bar{Y} - \bar{Y}_n)'\right] \\ &= \frac{1}{T}\left[\sum_{i=1}^{n} T_i(\bar{Y}_i - \bar{Y})(\bar{Y}_i - \bar{Y})'\right] \equiv \frac{1}{T}S_B, \end{aligned} \tag{3.15}$$

where S_B is the "between-group" sum of squares and cross-products matrix, with $\bar{Y} = \sum_{i=1}^{n} T_i\bar{Y}_i/T$. Let $\hat{V}_k$ denote the normalized eigenvector of $\tilde{\Sigma}_{\epsilon\epsilon}^{-1/2}\tilde{C}\hat{\Sigma}_{xx.z}\tilde{C}'\tilde{\Sigma}_{\epsilon\epsilon}^{-1/2} = \frac{1}{T}\tilde{\Sigma}_{\epsilon\epsilon}^{-1/2}S_B\tilde{\Sigma}_{\epsilon\epsilon}^{-1/2}$ corresponding to the kth largest eigenvalue $\hat{\lambda}_k^2$. Then $\hat{V}_k^* = \tilde{\Sigma}_{\epsilon\epsilon}^{-1/2}\hat{V}_k$ satisfies $(\hat{\lambda}_k^2\tilde{\Sigma}_{\epsilon\epsilon} - \frac{1}{T}S_B)\hat{V}_k^* = 0$ or $(\hat{\lambda}_k^2 S - S_B)\hat{V}_k^* = 0$, with $\hat{V}_k^*$ normalized so that $\hat{V}_k^{*\prime}\tilde{\Sigma}_{\epsilon\epsilon}\hat{V}_k^* = 1$. Let $\hat{V}_* = [\hat{V}_1^*, \ldots, \hat{V}_r^*] = \tilde{\Sigma}_{\epsilon\epsilon}^{-1/2}\hat{V}$ with $\hat{V} = [\hat{V}_1, \ldots, \hat{V}_r]$. Then from results of Section 3.1, the ML reduced-rank estimate of C is

$$\hat{C} = \hat{A}^{(r)}\hat{B}^{(r)} = \tilde{\Sigma}_{\epsilon\epsilon}^{1/2}\hat{V}\hat{V}'\tilde{\Sigma}_{\epsilon\epsilon}^{-1/2}\tilde{C} = \tilde{\Sigma}_{\epsilon\epsilon}\hat{V}_*\hat{V}_*'\tilde{C}. \tag{3.16}$$

The corresponding restricted estimate of $D = \mu_n$ is then

$$\begin{aligned}\hat{D} \equiv \hat{\mu}_n &= (\hat{\Sigma}_{yz} - \hat{C}\hat{\Sigma}_{xz})\hat{\Sigma}_{zz}^{-1} = \bar{Y} - \hat{C}\frac{1}{T}N \\ &= \bar{Y} - \tilde{\Sigma}_{\epsilon\epsilon}\hat{V}_*\hat{V}_*'\frac{1}{T}\sum_{i=1}^{n-1} T_i(\bar{Y}_i - \bar{Y}_n) = \bar{Y} + \tilde{\Sigma}_{\epsilon\epsilon}\hat{V}_*\hat{V}_*'(\bar{Y}_n - \bar{Y}). \end{aligned} \tag{3.17}$$

From (3.16) it is now easy to see that the ML reduced-rank estimates $\hat{\mu}_i$ of the individual population means μ_i are related as $\hat{\mu}_i - \hat{\mu}_n = \tilde{\Sigma}_{\epsilon\epsilon}\hat{V}_*\hat{V}_*'(\bar{Y}_i -$

$\bar{Y}_n$), so that using (3.17), the ML reduced-rank estimates of the μ_i are

$$\hat{\mu}_i = \bar{Y} + \tilde{\Sigma}_{\epsilon\epsilon}\hat{V}_*\hat{V}_*'(\bar{Y}_i - \bar{Y}), \qquad i = 1, \ldots, n. \tag{3.18}$$

The reduction in dimension for estimation of the $\mu_i - \mu_n$ appears through the use of only the r linear combinations $\hat{V}_*'(\bar{Y}_i - \bar{Y}_n)$ of the differences in sample mean vectors to estimate the $\mu_i - \mu_n$ under the reduced-rank restrictions. In addition, using the results in Sections 2.3 and 3.1, the ML estimate of the error covariance matrix $\Sigma_{\epsilon\epsilon}$ under the reduced-rank assumptions is given by

$$\hat{\Sigma}_{\epsilon\epsilon} = \tilde{\Sigma}_{\epsilon\epsilon} + (I_m - \tilde{\Sigma}_{\epsilon\epsilon}\hat{V}_*\hat{V}_*')\,\frac{1}{T}S_B\,(I_m - \hat{V}_*\hat{V}_*'\tilde{\Sigma}_{\epsilon\epsilon}) \tag{3.19}$$

and also note that $I_m - \tilde{\Sigma}_{\epsilon\epsilon}\hat{V}_*\hat{V}_*' = \tilde{\Sigma}_{\epsilon\epsilon}\hat{V}_-\hat{V}_-'$ with $\hat{V}_- = [\hat{V}_{r+1}^*, \ldots, \hat{V}_m^*]$.

Now we indicate the relation with linear discriminant analysis. The setting of linear discriminant analysis is basically that of the one-way multivariate ANOVA model. The objectives of discriminant analysis are to obtain representations of the data that best separate the n groups or populations, and based on this to construct a procedure to classify a new vector observation to one of the groups with reasonable accuracy. Discrimination is based on the separations among the m-dimensional mean vectors μ_i, $i = 1, \ldots, n$, of the n populations. If the mean vectors μ_i were to lie in some r-dimensional linear subspace, $r < \min(n-1, m)$, then only r linear functions of the μ_i (linear discriminant functions) would be needed to describe the separations among the n groups. The Fisher–Rao approach to discriminant analysis in the sample is to construct linear combinations (linear discriminants) $z_{ij} = a'Y_{ij}$ which successively maximize the ratio

$$\frac{a'S_B a}{a'Sa} = \sum_{i=1}^{n} T_i(\bar{z}_i - \bar{z})^2 \Big/ \sum_{i=1}^{n}\sum_{j=1}^{T_i}(z_{ij} - \bar{z}_i)^2. \tag{3.20}$$

Note that this is the ratio of the between-group (i.e., between means) to the within-group variation in the sample. Thus, the sample *first linear discriminant function* is the linear combination $a_1'Y_{ij}$ that maximizes the ratio in (3.20), the sample *second linear discriminant function* is the linear combination $a_2'Y_{ij}$ that maximizes the ratio subject to $a_2'\tilde{\Sigma}_{\epsilon\epsilon}a_1 = 0$ (i.e., the sample covariance between $a_1'Y_{ij}$ and $a_2'Y_{ij}$ is zero), and so on.

Now recall that the $\hat{V}_k^*$ in the solution to the above one-way ANOVA model reduced-rank estimation problem are the eigenvectors of the matrix $S^{-1}S_B$ corresponding to its nonzero eigenvalues $\hat{\lambda}_k^2$, for $k = 1, \ldots, n-1$, that is, they satisfy $(\hat{\lambda}_k^2 S - S_B)\hat{V}_k^* = 0$. From this it is known that the first eigenvector $\hat{V}_1^*$, for example, provides the linear combination $z_{ij} = a'Y_{ij}$ which maximizes the ratio (3.20), that is, $\hat{V}_1^*$ yields the sample first linear discriminant function in the linear discriminant analysis of Fisher–Rao, and $\hat{\lambda}_1^2$ represents the maximum value attained for this ratio. Then $\hat{V}_2^*$ gives the second linear discriminant function, in the sense of providing the linear combination which gives the maximum of the above ratio subject to the orthogonality condition $\hat{V}_1^{*\prime} S \hat{V}_2^* = 0$. In this same fashion, the collection of vectors $\hat{V}_k^*$ provides the set of $n-1$ sample linear discriminant functions. In linear discriminant analysis of n multivariate normal populations, these discriminant functions are generally applied to the sample mean vectors $\bar{Y}_i$, to consider $\hat{V}_k^{*\prime}(\bar{Y}_i - \bar{Y})$, in an attempt to provide information that best separates or discriminates between the means of the n groups or populations. In practice, it may be desirable to choose a subspace of rank $r < n-1$ and use only the corresponding first r linear discriminant functions that maximize the separation of the group means. From the above discussion, these would be obtained as the first r eigenvectors $\hat{V}_1^*, \dots, \hat{V}_r^*$, and they might contain nearly all the available information for discrimination. We thus see that restriction to consideration of only these r linear discriminant functions can be formalized by formulating the one-way ANOVA model with a *reduced-rank* model structure for the population mean vectors μ_i.

The above results on the ML estimates in the one-way ANOVA model under the reduced-rank structure were obtained by Anderson (1951), and later by Healy (1980), Campbell (1984), and Srivastava (1997) [also by Theobald (1975) with $\Sigma_{\epsilon\epsilon}$ assumed known], while the related problem in discriminant analysis was considered earlier by Fisher (1938). The connections between reduced-rank regression and discriminant analysis were also discussed by Campbell (1984), and a more direct formulation of the structure on the population means along the lines of (3.18), i.e., $\mu_i = \mu + \Sigma_{\epsilon\epsilon} V_* \xi_i \equiv \mu + \Sigma_{\epsilon\epsilon} V_* V_*'(\mu_i - \mu)$ was considered there. The reduced-rank linear discriminant analysis method forms the basis of recent extensions of discriminant analysis that are useful in analyzing more complex pattern recognition problems (Hastie and Tibshirani, 1996) and in analyzing longitudinal data through growth curve models (Albert and Kshirsagar, 1993).

3.3 Numerical Example Using Chemometrics Data

In the example of this section, we apply the multivariate reduced-rank regression methods of Chapter 2 and Section 3.1 to chemometrics data obtained from simulation of a low-density polyethylene tubular reactor. The data are taken from Skagerberg, MacGregor, and Kiparissides (1992), who used partial least squares (PLS) multivariate regression modeling applied to these data both for predicting properties of the produced polymer and for multivariate process control. The data were also considered by Breiman and Friedman (1997) to illustrate the relative performance of different multivariate prediction methods.

The data set consists of $T = 56$ multivariate observations, with $m = 6$ response variables and $n = 22$ predictor (or 'process') variables. The response variables are the following output properties of the polymer produced:

y_1, number-average molecular weight
y_2, weight-average molecular weight
y_3, frequency of long chain branching
y_4, frequency of short chain branching
y_5, content of vinyl groups in the polymer chain
y_6, content of vinylidene groups in the polymer chain

The process variable measurements employed consist of 20 different temperatures measured at equal distances along the reactor ($x_1 - x_{20}$), complemented with the wall temperature of the reactor (x_{21}) and the feed rate of the solvent (x_{22}) that also acts as a chain transfer agent. The temperature measurements in the temperature profile along the reactor are expected to be highly correlated and therefore reduced-rank methods may be especially useful to reduce the dimensionality and complexity of the input or predictor set of data. For interpretational convenience, the response variables y_3, y_5, and y_6 were re-scaled by the multiplicative factors 10^2, 10^3, and 10^2, respectively, before performing the analysis presented here, so that all six response variables would have variability of the same order of magnitude. The predictor variable x_{21} was also re-scaled by the factor 10^{-1}.

As usual, both the response and predictor variables were first centered by subtraction of appropriate sample means prior to the multivariate analysis, and we let $\boldsymbol{Y}$ and $\boldsymbol{X}$ denote the resulting data matrices of dimensions 6×56 and 22×56, respectively. To obtain a preliminary 'benchmark' model for comparison with subsequent analyses, a multivariate regression model was fit by LS to the vector of response variables using only the two predictor variables x_{21} and x_{22}, the wall temperature of the reactor and the solvent feed rate. The LS estimate of the regression coefficient matrix (with estimated standard errors given in parentheses) from this preliminary model is obtained as

$$\begin{aligned}\tilde{C}_1' &= (\boldsymbol{X}_1\boldsymbol{X}_1')^{-1}\boldsymbol{X}_1\boldsymbol{Y}' \\ &= \begin{bmatrix} \begin{matrix}0.1357\\(0.0475)\end{matrix} & \begin{matrix}1.0956\\(0.1230)\end{matrix} & \begin{matrix}-1.5618\\(0.1660)\end{matrix} & \begin{matrix}1.4234\\(0.2045)\end{matrix} & \begin{matrix}0.5741\\(0.0813)\end{matrix} & \begin{matrix}0.5339\\(0.0783)\end{matrix} \\ \begin{matrix}5.8376\\(0.1704)\end{matrix} & \begin{matrix}28.3353\\(0.4414)\end{matrix} & \begin{matrix}-0.1305\\(0.5953)\end{matrix} & \begin{matrix}0.1715\\(0.7336)\end{matrix} & \begin{matrix}0.1065\\(0.2917)\end{matrix} & \begin{matrix}0.0596\\(0.2809)\end{matrix} \end{bmatrix},\end{aligned}$$

where $\boldsymbol{X}_1$ is 2×56. The unbiased estimate of the covariance matrix of the errors ϵ_k from the multivariate regression model (with correlations shown above the diagonal) is given by

$$\overline{\Sigma}_{\epsilon\epsilon} = \frac{1}{56-3}\,\hat{\epsilon}\hat{\epsilon}'$$

$$= \begin{bmatrix} 0.15323 & 0.79248 & 0.26439 & 0.88877 & 0.89995 & 0.90231 \\ 0.31458 & 1.02835 & -0.10505 & 0.77472 & 0.75447 & 0.76567 \\ 0.14156 & -0.14571 & 1.87093 & 0.10493 & 0.14821 & 0.13324 \\ 0.58637 & 1.32412 & 0.24190 & 2.84066 & 0.95115 & 0.95700 \\ 0.23610 & 0.51276 & 0.13587 & 1.07438 & 0.44916 & 0.95458 \\ 0.22792 & 0.50104 & 0.11760 & 1.04082 & 0.41282 & 0.41640 \end{bmatrix},$$

where $\hat{\epsilon} = \boldsymbol{Y} - \tilde{C}_1\boldsymbol{X}_1$. We also have $\mathrm{tr}(\overline{\Sigma}_{\epsilon\epsilon}) = 6.75873$, and the determinant of the ML error covariance matrix estimate, $\tilde{\Sigma}_{\epsilon\epsilon} = \frac{1}{56}\hat{\epsilon}\hat{\epsilon}'$, is $|\tilde{\Sigma}_{\epsilon\epsilon}| = 0.25318 \times 10^{-4}$.

Clearly, from the results above we see that the response variables y_1 and y_2 are very highly related to x_{22}, and only more mildly related to x_{21}, whereas response variables $y_3 - y_6$ are more strongly related to x_{21} with little or no dependence on x_{22}. It is also found that the errors from the fitted regressions are very strongly positively correlated among the variables y_1, $y_4 - y_6$, the errors for y_2 are rather highly positively correlated with each of y_1 and $y_4 - y_6$, and the errors for y_3 show relatively small correlations with the errors for all the other response variables.

We now incorporate the 20 measurements of the temperature profile along the reactor into the multivariate regression analysis. The LS estimation of the response variables regressed on the entire set of 22 predictor variables ($x_1 - x_{22}$) is performed, and yields a ML estimate $\tilde{\Sigma}_{\epsilon\epsilon} = \frac{1}{56}\hat{\epsilon}\hat{\epsilon}'$ of the error covariance matrix with $|\tilde{\Sigma}_{\epsilon\epsilon}| = 0.97557 \times 10^{-8}$, where $\hat{\epsilon} = \boldsymbol{Y} - \tilde{C}\boldsymbol{X}$. A requisite LR test of the hypothesis that the regression coefficients for all temperature profile variables ($x_1 - x_{20}$) are equal to zero for all response variables gives a test statistic value, from (1.16), equal to $\mathcal{M} = -[56-23+(20\text{–}6\text{–}1)/2]\log(0.97557\times 10^{-8}/0.25318\times 10^{-4}) = 310.53$, with $6{*}20 = 120$ degrees of freedom. This leads to clear rejection of the null hypothesis and indicates that the set of 20 temperature variables is useful in explaining variations of the response variables. In addition, the "unbiased" estimate of the error covariance matrix, $\overline{\Sigma}_{\epsilon\epsilon} = \frac{1}{56-23}\hat{\epsilon}\hat{\epsilon}'$, has diagonal elements equal to $0.03297, 0.38851, 0.45825, 0.21962, 0.05836, 0.03717$, with trace equal to 1.19487. Comparison to the corresponding values for the simple two predictor variable model above indicates that the inclusion of the temperature variables has its most dramatic effect in reduction of the error variances for response variables y_4, y_5, and y_6. A rather high proportion of the t-statistics for the individual regression coefficients of the temperature variables in the LS fitted model are not 'significant', however, partly due to the high level of correlations among many of the 20 temperature profile variables. Because of this high degree of correlation among the temperature measurements, it is obviously desirable to search for alternatives to the ordinary least squares estimation results based on all 22 predictor variables. One common approach to resolve the problem of dealing with a large number of predictor variables might be to attempt to select a smaller subset of 'key variables' as predictors for the regression, by using various variable selection methods, principal components methods, or other more ad hoc procedures. Unfortunately, this type of approach could lead to discarding important information, which might never be detected, and also to loss of precision and quality of the regression model. Therefore, to avoid this possibility, we consider the more systematic approach of using reduced-rank regression methods.

Because of the nature of the set of predictor variables, as a 'natural' reduced-rank model for these data we consider the model (3.13) of Section 3.1, $Y_k = ABX_k + DZ_k + \epsilon_k$. Here, we take the 'primary' or full-rank set of predictor variables as $Z = (1, x_{21}, x_{22})'$, the wall temperature and solvent feed rate variables, and the reduced-rank set as $X = (x_1, \ldots, x_{20})'$, the 20 temperature measurements along the reactor. To determine an appropriate rank for the coefficient matrix $C = AB$ of the variables X_k, we perform a partial canonical correlation analysis of Y_k on X_k, after eliminating the effects of Z_k, and let $\hat{\rho}_{1j}$, $j = 1, \ldots, m$, denote the sample partial canonical correlations. The LR test statistic $\mathcal{M} = -[T-n+(n_1-m-1)/2] \sum_{j=r_1+1}^{m} \log(1-\hat{\rho}_{1j}^2)$, with $n = 23$ and $n_1 = 20$, discussed in Section 3.1 in relation to (3.12), is used for $r_1 = 0, 1, \ldots, 5$, to test the rank of C, $H_0 : \text{rank}(C) \le r_1$. The results are presented in Table 3.1, and they strongly indicate that the rank of C can be taken as equal to two.

Table 3.1 Summary of results for LR tests on rank of coefficient matrix for chemometrics data.

Eigenvalues $\hat{\lambda}_{1j}^2$	Partial Canonical Correlations $\hat{\rho}_{1j}$	Rank $\le r_1$	$\mathcal{M}$ = LR Statistic	d.f.	Critical Value (5%)
57.198	0.991		310.53	120	146.57
6.427	0.930	1	150.00	95	118.75
1.119	0.727	2	70.80	72	92.81
0.667	0.633	3	41.14	51	68.67
0.396	0.533	4	20.95	32	46.19
0.217	0.422	5	7.76	15	25.00

We then obtain the ML estimates for the model with rank $r_1 = 2$ by first finding the normalized eigenvectors of the matrix

$$\hat{R} = \tilde{\Sigma}_{\epsilon\epsilon}^{-1/2} \hat{\Sigma}_{yx.z} \hat{\Sigma}_{xx.z}^{-1} \hat{\Sigma}_{xy.z} \tilde{\Sigma}_{\epsilon\epsilon}^{-1/2}$$

associated with its two largest eigenvalues, $\hat{\lambda}_{11}^2$ and $\hat{\lambda}_{12}^2$, in Table 3.1. These normalized eigenvectors are found to be

$$\hat{V}_1 = (\,0.15799,\ 0.26926,\ -0.37958,\ 0.57403,\ 0.46378,\ 0.46244)',$$

$$\hat{V}_2 = (\,0.13435,\ 0.03954,\ 0.81760,\ 0.43827,\ -0.21363,\ 0.27241)'.$$

Maximum likelihood estimates of the matrix factors A and B are then computed as $\hat{A}^{(2)} = \tilde{\Sigma}_{\epsilon\epsilon}^{1/2}\hat{V}_{(2)}$ and $\hat{B}^{(2)} = \hat{V}_{(2)}'\tilde{\Sigma}_{\epsilon\epsilon}^{-1/2}\hat{\Sigma}_{yx.z}\hat{\Sigma}_{xx.z}^{-1}$, with $\hat{V}_{(2)} = [\hat{V}_1, \hat{V}_2]$, and the ML estimate of the coefficient matrix D is obtained from $\hat{D} = (\hat{\Sigma}_{yz} - \hat{A}^{(2)}\hat{B}^{(2)}\hat{\Sigma}_{xz})\hat{\Sigma}_{zz}^{-1}$. The ML estimates for A and D are found to be

$$\hat{A}' = \begin{bmatrix} 0.039611 & 0.108042 & -0.047137 & 0.196351 & 0.076314 & 0.074153 \\ 0.067271 & 0.037769 & 0.452448 & 0.229329 & 0.082096 & 0.091330 \end{bmatrix}$$

and

$$\hat{D}' = \begin{bmatrix} 24.5251 & 18.9297 & 109.5563 & 97.6315 & 41.2813 & 40.7216 \\ -0.3685 & -0.2129 & -1.2063 & -1.0284 & -0.3756 & -0.3943 \\ 5.4768 & 27.6520 & -0.8007 & -1.4021 & -0.4906 & -0.5444 \end{bmatrix}.$$

Essentially all of the coefficient estimates in $\hat{A}$ and $\hat{D}$ are highly significant. On examination of the two rows of the ML estimate $\hat{B}^{(2)}$, it is found that the two linear combinations in the 'transformed set of predictor variables' of reduced dimension two, $X_k^* = (x_{1k}^*, x_{2k}^*)' = \hat{B}^{(2)}X_k$, are roughly approximated by

$$\begin{aligned} x_1^* \;\approx\; & -114x_1 - 121x_2 + 154x_3 + 58x_{14} - 85x_{15} \\ & +124x_{16} - 245x_{17} - 108x_{18} + 191x_{19} + 52x_{20}, \end{aligned}$$

$$x_2^* \;\approx\; -175x_2 + 93x_3 - 42x_{11} - 63x_{13} + 94x_{14} - 59x_{15} + 49x_{19} - 108x_{20}.$$

Thus, it is found that certain of the temperature variables, particularly those in the range $x_5 - x_{10}$, do not play a prominent role in the modeling of the response variables, although a relatively large number among the 20 temperature variables do contribute substantially to the two linear combination predictor variables x_1^* and x_2^*. The total number of estimated coefficients in the regression model with the reduced-rank of two imposed on C is $6*3 + 6*2 + 2*20 - 4 = 66$ as compared to $6*23 = 138$ in the full-rank ordinary LS fitted model. The 'average' number of parameters per regression equation in the reduced-rank model is thus 11, as opposed to 23 in the full-rank model. An "approximate unbiased" estimator of the error covariance matrix $\Sigma_{\epsilon\epsilon}$ under the reduced-rank model may be constructed as

$$\begin{aligned} \overline{\Sigma}_{\epsilon\epsilon} &= \frac{1}{56-11}\,\hat{\epsilon}\hat{\epsilon}' \\ &= \begin{bmatrix} 0.03260 & 0.04556 & 0.05621 & 0.01363 & 0.01874 & 0.01023 \\ 0.04556 & 0.36888 & 0.05422 & -0.01976 & -0.00776 & -0.00774 \\ 0.05621 & 0.05422 & 0.40822 & 0.11388 & 0.11901 & 0.05683 \\ 0.01363 & -0.01976 & 0.11388 & 0.18082 & 0.04824 & 0.02197 \\ 0.01874 & -0.00776 & 0.11901 & 0.04824 & 0.06057 & 0.02345 \\ 0.01023 & -0.00774 & 0.05683 & 0.02197 & 0.02345 & 0.03232 \end{bmatrix}, \end{aligned}$$

with $\operatorname{tr}(\overline{\Sigma}_{\epsilon\epsilon}) = 1.08341$, where $\hat{\epsilon} = \boldsymbol{Y} - \hat{A}\hat{B}\boldsymbol{X} - \hat{D}\boldsymbol{Z}$. The diagonal elements of the estimate $\overline{\Sigma}_{\epsilon\epsilon}$ above, when compared to the corresponding entries of the unbiased estimate of $\Sigma_{\epsilon\epsilon}$ under full-rank LS as reported earlier, show substantial reductions in the estimated error variances for the response variables y_3, y_4, and y_6 (of around 14% on average), slight reductions for y_1 and y_2 (around 3% average), and only slight increase for y_5 (around 4%). The reduced-rank model (3.13) thus provides considerable simplification over the full-rank model fitted by LS, and generally provides a better fit of model as well. In the reduced-rank model results, it might also be of interest and instructive to examine in some detail the vectors $\hat{\ell}_j = \tilde{\Sigma}_{\epsilon\epsilon}^{-1/2}\hat{V}_j$, $j = 3, \ldots, 6$, which provide estimates for linear combinations $y^*_{jk} = \hat{\ell}'_j Y_k$ of the response variables that, supposedly, are little influenced by the set of temperature variables. For example, two such linear combinations are estimated to be, approximately, $y^*_6 \approx 1.51y_1 - 1.84y_2 + 2.96y_5$ and $y^*_5 \approx -4.98y_1 + 2.40y_4 - 2.94y_6$. Interpretations for these linear combinations require further knowledge of the subject matter which goes beyond the scope of this illustrative example.

An additional application of the reduced-rank methods in the content of this example is also suggested. One of the goals of the study by Skagerberg et al. (1992) was to 'monitor the performance of the process by means of multivariate control charts' over time, to enable quick detection of changes early in the process. In the case of a univariate measurement, this is typically accomplished by use of statistical process control charts. For this multivariate chemometrics data set, one might require as many as 22 charts for the predictor variables and 6 for the response variables. Because of the association established between the response and the predictor variables in the reduced-rank model, involving only two linear combinations of the temperature profile variables, a small number of linear combinations only might be used to effectively monitor the process. Skagerberg et al. (1992) suggested a similar idea based on linear combinations that result from partial least squares regression methods.

3.4 Both Regression Matrices of Lower Ranks – Model and Its Applications

The assumption that the coefficient matrix of the regressors Z_k in (3.1) is of full rank and cannot be collapsed may not hold in practice. We propose a model which assumes that there is a natural division of regressor variables into two sets, and each coefficient matrix is of reduced rank. This model seems to have wider applications not only for analyzing cross-sectional data where observations are assumed to be independent but also for analyzing chronological (time series) data. The usefulness of the extended model will be illustrated with an application in meteorology where the total ozone measurements taken at five European stations are related to temperature measurements taken at various pressure levels. This model will be shown also to lead to some intuitively simpler models which cannot be otherwise identified as applicable to this data set. Thus, the merit of reduced-rank regression techniques will become more apparent.

The extended reduced-rank model is given as in (3.1) by

$$Y_k = CX_k + DZ_k + \epsilon_k, \qquad k = 1, \ldots, T, \tag{3.21}$$

with the assumption that

$$\begin{aligned} \text{rank}(C) &= r_1 \leq \min(m, n_1) \\ \text{rank}(D) &= r_2 \leq \min(m, n_2) \end{aligned} \tag{3.22}$$

accommodating the possibility that the dimension reduction could be different for each set. Hence we can write $C = AB$ where A is an $m \times r_1$ matrix and B is an $r_1 \times n_1$ matrix. Similarly, we can write $D = EF$ where E is an $m \times r_2$ matrix and F is an $r_2 \times n_2$ matrix.

A situation that readily fits into this setup is given in the appendix of the paper by Jöreskog and Goldberger (1975). Suppose there are two latent variables y_1^* and y_2^* each determined by its own set of exogenous causes and its own disturbances so that

$$y_{1k}^* = \beta_1' X_k + V_{1k}, \qquad y_{2k}^* = \beta_2' Z_k + V_{2k}.$$

The observable indicators Y_k are assumed to be related to the latent variables linearly, giving $Y_k = \alpha_1 y_{1k}^* + \alpha_2 y_{2k}^* + U_k$. Therefore the reduced-rank form equations are

$$Y_k = \alpha_1 \beta_1' X_k + \alpha_2 \beta_2' Z_k + \epsilon_k = CX_k + DZ_k + \epsilon_k$$

where C and D are of unit rank. This example can be formally expanded into two blocks of latent variables each explained by their own causal set. This would result in a higher dimensional example for the extended model.

With the factorization for C and D, we can write the model (3.21) as

$$Y_k = ABX_k + EFZ_k + \epsilon_k, \qquad k = 1, \ldots, T. \tag{3.23}$$

Following Brillinger's interpretation, the n_1 component vector X_k and the n_2 component vector Z_k that arise from two different sources might be regarded as determinants of a "message" Y_k having m components, but there are restrictions on the number of channels through which the message can be transmitted. These restrictions are of different orders for X_k and Z_k. We assume that these vectors X_k and Z_k cannot be assembled at one place. This means that they cannot be collapsed jointly but have to be condensed individually when they arise from their sources. Thus BX_k and FZ_k act as codes from their sources of origin and on receipt they are premultiplied by A and E respectively and summed to yield Y_k, subject to error ϵ_k.

Estimation and inference aspects for the extended reduced-rank model (3.23) were considered by Velu (1991). While the focus of that paper and the remainder of this chapter is on the estimation of component matrices, the complementary feature of linear constraints which are of different dimensions and forms for the different coefficient matrices, has been shown to be useful in multidimensional scaling. Takane, Kiers, and de Leeuw (1995) consider a further extension of model (3.23), to allow for more than two sets of regressor variables, and provide a convergent alternating least squares algorithm to compute the least squares estimates of the component matrices of the reduced-rank coefficient matrices.

A model similar to (3.23) can be used to represent the time lag dependency of a multiple autoregressive time series $\{Y_t\}$. For illustration, let $X_t = Y_{t-1}$ and $Z_t = Y_{t-2}$. Model (3.23) for the time series $\{Y_t\}$ can then be written as $Y_t = ABY_{t-1} + EFY_{t-2} + \epsilon_t$, which provides a framework for dimension reduction for multiple time series data. An even more condensed model follows if, further, $B = F$ when $\text{rank}(C) = \text{rank}(D) = r$. Then $Y_t = A(BY_{t-1}) + E(BY_{t-2}) + \epsilon_t$, and this model has an interpretation in terms of index variables. In this case the lower-dimensional $Y_t^* = BY_t$ acts as an index vector driving the higher-dimensional time series Y_t. These dimension reducing aspects in a multiple autoregressive time series context are discussed in Reinsel (1983), Velu, Reinsel, and Wichern (1986), and Ahn and Reinsel (1988). The reduced-rank model discussed in Section 3.1 would also be sensible to apply in the multiple time series context. For example, we might consider $Y_t = DY_{t-1} + ABY_{t-2} + \epsilon_t$, where the coefficient matrix of the lag one variables Y_{t-1} may be of full rank. An interpretation is that because of the closer ordering in time to Y_t, all variables in Y_{t-1} are needed for Y_t, but some (linear combinations of) variables in the more distant lagged term Y_{t-2} are not needed. This provides an example of a "nested reduced-rank" autoregressive model, and a general class of models of this type was considered by Ahn and Reinsel (1988). Autoregressive models with reduced rank are discussed in detail in Chapter 5.

3.5 Estimation and Inference for the Model

Before we present the efficient estimators of the coefficient matrices for model (3.23), it is instructive to note that combining X_k and Z_k through blocking, as in the conventional reduced-rank model considered in Chapter 2, may not produce a reduced-rank structure. The extended model (3.23) can be written as

$$Y_k = C^* \begin{bmatrix} X_k \\ Z_k \end{bmatrix} + \epsilon_k$$

where $C^* = [C, D]$ is an $m \times (n_1 + n_2)$ matrix. From matrix theory, we have

$$\min(m, r_1 + r_2) \geq \operatorname{rank}(C^*) \geq \max(r_1, r_2).$$

Treating the above model as a classical reduced-rank regression model, when $r = \operatorname{rank}(C^*)$, would yield the factorization

$$C^* = GH = [GH_1, GH_2]$$

where G is an $m \times r$ matrix and H is an $r \times (n_1 + n_2)$ matrix with partitions H_1 and H_2 of dimensions $r \times n_1$ and $r \times n_2$ respectively. Note that both submatrices H_1 and H_2 in the partitioned matrix H have (at most) rank r but this formulation does not yield the desired factorization. Even if the individual matrices H_1 and H_2 are of lower ranks, when put in partitioned form, the partitioned matrix H may be of full rank (i.e., $r = m$) and the model may have no particular reduced-rank structure at all. When m is assumed to be greater than $r_1 + r_2$, the above blocking amounts to overestimating the rank of each individual matrix by $(r_1 + r_2) - \max(r_1, r_2)$. There is also a conceptual difference between the above model and the model in (3.23). This is best seen through the simple case $r = r_1 + r_2$ which we now consider.

The "correct" factorization we are seeking under model (3.23) can be written as

$$[C, D] = [AB, EF] = [A, E] \begin{bmatrix} B & 0 \\ 0 & F \end{bmatrix}$$

where the orders of A, B, E and F are defined as before. The factorization we would arrive at if we worked through a "blocked" version of model (3.23) is

$$[C, D] = GH = [G_1, G_2] \begin{bmatrix} H_{11} & H_{12} \\ H_{21} & H_{22} \end{bmatrix}$$

where G_1 is an $m \times r_1$ matrix, G_2 is an $m \times r_2$ matrix, H_{11} is $r_1 \times n_1$, H_{12} is $r_1 \times n_2$, H_{21} is $r_2 \times n_1$, and H_{22} is $r_2 \times n_2$. Recall the right-hand side matrix in the factorization is used to form codes out of the regressors and in the original formulation of the model (3.23) each regressor set, X_k and Z_k, has its own codes. This feature is violated if we work with the blocked model above.

3.5.1 *Efficient Estimator*

To identify the individual components in the extended reduced-rank model (3.23), e.g., A and B from C, it is necessary to impose certain normalization conditions on them. Normalizations similar to those used in Chapter 2 for the reduced-rank regression model with one set of regressor variables are chosen for this model. For the sample situation, these normalizations are

$$\begin{aligned} A'\Gamma A &= I_{r_1}; \quad B\left(\tfrac{1}{T}\boldsymbol{X}\boldsymbol{X}'\right)B' = \Lambda_{r_1}^2 \\ E'\Gamma E &= I_{r_2}; \quad F\left(\tfrac{1}{T}\boldsymbol{Z}\boldsymbol{Z}'\right)F' = \Lambda_{r_2}^{*2} \end{aligned} \tag{3.24}$$

where $\boldsymbol{X} = [X_1, X_2, \dots, X_T]$ and $\boldsymbol{Z} = [Z_1, Z_2, \dots, Z_T]$ are data matrices, Γ is a positive-definite matrix and $\Lambda_{r_1}^2$ and $\Lambda_{r_2}^{*2}$ are diagonal matrices. Γ can be specified by the analyst whereas the diagonal elements of Λ_{r_1} and $\Lambda_{r_2}^*$ are dependent on C and D respectively. Let $\theta = (\alpha', \beta', \gamma', \delta')'$ denote the vector of unknown regression parameters in model (3.23), where $\alpha = \text{vec}(A')$, $\beta = \text{vec}(B')$, $\gamma = \text{vec}(E')$, and $\delta = \text{vec}(F')$. To arrive at the "efficient" estimator of θ, consider the criterion

$$S_T(\theta) = \frac{1}{2T}\text{tr}\{\Gamma(\boldsymbol{Y} - AB\boldsymbol{X} - EF\boldsymbol{Z})(\boldsymbol{Y} - AB\boldsymbol{X} - EF\boldsymbol{Z})'\} \tag{3.25}$$

where $\boldsymbol{Y} = [Y_1, Y_2, \dots, Y_T]$ is the data matrix of the dependent variables. The estimates are obtained by minimizing criterion (3.25) subject to the normalization conditions (3.24). Criterion (3.25) is the likelihood criterion when the errors ϵ_k are assumed to be normal, and $\Gamma = \hat{\Sigma}_{\epsilon\epsilon}^{-1}$ where $\hat{\Sigma}_{\epsilon\epsilon} = \frac{1}{T}(\boldsymbol{Y} - \hat{A}\hat{B}\boldsymbol{X} - \hat{E}\hat{F}\boldsymbol{Z})(\boldsymbol{Y} - \hat{A}\hat{B}\boldsymbol{X} - \hat{E}\hat{F}\boldsymbol{Z})'$ is the maximum likelihood estimate of the error covariance matrix. For this reason, the resulting estimators can be called efficient. However, for the asymptotic theory we shall initially take $\Gamma = \Sigma_{\epsilon\epsilon}^{-1}$ as known in criterion (3.25), for convenience. The asymptotic theory of the proposed estimator follows from Robinson (1973) or Silvey (1959) and will be presented later.

Several procedures are available for computing the efficient estimators. The first order equations that result from the criterion (3.25) refer only to the necessary conditions for the minimum to occur and so do not guarantee the global minimum. Therefore, we suggest various computational schemes to check on the optimality of a solution found by an alternate method.

The first order equations resulting from (3.25) are

$$\frac{\partial S_T(\theta)}{\partial A'} = -\frac{1}{T}B\boldsymbol{X}(\boldsymbol{Y} - AB\boldsymbol{X} - EF\boldsymbol{Z})'\Gamma \tag{3.26a}$$

$$\frac{\partial S_T(\theta)}{\partial B'} = -\frac{1}{T}\boldsymbol{X}(\boldsymbol{Y} - AB\boldsymbol{X} - EF\boldsymbol{Z})'\Gamma A \tag{3.26b}$$

$$\frac{\partial S_T(\theta)}{\partial E'} = -\frac{1}{T}F\boldsymbol{Z}(\boldsymbol{Y} - AB\boldsymbol{X} - EF\boldsymbol{Z})'\Gamma \tag{3.26c}$$

$$\frac{\partial S_T(\theta)}{\partial F'} = -\frac{1}{T}\boldsymbol{Z}(\boldsymbol{Y} - AB\boldsymbol{X} - EF\boldsymbol{Z})'\Gamma E \tag{3.26d}$$

An iterative procedure that utilizes only the first order equations is the partial least squares procedure [see Gabriel and Zamir (1979)]. The procedure is easy to use when compared with other suggested procedures, and is now described. To start assume that the matrices B and F in model (3.23) are known. The estimates of A and E are easily obtained by least squares regression of Y_k on $[BX_k, FZ_k]$. Thus, starting with estimates $\hat{B}$ and $\hat{F}$ that satisfy the constraints (3.24), $\hat{A}$ and $\hat{E}$ can be obtained by the usual multivariate regression procedure. The resulting $\hat{A}$ and $\hat{E}$ are then adjusted to satisfy the normalization constraints (3.24) before moving to the next step of obtaining new estimates $\hat{B}$ and $\hat{F}$ of B and F.

The first order equations $\partial S_T(\theta)/\partial B = 0$ lead to the solution for B as

$$\hat{B} = \hat{A}'\Gamma[\frac{1}{T}(\boldsymbol{Y} - \hat{E}\hat{F}\boldsymbol{Z})\boldsymbol{X}'][\frac{1}{T}\boldsymbol{X}\boldsymbol{X}']^{-1}. \tag{3.27}$$

Substituting this $\hat{B}$ into the first order equation $\partial S_T(\theta)/\partial F = 0$, we arrive at the equations

$$\hat{F}(\frac{1}{T}\boldsymbol{Z}\boldsymbol{Z}') - Q\hat{F}R = P$$

where $R = \frac{1}{T}\boldsymbol{Z}\boldsymbol{X}'(\frac{1}{T}\boldsymbol{X}\boldsymbol{X}')^{-1}\frac{1}{T}\boldsymbol{X}\boldsymbol{Z}'$, $Q = (\hat{E}'\Gamma\hat{A})(\hat{A}'\Gamma\hat{E})$, and

$$P = \hat{E}'\Gamma(\frac{1}{T}\boldsymbol{Y}\boldsymbol{Z}') - (\hat{E}'\Gamma\hat{A})\hat{A}'\Gamma(\frac{1}{T}\boldsymbol{Y}\boldsymbol{X}')(\frac{1}{T}\boldsymbol{X}\boldsymbol{X}')^{-1}(\frac{1}{T}\boldsymbol{X}\boldsymbol{Z}').$$

We can readily solve for $\hat{F}$ by conveniently stacking the columns of $\hat{F}$, using the "vec" operator and Kronecker products, from

$$[(\frac{1}{T}\boldsymbol{Z}\boldsymbol{Z}' \otimes I) - (R \otimes Q)]\text{vec}(\hat{F}) = \text{vec}(P). \tag{3.28}$$

Observe that with the regression estimates $\hat{A}$ and $\hat{E}$, using (3.28) we can solve for $\hat{F}$ and then use all of these estimates in (3.27) to solve for $\hat{B}$. The normalizations implied by the conditions (3.24) can be carried out using the spectral decomposition of matrices, e.g., Rao (1973, p. 39).

An alternative procedure that can be formulated is along the lines of the modification of the Newton–Raphson method due to Aitchison and Silvey (1958). The details of this procedure will be made clearer when the asymptotic inference results are presented later, but the essential steps can be described as follows. Let $h(\theta)$ be an $r^* = r_1^2 + r_2^2$ vector obtained by stacking the normalization conditions (3.24) without duplication, and let the matrix of first partial derivatives of these constraints be $H_\theta = (\partial h_j(\theta)/\partial\theta_i)$. Note that H_θ is of dimension $s \times r^*$ where $s = r_1(m+n_1) + r_2(m+n_2)$. The first order equations that are obtained by minimizing the criterion $S_T(\theta) - h(\theta)'\lambda$, where λ is a vector of Lagrange multipliers, can be expanded through Taylor's Theorem to arrive at

$$\begin{bmatrix} B_\theta & -H_\theta \\ -H_\theta' & 0 \end{bmatrix} \begin{bmatrix} \sqrt{T}(\hat{\theta} - \theta) \\ \sqrt{T}\hat{\lambda} \end{bmatrix} = \begin{bmatrix} -\sqrt{T}\partial S_T(\theta)/\partial\theta \\ 0 \end{bmatrix} \tag{3.29}$$

In the above expression, $\partial S_T(\theta)/\partial\theta$ is essentially (3.26) arranged in a vector form and B_θ is the expected value of the matrix of second-order derivatives, $\{\partial^2 S_T(\theta)/\partial\theta_i\partial\theta_j\}$.

To obtain explicit expressions for these quantities, we write the matrix form of model (3.23), $\boldsymbol{Y} = C\boldsymbol{X} + D\boldsymbol{Z} + \epsilon = AB\boldsymbol{X} + EF\boldsymbol{Z} + \epsilon$, in vector form as

$$\begin{aligned} y = \mathrm{vec}(\boldsymbol{Y}') &= (I_m \otimes \boldsymbol{X}')\mathrm{vec}(C') + (I_m \otimes \boldsymbol{Z}')\mathrm{vec}(D') + e \\ &= (I_m \otimes \boldsymbol{X}'B')\alpha + (I_m \otimes \boldsymbol{Z}'F')\gamma + e \\ &= (A \otimes \boldsymbol{X}')\beta + (E \otimes \boldsymbol{Z}')\delta + e, \end{aligned} \tag{3.30}$$

where $e = \mathrm{vec}(\boldsymbol{\epsilon}')$, and we have used the relations $\mathrm{vec}(C') = \mathrm{vec}(B'A') = (I_m \otimes B')\mathrm{vec}(A') = (A \otimes I_{n_1})\mathrm{vec}(B')$ and similarly for $\mathrm{vec}(D')$. Then the criterion in (3.25) can be written as $S_T(\theta) = \frac{1}{2T}e'(\Gamma \otimes I_T)e$ with e as in (3.30), and so we find that $\partial S_T(\theta)/\partial\theta$ can be expressed as

$$\frac{\partial S_T(\theta)}{\partial\theta} = \frac{1}{T}\frac{\partial e'}{\partial\theta}(\Gamma \otimes I_T)e = -\frac{1}{T}M'\begin{bmatrix} \Gamma \otimes \boldsymbol{X} \\ \Gamma \otimes \boldsymbol{Z} \end{bmatrix} e, \tag{3.31}$$

where $M' = \mathrm{Diag}[M_1', M_2']$ is block diagonal with the matrices M_1' and M_2' in its main diagonals,

$$M_1 = [I_m \otimes B', A \otimes I_{n_1}] \equiv \left(\frac{\partial \mathrm{vec}(C')}{\partial\alpha'}, \frac{\partial \mathrm{vec}(C')}{\partial\beta'}\right)$$

and

$$M_2 = [I_m \otimes F', E \otimes I_{n_2}] \equiv \left(\frac{\partial \mathrm{vec}(D')}{\partial\gamma'}, \frac{\partial \mathrm{vec}(D')}{\partial\delta'}\right).$$

Notice that the vector of partial derivatives $\partial e/\partial\theta'$ that is needed in (3.31) is easily obtained from (3.30) as $\partial e/\partial\theta' = -[I_m \otimes \boldsymbol{X}'B', A \otimes \boldsymbol{X}', I_m \otimes \boldsymbol{Z}'F', E \otimes \boldsymbol{Z}']$. It also follows, noting that $E(e) = 0$, that the matrix B_θ can be written as

$$\begin{aligned} B_\theta \equiv E\left\{\frac{\partial^2 S_T(\theta)}{\partial\theta\,\partial\theta'}\right\} &= \frac{1}{T}E\left\{\frac{\partial e'}{\partial\theta}(\Gamma \otimes I_T)\frac{\partial e}{\partial\theta'}\right\} \\ &= M'\frac{1}{T}E\begin{bmatrix} \Gamma \otimes \boldsymbol{XX}' & \Gamma \otimes \boldsymbol{XZ}' \\ \Gamma \otimes \boldsymbol{ZX}' & \Gamma \otimes \boldsymbol{ZZ}' \end{bmatrix} M \\ &\equiv M'\Sigma^* M, \end{aligned} \tag{3.32}$$

where

$$\Sigma^* = \begin{bmatrix} \Gamma \otimes \Sigma_{xx} & \Gamma \otimes \Sigma_{xz} \\ \Gamma \otimes \Sigma_{zx} & \Gamma \otimes \Sigma_{zz} \end{bmatrix}.$$

The basic idea in this procedure is to solve for the unknown quantities using equation (3.29), where in practice the sample version of B_θ (that is, the sample version of Σ^*) is used in which the expected value matrices Σ_{xx}, Σ_{xz}, and Σ_{zz} are replaced by their corresponding sample quantities $\hat{\Sigma}_{xx} = \frac{1}{T}\boldsymbol{XX}'$, $\hat{\Sigma}_{xz} = \frac{1}{T}\boldsymbol{XZ}'$, and $\hat{\Sigma}_{zz} = \frac{1}{T}\boldsymbol{ZZ}'$, respectively. There are some essential differences between this procedure and the partial least squares procedure. This procedure uses the second-order derivatives of $S_T(\theta)$ in addition to the first-order derivatives. In addition, the estimates and also the constraints are simultaneously iterated. Because of these differences, the dimensions of the estimators are much larger. An added problem to this procedure is that we cannot use the results on inverting the partitioned matrix, because the rank of B_θ is the same as the rank of M' (which equals $r_1(m+n_1-r_1)+r_2(m+n_2-r_2)$) and therefore B_θ is singular. But the singularity can be removed using a modification suggested by Silvey (1959, p. 399). Observe that the matrix H_θ is of rank $r_1^2+r_2^2$ and it can be demonstrated that $B_\theta + H_\theta H_\theta'$ is of full rank $r_1(m+n_1)+r_2(m+n_2)$. Using this fact, we can rewrite the equation (3.29) as

$$\begin{bmatrix} B_\theta + H_\theta H_\theta' & -H_\theta \\ -H_\theta' & 0 \end{bmatrix} \begin{bmatrix} \sqrt{T}(\hat{\theta}-\theta) \\ \sqrt{T}\hat{\lambda} \end{bmatrix} = \begin{bmatrix} -\sqrt{T}\partial S_T(\theta)/\partial\theta \\ 0 \end{bmatrix}.$$

The iterated estimate of θ at the ith step is given as

$$\begin{bmatrix} \hat{\theta}_i \\ \hat{\lambda}_i \end{bmatrix} = \begin{bmatrix} \hat{\theta}_{i-1} \\ 0 \end{bmatrix} + \begin{bmatrix} B_{\hat{\theta}_{i-1}} + H_{\hat{\theta}_{i-1}} H'_{\hat{\theta}_{i-1}} & -H_{\hat{\theta}_{i-1}} \\ -H'_{\hat{\theta}_{i-1}} & 0 \end{bmatrix}^{-1} \begin{bmatrix} -\partial S_T(\hat{\theta}_{i-1})/\partial\theta \\ 0 \end{bmatrix} \tag{3.33}$$

and using a stopping rule based on either the successive differences between the $\hat{\theta}_i$ or the magnitudes of the Lagrange multipliers $\hat{\lambda}_i$ or $\partial S_T(\hat{\theta}_{i-1})/\partial\theta$, the estimates can be calculated. In general we would expect this procedure to perform better numerically than the partial least squares procedure, because it uses more information. However, in practical applications, the size of the matrix B_θ can be quite large.

A third estimation procedure, which is not (statistically) efficient relative to the above efficient estimator, is presented in the discussion below as an alternative estimator.

3.5.2 An Alternative Estimator

The proposed alternative estimator is similar to the estimator of the parameters in the model (3.3) of Section 3.1. It is inefficient but intuitive and it is based on the full-rank coefficient estimates. Recall that in the model (3.1) or (3.3), D was assumed to be full rank but C could be of reduced rank. It was noted at the end of Section 3.1 that, in the population, solutions for the components A and B of the matrix $C \equiv AB$ under model (3.3) can be obtained as follows: the columns of $\Sigma_{xx.z}^{1/2}B'$ are chosen to be the eigenvectors corresponding to the largest r_1 eigenvalues of $\Sigma_{xx.z}^{-1/2}\Sigma_{xy.z}\Gamma\Sigma_{yx.z}\Sigma_{xx.z}^{-1/2}$ and $A = \Sigma_{yx.z}B'(B\Sigma_{xx.z}B')^{-1}$. The matrix D then is obtained as $D = \Sigma_{yz}\Sigma_{zz}^{-1} - AB\Sigma_{xz}\Sigma_{zz}^{-1}$. The r_1 eigenvalues can be related to partial canonical correlations between Y_k and X_k eliminating Z_k and are used to estimate the rank of C (see Section 3.1). The corresponding estimates in the sample are easily computable using the output from the standard documented computer programs as in the case of the reduced-rank model discussed in Chapter 2.

The alternative estimator is in the same spirit as the estimator in model (3.1) but differs only in the estimation of the matrix C. In Theorem 3.1 the criterion was decomposed into two parts:

$$\operatorname{tr}\{E[\Gamma^{1/2}(Y - ABX^*)(Y - ABX^*)'\Gamma^{1/2}]\}$$
$$+ \operatorname{tr}\{E[\Gamma^{1/2}(Y - D^*Z)(Y - D^*Z)'\Gamma^{1/2}]\}.$$

Here we modify the second part as $\operatorname{tr}\{E[\Gamma^{1/2}(Y - DZ^*)(Y - DZ^*)'\Gamma^{1/2}]\}$ where $D = EF$ and $Z^* = Z - \Sigma_{zx}\Sigma_{xx}^{-1}X$, leaving the first part as it is in the procedure of Section 3.1. This is equivalent to assuming that C is of full rank when estimating the components of the matrix D. Therefore the columns of $\Sigma_{zz.x}^{1/2}F'$ are given by the eigenvectors of $\Sigma_{zz.x}^{-1/2}\Sigma_{zy.x}\Gamma\Sigma_{yz.x}\Sigma_{zz.x}^{-1/2}$ and $E = \Sigma_{yz.x}F'(F\Sigma_{zz.x}F')^{-1}$. Stated differently, in each case (either C or D), in the alternative procedure we first compute the full-rank estimator and then project it to a lower dimension (reduced rank) individually, ignoring the simultaneous nature of the reduced-rank problem. The alternative estimators of the components A, B and E, F are simply the corresponding sample versions of the above expressions.

Implicit in the solution is the fact that the matrices B and F are normalized with respect to partial covariance matrices $\Sigma_{xx.z}$ and $\Sigma_{zz.x}$ respectively, rather than the covariance matrices Σ_{xx} and Σ_{zz} as for the efficient estimator [see equation (3.8)]. For different choices of normalizations, it is assumed that the true parameter belongs to different regions of the parameter space. However, through the transformations from one set of true parameters to the other using the spectral decomposition of matrices, the estimators are made comparable to each other. Thus we provide yet another procedure to arrive at estimates of the extended model (3.23).

3.5.3 Asymptotic Inference

The asymptotic distribution of the maximum likelihood estimator and the alternative estimator can be derived using the theoretical results from Robinson (1973) or Silvey (1959). Before we present the main results we observe the following points. It must be noted that with the choice of $\Gamma = \hat{\Sigma}_{\epsilon\epsilon}^{-1}$, the normalization conditions (3.24) imposed on θ vary with sample quantities. To apply Silvey's (1959) results directly for estimators of θ, the constraints are required to be fixed and not vary with sample quantities. The modifications to account for the randomness could be accommodated by considering the asymptotic theory of Silvey (1959) for the combined set of ML estimators $\hat{\theta}$ and $\hat{\Sigma}_{\epsilon\epsilon}$ jointly, but explicit details on the asymptotic distribution of the ML estimator $\hat{\theta}$ would then become more complicated. Hence, we do not consider results for this case explicitly. As in the case of the asymptotic results of Theorem 2.4 for the single regressor model, we use $\Gamma = \Sigma_{\epsilon\epsilon}^{-1}$ as known instead of $\Gamma = \hat{\Sigma}_{\epsilon\epsilon}^{-1}$, to obtain more convenient explicit results on the asymptotic covariance matrix of the estimator of θ. It is expected that the use of $\Gamma = \Sigma_{\epsilon\epsilon}^{-1}$ instead of $\Gamma = \hat{\Sigma}_{\epsilon\epsilon}^{-1}$ will not have much impact on the asymptotic distribution. Suppose we let (3.25) stand for the criterion with the choice $\Gamma = \hat{\Sigma}_{\epsilon\epsilon}^{-1}$ and let the same criterion and the constraints in (3.24) for the choice of fixed $\Gamma = \Sigma_{\epsilon\epsilon}^{-1}$ be denoted as $S_T^*(\theta)$.

Theorem 3.2 Assume the limits

$$\lim_{T\to\infty}\left(\tfrac{1}{T}\textstyle\sum_k X_k X_k'\right) = \Sigma_{xx}, \qquad \lim_{T\to\infty}\left(\tfrac{1}{T}\textstyle\sum_k Z_k Z_k'\right) = \Sigma_{zz},$$

and $\lim_{T\to\infty}\left(\tfrac{1}{T}\sum_k X_k Z_k'\right) = \Sigma_{xz}$ exist almost surely and $\begin{pmatrix}\Sigma_{xx} & \Sigma_{xz}\\ \Sigma_{zx} & \Sigma_{zz}\end{pmatrix}$

and $\Sigma_{\epsilon\epsilon}$ are positive-definite. Let $\hat{\theta}$ denote the estimator that absolutely minimizes the criterion (3.25) subject to the normalizations (3.24), with the choice $\Gamma = \Sigma_{\epsilon\epsilon}^{-1}$ known. Then as $T \to \infty$, $\hat{\theta} \to \theta$ almost surely and $\sqrt{T}(\hat{\theta} - \theta)$ has a limiting multivariate normal distribution with null mean vector and singular covariance matrix

$$W = (B_\theta + H_\theta H_\theta')^{-1} B_\theta (B_\theta + H_\theta H_\theta')^{-1}, \tag{3.34}$$

where $B_\theta = M'\Sigma^* M$ is given in (3.32) and $H_\theta' = \partial h(\theta)/\partial\theta'$.

Proof: With assumptions stated in the theorem and with an assumption that the parameter space Θ is compact and if the true parameter is taken to be an interior point of Θ, the almost sure convergence follows from Silvey's (1959) results. To prove the limiting distribution part of the theorem, we follow the standard steps. The first-order equations are

$$\begin{aligned} \frac{\partial S_T^*(\theta)}{\partial\theta} - H_\theta\lambda &= 0 \\ h(\theta) &= 0 \end{aligned} \tag{3.35}$$

Let $\hat{\theta}$ and $\hat{\lambda}$ be solutions of the above equations. Because the partial derivatives of $h_i(\theta)$ are continuous,

$$H_{\hat{\theta}} = H_\theta + o(1) \tag{3.36a}$$

and

$$h(\hat{\theta}) = [H_\theta' + o(1)](\hat{\theta} - \theta). \tag{3.36b}$$

By Taylor's theorem,

$$\frac{\partial S_T^*(\theta)}{\partial\theta}\Big|_{\hat{\theta}} = \frac{\partial S_T^*(\theta)}{\partial\theta} + \left\{\frac{\partial^2 S_T^*(\theta)}{\partial\theta\partial\theta'} + o(1)\right\}(\hat{\theta} - \theta) \tag{3.37}$$

Observe that $\frac{\partial^2 S_T^*(\theta)}{\partial\theta\partial\theta'} \to B_\theta$ almost surely. Substituting (3.36) and (3.37) in the equations (3.35) evaluated at $\hat{\theta}$, $\hat{\lambda}$, we arrive at (3.29) which forms the basic components of the asymptotic distribution of $\sqrt{T}(\hat{\theta} - \theta)$.

If we let $\boldsymbol{\epsilon} = \boldsymbol{Y} - AB\boldsymbol{X} - EF\boldsymbol{Z}$, the first-order quantities $\partial S_T^*(\theta)/\partial\theta$ can be written compactly as in (3.31). Observe that the same quantities written in the matrix form in (3.26) are

$$\frac{\partial S_T^*(\theta)}{\partial A'} = -\, B(\frac{1}{T}\boldsymbol{X}\boldsymbol{\epsilon}')\Gamma$$

$$\frac{\partial S_T^*(\theta)}{\partial B'} = -\, (\frac{1}{T}\boldsymbol{X}\boldsymbol{\epsilon}')\Gamma A$$

$$\frac{\partial S_T^*(\theta)}{\partial E'} = -\, F(\frac{1}{T}\boldsymbol{Z}\boldsymbol{\epsilon}')\Gamma$$

$$\frac{\partial S_T^*(\theta)}{\partial F'} = -\, (\frac{1}{T}\boldsymbol{Z}\boldsymbol{\epsilon}')\Gamma E$$

The above can be reexpressed, as in (3.31), using the vec operator and writing $(\Gamma \otimes \boldsymbol{X})e = \mathrm{vec}(\boldsymbol{X}\boldsymbol{\epsilon}'\Gamma)$ and $(\Gamma \otimes \boldsymbol{Z})e = \mathrm{vec}(\boldsymbol{Z}\boldsymbol{\epsilon}'\Gamma)$, as

$$\sqrt{T}\frac{\partial S_T^*(\theta)}{\partial\theta} = -\, M'\sqrt{T}\begin{pmatrix} \mathrm{vec}(\frac{1}{T}\boldsymbol{X}\boldsymbol{\epsilon}'\Gamma) \\ \mathrm{vec}(\frac{1}{T}\boldsymbol{Z}\boldsymbol{\epsilon}'\Gamma) \end{pmatrix}.$$

A simple interpretation for this relation is as

$$\partial S_T^*(\theta)/\partial\theta = (\,\partial\phi'/\partial\theta\,)\,(\,\partial S_T^*(\theta)/\partial\phi\,)$$

where $\phi = \mathrm{vec}(C', D') = (\mathrm{vec}(C')', \mathrm{vec}(D')')'$ denotes the vector of "unrestricted" regression coefficient parameters, with $\partial\phi'/\partial\theta = M'$ and

$$\partial S_T^*(\theta)/\partial\phi = -\,\mathrm{vec}(\frac{1}{T}\boldsymbol{X}\boldsymbol{\epsilon}'\Gamma, \frac{1}{T}\boldsymbol{Z}\boldsymbol{\epsilon}'\Gamma) \equiv -\,\Gamma^{*\prime}\mathrm{vec}(\frac{1}{T}\boldsymbol{X}\boldsymbol{\epsilon}', \frac{1}{T}\boldsymbol{Z}\boldsymbol{\epsilon}'),$$

with $\Gamma^{*\prime} = \mathrm{Diag}[\Gamma \otimes I_{n_1}, \Gamma \otimes I_{n_2}]$. From the multivariate linear regression theory, we know that

$$\sqrt{T}\begin{pmatrix} \mathrm{vec}(\frac{1}{T}\boldsymbol{X}\boldsymbol{\epsilon}') \\ \mathrm{vec}(\frac{1}{T}\boldsymbol{Z}\boldsymbol{\epsilon}') \end{pmatrix} \xrightarrow{d} N(0, \Sigma^{**})$$

where

$$\Sigma^{**} = \begin{bmatrix} \Sigma_{\epsilon\epsilon} \otimes \Sigma_{xx} & \Sigma_{\epsilon\epsilon} \otimes \Sigma_{xz} \\ \Sigma_{\epsilon\epsilon} \otimes \Sigma_{zx} & \Sigma_{\epsilon\epsilon} \otimes \Sigma_{zz} \end{bmatrix}$$

and, therefore, with $\Gamma = \Sigma_{\epsilon\epsilon}^{-1}$, $\sqrt{T}\frac{\partial S_T^*(\theta)}{\partial\theta} \xrightarrow{d} N(0, B_\theta)$, where $B_\theta = M'\Gamma^{*\prime}\Sigma^{**}\Gamma^* M = M'\Sigma^* M$. The asymptotic covariance matrix of $\hat{\theta}$ is the relevant partition (upper left block) of the following matrix:

$$\begin{bmatrix} B_\theta & -H_\theta \\ -H_\theta' & 0 \end{bmatrix}^{-1} \begin{bmatrix} B_\theta & 0 \\ 0 & 0 \end{bmatrix} \begin{bmatrix} B_\theta & -H_\theta \\ -H_\theta' & 0 \end{bmatrix}^{-1} \tag{3.38}$$

The above expression follows from (3.29) and the upper left partitioned portion of the resulting matrix product will yield the asymptotic covariance matrix of $\hat{\theta}$. In order to arrive at the explicit analytical expression W, we can use Silvey's (1959) modification, where we replace B_θ by $B_\theta + H_\theta H_\theta'$ in (3.38) in the matrices that require inverses. The standard results for finding the inverse of a partitioned matrix [e.g., see Rao (1973, p. 33)] will yield the asymptotic covariance matrix W.

For the numerical example that will be illustrated in the last section of this chapter, we compute estimates using the partial least squares procedure and their standard errors are calculated via the matrix W in the above theorem. Although the quantity H_θ does not appear to be easily written in a compact form, it can be computed using standard results on matrix differentiation of the quadratic functions of the elements of A, B, E, and F involved in the normalization conditions (3.24).

The asymptotic distribution of the alternative estimator, which was motivated by the model (3.1) of Anderson and which was discussed earlier, can also be considered. A limiting multivariate normal distribution result can be established for this estimator by following the same line of argument as in the proof of Theorem 3.2, but we omit the details. A main feature of the results is that the alternative estimator is asymptotically inefficient relative to the estimator of Theorem 3.2, although the expression for the asymptotic covariance matrix of the alternative estimator is somewhat cumbersome and so is not easily compared with the asymptotic covariance matrix (3.34) in Theorem 3.2. The proposed alternative estimator, it is again emphasized, does not have any overall optimality properties corresponding to the extended reduced-rank model (3.23), but is mainly motivated by the model (3.3) in Section 3.1. The usefulness of this estimator is that it provides initial values in the iterative procedures for the efficient estimator and more importantly that the associated partial canonical correlations can be used to specify (identify) the ranks of the coefficient matrices C and D in model (3.23). This topic is taken up in the following section.

Comments. For the reduced-rank model (2.11) with a single regressor set X_k considered in Chapter 2, the asymptotic distribution results for the corresponding reduced-rank estimators $\hat{A}$ and $\hat{B}$ can be obtained as a special case of the approach and results in Theorem 3.2 simply by eliminating aspects related to the second regressor set Z_k. In Theorem 2.4 of Section 2.5, the asymptotic results for the reduced-rank estimators in the single regressor model (2.11) were stated in more explicit detail in terms of eigenvalues and eigenvectors of relevant matrices, and the estimators were obtained in more explicit noniterative form as compared to results for the extended model (3.23). Nevertheless, the final asymptotic distribution results are the same in both approaches. In addition, as in the single regressor model, the results of Theorem 3.2 enable the asymptotic distribution of the overall coefficient matrix estimators $\hat{C} = \hat{A}\hat{B}$ and $\hat{D} = \hat{E}\hat{F}$, obtained as functions of the reduced-rank estimators in the extended model, to be recovered. Specifically, from the relation

$$
\begin{aligned}
T^{1/2}\,\mathrm{vec}(\hat{C}' - C') &= T^{1/2}[I_m \otimes B',\, A \otimes I_{n_1}] \begin{bmatrix} \mathrm{vec}(\hat{A}' - A') \\ \mathrm{vec}(\hat{B}' - B') \end{bmatrix} + o_p(1) \\
&\equiv T^{1/2}\, M_1 \begin{bmatrix} \mathrm{vec}(\hat{A}' - A') \\ \mathrm{vec}(\hat{B}' - B') \end{bmatrix} + o_p(1)
\end{aligned}
$$

and a similar relation for $T^{1/2}\mathrm{vec}(\hat{D}' - D')$ in terms of $T^{1/2}\mathrm{vec}(\hat{E}' - E')$ and $T^{1/2}\mathrm{vec}(\hat{F}' - F')$, the asymptotic distribution result $T^{1/2}\,\mathrm{vec}(\hat{C}' - C', \hat{D}' - D') \xrightarrow{d} N(0, MWM')$ readily follows, where M is defined below equation (3.31) and W is the asymptotic covariance matrix of Theorem 3.2 as defined in (3.34). Also notice that the covariance matrix W is singular, since the matrix $B_\theta = M'\Sigma^* M$ is singular. However, it can be verified that W is a (reflexive) generalized inverse of B_θ (that is, the relations $B_\theta W B_\theta = B_\theta$ and $W B_\theta W = W$ both hold), which we denote as $W = B_\theta^-$. Therefore, the above asymptotic covariance matrix for $T^{1/2}\,\mathrm{vec}(\hat{C}' - C', \hat{D}' - D')$ can be expressed as $MB_\theta^- M' = M\{M'\Sigma^* M\}^- M'$. This form can be compared to the asymptotic covariance matrix for the full-rank least squares estimators $\tilde{C}$ and $\tilde{D}$ in the extended model, which ignore the reduced-rank restrictions. The asymptotic covariance matrix of this estimator, $T^{1/2}\,\mathrm{vec}(\tilde{C}' - C', \tilde{D}' - D')$, is given by Σ^{*-1}.

3.6 Identification of Ranks of Coefficient Matrices

It has been assumed up to this point that the ranks of the coefficient matrices C and D in model (3.21) are known. Usually in practice these ranks are unknown a priori and have to be estimated from the sample data. A formal testing procedure is presented, but the effectiveness of the procedure has not been fully investigated.

In the extended reduced-rank model (3.21) if the structure on the coefficient matrix C is ignored, then the rank of D can be estimated from partial canonical correlation analysis between Y_k and Z_k eliminating X_k (see Section 3.1). The rank of D is taken to be the number of eigenvalues $\hat{\rho}_{2j}^2$ of $\hat{\Sigma}_{yy.x}^{-1/2}\hat{\Sigma}_{yz.x}\hat{\Sigma}_{zz.x}^{-1}\hat{\Sigma}_{zy.x}\hat{\Sigma}_{yy.x}^{-1/2}$ which are not significantly different from zero. The likelihood ratio criterion for testing whether the number of corresponding population eigenvalues which are zero is equal to $m-r_2$ is given by the test statistic similar to (3.12), namely $-T\sum_{j=r_2+1}^{m}\log(1-\hat{\rho}_{2j}^2)$, where the $\hat{\rho}_{2j}$ are the sample partial canonical correlations between Y_k and Z_k eliminating X_k. For large samples, this statistic is approximately the same as $T\sum_{j=r_2+1}^{m}\hat{\rho}_{2j}^2$. These criteria for testing the hypothesis that $\text{rank}(D) = r_2$ against the alternative that D is of full rank are shown to be asymptotically distributed as χ^2 with $(m-r_2)(n_2-r_2)$ degrees of freedom (e.g., Anderson, 1951). We can use similar criteria to test for the rank of C assuming that D is full rank.

Convenient formal test procedures for the ranks when both coefficient matrices are assumed to be rank deficient are difficult to formulate because of the simultaneous nature of the problem. Given $X_k^* = BX_k$, model (3.23) is in the framework of the model (3.3) studied in Section 3.1, and hence the rank of D is precisely estimated by partial canonical correlation analysis between Y_k and Z_k eliminating X_k^*. Similarly partial canonical correlation analysis between Y_k and X_k eliminating $Z_k^* = FZ_k$ can be used to test for the rank of C, if the matrix F were given. In these computations, it is assumed that B or F is known, but in practice these have to be estimated by the corresponding sample quantities. Hence, objective criteria that do not depend on the values of B or F are not immediately apparent. However, we argue that, in the population, conditioning on the total information X_k is equivalent to conditioning on the partial information $X_k^* = BX_k$.

We can write model (3.23) as $Y_k = [A,\ 0]V_k + EFZ_k + \epsilon_k$, where $V_k = [B',\ B_1']'X_k$ and B_1 is a matrix of dimension $(n_1 - r_1) \times n_1$. Assuming the population version of the normalization conditions (3.24) we have $B\Sigma_{xx}B' = \Lambda_{r_1}^2$ and we can complete the orthogonalization procedure by finding B_1 such that $B_1\Sigma_{xx}B_1' = \Lambda_{n_1-r_1}^2$ and $B\Sigma_{xx}B_1' = 0$. The additional variables B_1X_k included in V_k are intended to capture any information from X_k that exists beyond the linear combinations BX_k that are assumed to be relevant in the model (3.23). Hence, we have a one-to-one transformation from X_k to V_k, and for a known $[B',\ B_1']'$ conditioning on V_k is equivalent to conditioning on X_k. Thus, in the population the rank of the matrix $D = EF$ can be determined as the number of nonzero partial canonical correlations between Y_k and Z_k given X_k as in the model of Section 3.1. A more general population result that throws some light on relative efficiencies of the two procedures, i.e., canonical correlation analysis between Y_k and Z_k eliminating $X_k^* = BX_k$ versus eliminating X_k, in the sample is presented below. An analogous sample version of the result indicates that using sample partial canonical correlations, eliminating X_k, to test whether the hypothesis that $\text{rank}(D) = r_2$ is true, is less powerful than using those eliminating only BX_k.

Lemma 3.1. Let the true model be $Y = AX^* + DZ + \epsilon$ where $X^* = BX$. Also let ρ_j^{*2} and ρ_j^2 $(j = 1, 2, \ldots, m)$ be the (population) squared partial canonical correlations between Y and Z obtained by eliminating X^* and X, respectively. Suppose that $\text{rank}(D) = r_2$; then (a) $\rho_j^{*2} \geq \rho_j^2$, for all j, and (b) $\rho_j^* = \rho_j = 0$, $j = r_2 + 1, \ldots, m$.

Proof: Assume without loss of generality that the covariance matrices Σ_{zz}, Σ_{xx} and $\Sigma_{\epsilon\epsilon}$ are identity matrices. Then, since $\Sigma_{yz} = [A,\ 0]\Sigma_{xz} + D\Sigma_{zz} = [A,\ 0]\Sigma_{xz} + D$, $\Sigma_{yx} = [A,\ 0]\Sigma_{xx} + D\Sigma_{zx} = [A,\ 0] + D\Sigma_{zx}$, and $\Sigma_{yy} = AA' + DD' + I + [A,\ 0]\Sigma_{xz}D' + D\Sigma_{zx}[A,\ 0]'$, we find that the relevant partial covariance matrices needed in the partial canonical correlation analysis of Y and Z eliminating X are given by $\Sigma_{zz.x} = \Sigma_{zz} - \Sigma_{zx}\Sigma_{xx}^{-1}\Sigma_{xz} = I - \Sigma_{zx}\Sigma_{xz}$, $\Sigma_{yz.x} = \Sigma_{yz} - \Sigma_{yx}\Sigma_{xx}^{-1}\Sigma_{xz} = D(I - \Sigma_{zx}\Sigma_{xz})$, and $\Sigma_{yy.x} = \Sigma_{yy} - \Sigma_{yx}\Sigma_{xx}^{-1}\Sigma_{xy} = I + D(I - \Sigma_{zx}\Sigma_{xz})D'$. The partial covariance matrices when eliminating only X^* can be obtained by simply replacing X by X^* in these expressions. Thus it follows that the squared partial canonical correlations between Y and Z eliminating X are the roots of the determinantal equation $|D(I - \Sigma_{zx}\Sigma_{xz})D' - \rho^2/(1 - \rho^2)I| = 0$. Similarly, the squared partial canonical correlations between Y and Z eliminating X^* are the roots of the determinantal equation $|D(I - \Sigma_{zx^*}\Sigma_{x^*z})D' - \rho^{*2}/(1 - \rho^{*2})I| = 0$. Since $I - \Sigma_{zx}\Sigma_{xz} \geq 0$, the number of nonzero eigenvalues (roots) in either equation is the same as the rank of D.

To prove part (a) note that $D(I - \Sigma_{zx^*}\Sigma_{x^*z})D' \geq D(I - \Sigma_{zx}\Sigma_{xz})D'$ since the matrices in parentheses are respectively the residual covariance matrices obtained by regressing Z on X and Z on X^*, a subset of X. This implies that $\rho_j^{*2}/(1 - \rho_j^{*2}) \geq \rho_j^2/(1 - \rho_j^2)$ for all j, and hence, $\rho_j^{*2} \geq \rho_j^2$, $j = 1, 2, \ldots, m$.

The above lemma indicates that in the population, the rank of D for the extended model (3.23) can be determined using the partial canonical correlations similar to methods for the model (3.3) of Section 3.1. But in the sample, an analogous version of the above result implies that the procedure is not efficient. To emphasize that the quantity ρ^* depends on the choice of the matrix B, we write it as $\rho^*(B)$. Under the null hypothesis that $\text{rank}(D) = r_2$, both statistics $\hat{\psi} = -T\sum_{j=r_2+1}^{m}\log(1 - \hat{\rho}_j^2)$ and $\hat{\psi}^* = -T\sum_{j=r_2+1}^{m}\log(1 - \hat{\rho}_j^{*2}(B))$ follow χ^2 distributions with $(m - r_2)(n_2 - r_2)$ degrees of freedom (when $\hat{\psi}^*$ uses the true value of B). Since $\hat{\psi}^* \geq \hat{\psi}$, which follows from the result $\hat{\rho}_j^{*2} \geq \hat{\rho}_j^2$ for all $j = 1, \ldots, m$ that is implied by a sample analogue of the above lemma, use of (sample) partial canonical correlation analysis with the statistic $\hat{\psi}$ to test the null hypothesis that $\text{rank}(D) = r_2$ is less powerful than with the statistic $\hat{\psi}^*$.

An alternative data-based method for estimating the ranks of the coefficient matrices is the rank trace method given by Izenman (1980). This calls for examining the residual covariance matrix for every possible rank combination, and so it involves a large amount of computational effort. Our limited experience suggests that the use of partial canonical correlations, which represents a much easier computation, is a good way to identify the ranks.

3.7 An Example on Ozone Data

Ozone is found in the atmosphere between about 5 km and 60 km above the earth's surface and plays an important role in the life-cycle on earth. It is well known that stratospheric ozone is related to other meteorological variables, such as stratospheric temperatures. Reinsel et al. (1981), for example, investigated the correlation between total column ozone measured from Dobson ground stations and measurements of atmospheric temperature at various pressure levels taken near the Dobson stations. A typical pattern of correlations between total ozone and temperature as a function of pressure level of the temperature reading was given by Reinsel et al. (1981) for the data from Edmonton, Canada.

Temperature measurements taken at pressure levels of 300 mbar and below are negatively correlated with total ozone measurements, and temperature measurements taken at pressure levels of 200 mbar and above are positively correlated with ozone. For the observations from Edmonton, a maximum negative correlation between total ozone and temperature occurs in the middle troposphere at 500 mbar ($\simeq$ 5 km), and a maximum positive correlation occurs in the stratosphere at 100 mbar ($\simeq$ 16 km). However, these patterns are known to change somewhat with location, especially with latitude.

We consider here the monthly averages of total ozone measurements taken from January 1968 to December 1977 for five stations in Europe. The stations are Arosa (Switzerland), Vigna Di Valle (Italy), Mont Louis (France), Hradec Kralove (Czech Republic), and Hohenpeissenberg (Germany). The ozone observations, which represent the time series Y_t, are to be related to temperature measurements taken at 10 different pressure levels. The temperature measurement at each pressure level is an average of the temperature measurements at that pressure level taken around the three stations Arosa, Vigna Di Valle, and Hohenpeissenberg. The series are seasonally adjusted using regression on sinusoidal functions with periods 12 and 6 months, before the multivariate regression analysis is carried out.

The correlations, averaged over the five ozone stations, between the ozone measurements and the temperature data for the 10 pressure levels are displayed in Figure 3.1. As can be seen from Figure 3.1, the average cor-

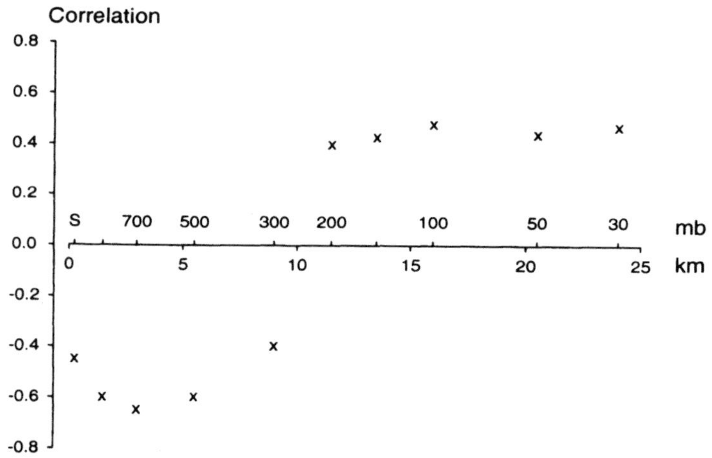

Figure 3.1. Average correlation between ozone and temperatures at European stations versus pressure level (and average height)

relations confirm the pattern mentioned earlier. But, in general, negative correlations are of greater magnitude than positive correlations. The zone between the troposphere and stratosphere known as the tropopause provides a natural division between the temperature measurements in influencing ozone measurements. The temperature measurements in the troposphere, at the first five pressure levels (the pressure levels below 10 km), form the first set of time series regressor variables X_t, and the temperature measurements at the next five pressure levels (above 10 km) in the stratosphere form the second set of regressors Z_t. As mentioned earlier, all series are seasonally adjusted prior to the multivariate analysis. The ozone series exhibit weak autocorrelations of magnitude about 0.2 which have been ignored in the analysis reported here. Because the ozone stations are relatively close to each other, and the temperature series are averages representing the region, with substantial correlations in neighboring pressure levels, some redundancy in the parameters can be expected and thus the extended reduced-rank multivariate regression model may be appropriate.

A standard multivariate regression model $Y_t = CX_t + DZ_t + \epsilon_t$ was initially fitted to the data, and the least squares estimates of the coefficient matrices were obtained as

$$\tilde{C} = \begin{bmatrix} -0.99 & 1.28 & -4.51 & -5.09* & 2.66 \\ -0.95 & -0.61 & -3.70 & 0.48 & -1.87 \\ 0.33 & 1.77 & -1.51 & -2.02 & -3.35* \\ 0.59 & 0.65 & -5.74 & -7.29* & 6.15* \\ -1.12 & 2.20 & -6.14* & -5.53* & 3.95* \end{bmatrix}$$

$$\tilde{D} = \begin{bmatrix} -0.75 & -0.48 & 1.62 & -2.55 & 4.62* \\ 0.89 & -3.30 & 4.12* & -4.14* & 3.88* \\ 1.39 & -1.45 & 2.93 & -2.39 & 3.22* \\ -3.30 & 1.11 & 1.13 & -3.48 & 5.09* \\ -1.76 & -0.05 & 1.13 & -0.65 & 3.39* \end{bmatrix}$$

The asterisks indicate that the individual coefficients have t-ratios greater than 2. Taking that as a rule of thumb for their significance, it follows that the temperature measurements in the last three pressure levels of both the troposphere and the stratosphere have greater influence than the measurements in the first two pressure levels when all are considered together. The mere nonsignificance of the elements of the first two columns in both matrices $\tilde{C}$ and $\tilde{D}$ probably indicates that these coefficient matrices may be of reduced rank.

As argued in Section 3.6, formal test procedures for the ranks when both C and D are rank deficient are difficult to formulate, but we still may use partial canonical correlations to identify the ranks as in the testing procedure for the model considered in Section 3.1. Though this procedure of using partial canonical correlations is not efficient, it is a good practical tool, as illustrated by this example. The sample partial canonical correlations $\hat{\rho}_{1j}$ between Y_t and X_t eliminating Z_t are found to be 0.71, 0.43, 0.34, 0.11, and 0.07. Using Bartlett's test statistic (i.e., the asymptotically equivalent form of the LR test statistic given in (3.12) of Section 3.1) to test the hypothesis $H_0\colon \text{rank}(C) = r_1$ versus $H_1\colon \text{rank}(C) = 5$ for $r_1 = 1, 2$, the rejection tail probabilities (p-values) corresponding to the two null hypotheses are found to be 0.001 and 0.076. If the conventional rule of fixing the level of significance as either 0.05 or 0.01 is followed, the hypothesis that $\text{rank}(C) = 2$ cannot be rejected. Since this procedure is inefficient, the possibility that $\text{rank}(C) = 3$ (the corresponding tail probability is 0.076) may also be entertained. When a similar canonical correlation analysis between Y_t and Z_t eliminating X_t is conducted, the partial canonical correlations are 0.53, 0.35, 0.28, 0.12, and 0.08. Using Bartlett's test statistic to test the hypothesis $H_0 : \text{rank}(D) = r_2$ versus $H_1 : \text{rank}(D) = 5$ for $r_2 = 1, 2$, the rejection tail probabilities are found to be 0.04 and 0.24. Here again, the possibility that the rank of D is either 1 or 2 is entertained.

Comparison of various model fittings are based on the information criterion (BIC) as discussed by Schwarz (1978), and a summary of the results is presented in Table 3.2. Of the two-regressor-sets reduced-rank models, the model fitted with ranks 2 and 1 for C and D, respectively, has the smallest BIC value, which suggests that the partial canonical correlations may be useful tools for identifying the ranks of C and D. (If a stringent significance level of 0.01 had been used, the tests based on partial canonical correlations could have concluded that the ranks are 2 and 1.) The model with ranks 1 and 1 also comes close to attaining the minimum BIC value. The estimated components of the best fitting model, obtained using the partial least squares computational approach, are as follows (the entries in parentheses are the standard errors computed using equation (3.34) from the results in Theorem 3.2):

$$\hat{A}' = \begin{bmatrix} \begin{matrix} 8.28 \\ (0.39) \end{matrix} & \begin{matrix} 5.86 \\ (0.43) \end{matrix} & \begin{matrix} 4.59 \\ (0.56) \end{matrix} & \begin{matrix} 8.48 \\ (0.63) \end{matrix} & \begin{matrix} 8.55 \\ (0.39) \end{matrix} \\ \begin{matrix} 1.03 \\ (1.16) \end{matrix} & \begin{matrix} 0.21 \\ (1.10) \end{matrix} & \begin{matrix} -5.77 \\ (0.49) \end{matrix} & \begin{matrix} 4.12 \\ (1.35) \end{matrix} & \begin{matrix} 1.85 \\ (1.13) \end{matrix} \end{bmatrix}$$

$$\hat{B} = \begin{bmatrix} \begin{matrix} -0.06 \\ (0.13) \end{matrix} & \begin{matrix} 0.15 \\ (0.22) \end{matrix} & \begin{matrix} -0.61 \\ (0.24) \end{matrix} & \begin{matrix} -0.52 \\ (0.24) \end{matrix} & \begin{matrix} 0.16 \\ (0.17) \end{matrix} \\ \begin{matrix} 0.00 \\ (0.12) \end{matrix} & \begin{matrix} -0.38 \\ (0.21) \end{matrix} & \begin{matrix} -0.25 \\ (0.23) \end{matrix} & \begin{matrix} 0.46 \\ (0.19) \end{matrix} & \begin{matrix} 0.55 \\ (0.14) \end{matrix} \end{bmatrix}$$

$$\hat{E}' = \begin{bmatrix} \begin{matrix} 8.07 \\ (0.84) \end{matrix} & \begin{matrix} 5.74 \\ (0.89) \end{matrix} & \begin{matrix} 4.65 \\ (0.85) \end{matrix} & \begin{matrix} 9.01 \\ (1.09) \end{matrix} & \begin{matrix} 8.25 \\ (0.84) \end{matrix} \end{bmatrix}$$

$$\hat{F} = \begin{bmatrix} \begin{matrix} -0.09 \\ (0.14) \end{matrix} & \begin{matrix} -0.13 \\ (0.22) \end{matrix} & \begin{matrix} 0.32 \\ (0.22) \end{matrix} & \begin{matrix} -0.38 \\ (0.18) \end{matrix} & \begin{matrix} 0.57 \\ (0.13) \end{matrix} \end{bmatrix}$$

In this model the temperature measurements at the first two pressure levels do not play a significant role, as can be seen from the standard errors of the corresponding coefficients in both $\hat{B}$ and $\hat{F}$. This aspect was briefly noted in the discussion of the full-rank coefficient estimates. The reduced-rank model also has an interpretation that is consistent with the correlation pattern mentioned earlier. Specifically, the first linear combination of X_t has large negative values for the temperature measurements at 700 mbar and 500 mbar, and the linear combination of Z_t mainly consists of the last measurement in the stratosphere with a predominantly positive influence.

Table 3.2 Summary of results for reduced-rank regression models fitted to ozone-temperature data.

Type of regression model	Rank of C	Rank of D	Number of parameters, n^*	BIC $= \log\lvert\hat{\Sigma}\rvert$ $+(n^* \log T)/T$
–	–	–	0	$\log\lvert\hat{\Sigma}_{yy}\rvert = 22.3591$
Full-rank	5	5	50	22.4683
Two regressors	1	1	18	21.8511
	2	1	25	21.8466
	2	2	32	21.9704
	3	2	37	22.0670

The temperature series, it may be recalled, are averages taken over three stations and thus are meant to represent the average temperature across the region, representative for all the stations. Because of the average nature of the temperature series and the proximity of the Dobson ozone stations under study, it may be expected that the most significant linear combinations of the temperature measurements will have a similar influence on ozone in all five stations. This is broadly confirmed by the first row elements of the component $\hat{A}$ and by the elements of the component $\hat{E}$ in the model. The first linear combination of $\hat{B}X_t$ and $\hat{F}Z_t$ may have the same influence on ozone measurements of all five stations. This is tested by fitting a model where individual ranks of C and D are assumed to be unity and further constraining the elements within A and within E to be equal. This is equivalent to a multiple regression procedure where the collection of all ozone measurements (combined over all stations) is regressed on the 10 temperature measurements, taking account of contemporaneous correlations across the five stations. This will be called the 'constrained' model,

$$Y_t = \mathbf{1}BX_t + \mathbf{1}FZ_t + \epsilon_t \qquad \text{or} \qquad \boldsymbol{Y} = \mathbf{1}\,[B,\ F][\boldsymbol{X}',\ \boldsymbol{Z}']' + \boldsymbol{\epsilon},$$

where $\mathbf{1}$ denotes an $m \times 1$ vector of ones, and the following estimation results were obtained:

$$\hat{B} = 7.5\,\begin{bmatrix} -0.06 & 0.14 & -0.58* & -0.52 & 0.20 \end{bmatrix}$$

$$\hat{F} = 7.0\,\begin{bmatrix} -0.10 & -0.12 & 0.31* & -0.38* & 0.58* \end{bmatrix}$$

Notice that this constrained model can be viewed as a special case of a conventional reduced-rank model of rank one in which the left factor in the regression coefficient matrix is constrained to be equal to $\mathbf{1}$.

It is interesting to see that the estimated combinations of temperature measurements using the constrained model are roughly proportional to the first rows of $\hat{B}$ and $\hat{F}$. The merit of the reduced-rank regression techniques is apparent when we consider that we might not have identified the constrained model structure from the data by merely looking at the full-rank coefficient estimates $\tilde{C}$ and $\tilde{D}$. As a contrast to the reduced-rank model and the constrained model, a 'simplified' model can be fitted where a temperature measurement at one fixed pressure level from the tropospheric region and one from the stratospheric region are selected as the potential explanatory variables. The two particular levels may vary from station to station, and we select the temperature level whose estimated regression coefficient has the largest negative t-ratio in the troposphere and the level whose estimate has the largest positive t-ratio in the stratosphere, in the full-rank regression analysis, as potential variables.

The performance of the three models is compared in terms of residual variances of individual series rather than using an overall measure such as BIC. The results are given in Table 3.3. It is clear from Table 3.3 that the results for the reduced-rank model and the constrained model are fairly close to those of the full-rank model. This indicates that we do not lose too much information by reducing the dimension of the temperature measurements. Even the simplified model performs close to the other two models for the three stations Vigna Di Valle, Mont Louis, and Hohenpeissenberg.

Table 3.3 Summary of the analysis of the ozone-temperature data residual variances.

		Residual Variance			
Station	Total Variance	Full rank model	Reduced rank model	Simplified model	Constrained model
Arosa	259.08	99.84	100.67	114.13	104.04
Vigna Di Valle	159.97	73.35	80.27	84.09	85.74
Mont Louis	124.10	56.85	60.72	68.95	88.92
Hradec Kralove	323.94	139.59	147.35	182.14	158.76
Hohenpeissenberg	274.78	103.28	105.99	112.36	112.36

4

Reduced-Rank Regression Model With Autoregressive Errors

4.1 Introduction and the Model

The classical multivariate regression methods are based on the assumptions that (i) the regression coefficient matrix is of full rank and (ii) the error terms in the model are independent. In Chapters 2 and 3, we have presented regression models that describe the linear relationships between two or more large sets of variables with a fewer number of parameters than that posited by the classical model. The assumption (i) of full rank of the coefficient matrix was relaxed and the possibility of reduced rank for the coefficient matrix has produced a rich class of models. In this chapter we also weaken the assumption (ii) that the errors are independent, to allow for possible correlation in the errors which may be likely with time series data. For the ozone/temperature time series data considered in Chapter 3, the assumption of independence of errors appears to hold.

When the data are in the form of time series, as we consider in this chapter, the assumption of serial independence of errors is often not appropriate. In these circumstances, it is important to account for serial correlation in the errors. Robinson (1973) considered the reduced-rank model for time series data in the frequency domain with a general structure for errors. In particular, assuming the errors follow a general stationary process, Robinson studied the asymptotic properties of both the least squares and other more efficient estimators of the regression coefficient matrix. However, often the serially correlated disturbances will satisfy more restrictive assumptions and, in these circumstances, it would seem possible to exploit this structure to obtain estimates which may have better small sample properties than the estimators proposed by Robinson. [For a related argument, see Gallant and Goebel (1976).] This approach was taken by Velu and Reinsel (1987), and in this chapter we present the relevant methodology and results.

The assumption we make on the errors is that they follow a (first-order) multivariate autoregressive process. We consider the estimation problem with the autoregressive error assumption in the time domain. The asymptotic distribution theory for the proposed estimator is derived, and computational algorithms for the estimation procedure are also briefly discussed. Discussion of methods for the initial specification and testing of the rank of the coefficient matrix is presented, and an alternative estimator, which is motivated by the need for initial identification of the rank of the model, is also suggested. As an example, we present an analysis of some macroeconomic data of the United Kingdom which was previously considered by Gudmundsson (1977), and which has been discussed earlier in Example 2.2 of Section 2.2.

Consider the multivariate regression model for the vector time series $\{Y_t\}$,

$$Y_t = CX_t + U_t, \qquad t = 1, 2, \ldots, T, \tag{4.1}$$

where Y_t is an $m \times 1$ vector of response variables, X_t is an $n \times 1$ vector of predictor variables, and C is an $m \times n$ regression coefficient matrix. As in Chapter 2, we assume that C is of reduced rank,

$$\text{rank}(C) = r \leq \min(m, n). \tag{4.2}$$

We assume further that the error terms U_t satisfy the stochastic difference equation

$$U_t = \Phi U_{t-1} + \epsilon_t \tag{4.3}$$

where U_0 is taken to be fixed. Here the ϵ_t are independently distributed random vectors with mean zero and positive-definite covariance matrix $\Sigma_{\epsilon\epsilon}$, and $\Phi = (\phi_{ij})$ is an $m \times m$ matrix of unknown parameters, with all eigenvalues of Φ less than one in absolute value. Hendry (1971), Hatanaka (1976) and Reinsel (1979), among others, have considered maximum likelihood estimation of the model parameters for the specification (4.1) and (4.3). When the regression matrix C is assumed to be of lower rank r, following the discussion in Chapter 2, write $C = AB$ with normalization conditions as in (2.14),

$$A'\Gamma A = I, \qquad \text{and} \qquad B\Sigma_{xx}B' = \Lambda^2, \tag{4.4}$$

where $\Sigma_{xx} = \lim_{T\to\infty} \frac{1}{T}\sum_{t=1}^{T} X_t X_t'$, and Γ is a specified positive-definite matrix.

The main focus in this chapter is on the estimation of A and B based on a sample of T observations. Although we do not consider the complementary result of (4.2) that there are $(m - r)$ linear constraints (assuming $m < n$)

in detail here, the implications for the extended model (4.1)-(4.3) are the same as the model discussed in Chapter 2. The linear combinations $\ell_i' Y_t$, $i = 1, 2, \ldots, (m - r)$, such that $\ell_i' C = 0$, do not depend on the predictor variables X_t, but they are no longer white noise processes.

4.2 Example on U.K. Economy – Basic Data and Their Descriptions

To illustrate the procedures to be developed in this chapter, we consider the macroeconomic data of the United Kingdom which has been previously analyzed by Klein, Ball, Hazlewood and Vandome (1961) and by Gudmundsson (1977) (see Section 2.2). Klein et al. present a detailed description of an econometric model based on an extensive data base including the data used in our calculations. Quarterly observations, starting from the first quarter of 1948 to the last quarter of 1956, are used in the following analysis. The endogenous (response) variables are y_1 = index of industrial production, y_2 = consumption of food, drinks and tobacco at constant prices, y_3 = total unemployment, y_4 = index of volume of total imports, and y_5 = index of volume of total exports, and the exogenous (predictor) variable are x_1 = total civilian labor force, x_2 = index of weekly wage rates, x_3 = price index of total imports, x_4 = price index of total exports, and x_5 = price index of total consumption. The relationship between these two sets of variables is taken to reflect the demand side of the macrosystem of the economy of the United Kingdom. The first three series of the set of endogenous variables are seasonally adjusted before the analysis is carried out. Time series plots of the resulting data are given in Figure 4.1. These data are also presented in Table A.2 of the Appendix.

Initially the least squares regression of $Y_t = (y_{1t}, y_{2t}, y_{3t}, y_{4t}, y_{5t})'$ on $X_t = (x_{1t}, x_{2t}, x_{3t}, x_{4t}, x_{5t})'$ was performed without taking into account any autocorrelation structure on the errors. In order to get a rough view on the rank of the regression coefficient matrix, canonical correlation analysis was also performed. The squared sample canonical correlations between Y_t and X_t are 0.95, 0.67, 0.29, 0.14 and 0.01, which suggest a possible deficiency in the rank of C.

However, the residuals resulting from the least squares regression indicated a strong serial correlation in the errors. While commenting on the model of Klein et al. (1961), Gudmundsson (1977, p. 51) had a similar view that some allowance should be made for the serial correlation in the residuals of many of their equations. Therefore, we consider incorporating an autoregressive errors structure in the analysis.

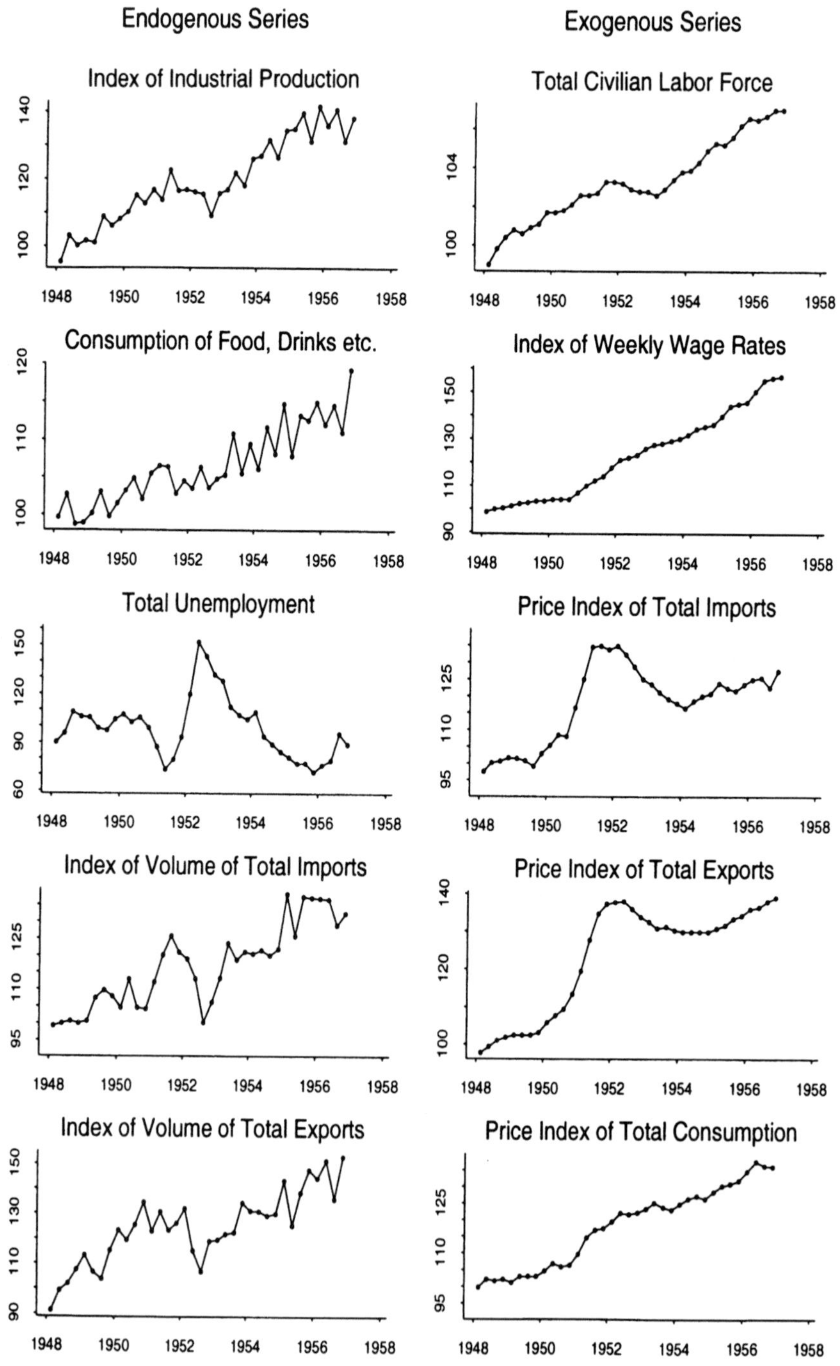

Figure 4.1. Quarterly macroeconomic data for the U.K., 1948–1956

4.3 Maximum Likelihood Estimators for the Model

We now consider an estimation procedure that will yield an efficient estimator, in the sense that it has the same asymptotic distribution as the maximum likelihood estimator under the normality assumption for the errors. From the model (4.1) and (4.3) we have

$$Y_t - \Phi Y_{t-1} = CX_t - \Phi CX_{t-1} + \epsilon_t, \tag{4.5}$$

where the error term ϵ_t in (4.5) satisfies the usual independence assumptions. Let $\boldsymbol{\epsilon} = [\epsilon_1, \ldots, \epsilon_T]$ and note that (4.5) leads to

$$\boldsymbol{Y} - \Phi \boldsymbol{Y}_{-1} = C\boldsymbol{X} - \Phi C\boldsymbol{X}_{-1} + \boldsymbol{\epsilon} \tag{4.6}$$

where $\boldsymbol{Y} = [Y_1, \ldots, Y_T]$ and $\boldsymbol{X} = [X_1, \ldots, X_T]$ are the $m \times T$ and $n \times T$ data matrices of observations on the response and predictor variables, and $\boldsymbol{Y}_{-1} = [Y_0, \ldots, Y_{T-1}]$ and $\boldsymbol{X}_{-1} = [X_0, \ldots, X_{T-1}]$ are the lagged data matrices. Before we examine the reduced-rank situation, we first briefly consider estimation of parameters C and Φ in the full-rank version of the model, that is, without the rank restriction on C. The full-rank estimator is useful, in particular, as a basis for the alternative reduced-rank estimator which is discussed in Section 4.5 and which is used in the identification procedure described in Section 4.6 to specify the rank of the coefficient matrix C.

We consider minimization of the criterion $S_T(C, \Phi) = \frac{1}{2T}\mathrm{tr}(\boldsymbol{\epsilon}\boldsymbol{\epsilon}'\Gamma) = \frac{1}{2T}e'(\Gamma \otimes I_T)e$, where $\boldsymbol{\epsilon} = \boldsymbol{Y} - \Phi\boldsymbol{Y}_{-1} - C\boldsymbol{X} + \Phi C\boldsymbol{X}_{-1}$, $e = \mathrm{vec}(\boldsymbol{\epsilon}')$, and the choice $\Gamma = \Sigma_{\epsilon\epsilon}^{-1}$ is taken. From (4.6) we have

$$\mathrm{vec}(\boldsymbol{Y}^{*\prime}) = \boldsymbol{X}^*\mathrm{vec}(C') + \mathrm{vec}(\boldsymbol{\epsilon}')$$

where $\boldsymbol{Y}^* = \boldsymbol{Y} - \Phi\boldsymbol{Y}_{-1}$ and $\boldsymbol{X}^* = (I_m \otimes \boldsymbol{X}') - (\Phi \otimes \boldsymbol{X}'_{-1})$ is an $mT \times mn$ matrix. It follows from standard results for the linear regression model that, for given values (estimates) $\tilde{\Phi}$ and $\tilde{\Sigma}_{\epsilon\epsilon}$ of Φ and $\Sigma_{\epsilon\epsilon}$, the full-rank estimate of C is obtained as the (estimated) generalized least squares (GLS) estimate, given by

$$\mathrm{vec}(\tilde{C}') = \tilde{G}^{-1}\tilde{\boldsymbol{X}}^{*\prime}(\tilde{\Sigma}_{\epsilon\epsilon}^{-1} \otimes I_T)\mathrm{vec}(\tilde{\boldsymbol{Y}}^{*\prime})/T, \tag{4.7}$$

where $\tilde{G} = \tilde{\boldsymbol{X}}^{*\prime}(\tilde{\Sigma}_{\epsilon\epsilon}^{-1} \otimes I_T)\tilde{\boldsymbol{X}}^*/T$, with $\tilde{\boldsymbol{Y}}^* = \boldsymbol{Y} - \tilde{\Phi}\boldsymbol{Y}_{-1}$ and $\tilde{\boldsymbol{X}}^* = (I_m \otimes \boldsymbol{X}') - (\tilde{\Phi} \otimes \boldsymbol{X}'_{-1})$. The covariance matrix of the full-rank estimate is estimated by $\hat{\mathrm{Cov}}[\mathrm{vec}(\tilde{C}')] = \frac{1}{T}\tilde{G}^{-1}$. Conversely, for given value (estimate) $\tilde{C}$ of C, the corresponding estimate of Φ to minimize the above criterion $S_T(C, \Phi)$ is given by

$$\tilde{\Phi} = \tilde{\boldsymbol{U}}\tilde{\boldsymbol{U}}'_{-1}(\tilde{\boldsymbol{U}}_{-1}\tilde{\boldsymbol{U}}'_{-1})^{-1}$$

where $\tilde{\boldsymbol{U}} = \boldsymbol{Y} - \tilde{C}\boldsymbol{X} = [\tilde{U}_1, \ldots, \tilde{U}_T]$, $\tilde{\boldsymbol{U}}_{-1} = [\tilde{U}_0, \ldots, \tilde{U}_{T-1}]$, with $\tilde{U}_t = Y_t - \tilde{C}X_t$, $t = 1, \ldots, T$, and $\Sigma_{\epsilon\epsilon}$ is estimated as $\tilde{\Sigma}_{\epsilon\epsilon} = \tilde{\boldsymbol{\epsilon}}\tilde{\boldsymbol{\epsilon}}'/T$ where $\tilde{\boldsymbol{\epsilon}} =$

$\tilde{\boldsymbol{U}} - \tilde{\Phi}\tilde{\boldsymbol{U}}_{-1}$. The estimation scheme is thus performed most conveniently by an iterative procedure which alternates between estimation of C through (4.7), for given estimate $\tilde{\Phi}$, and estimation of Φ as indicated for given estimate $\tilde{C}$. The most common starting value is the least squares estimate of C, given in (1.7), that ignores the AR(1) errors structure.

Now for estimation of the reduced-rank model with $C = AB$, let $\theta = (\alpha', \beta', \phi')'$, where $\alpha = \operatorname{vec}(A')$, $\beta = \operatorname{vec}(B')$, and $\phi = \operatorname{vec}(\Phi')$. We again consider minimization of the criterion

$$S_{1T}(\theta) = \frac{1}{2T}\operatorname{tr}(\boldsymbol{\epsilon}\boldsymbol{\epsilon}'\Gamma), \tag{4.8}$$

subject to normalization conditions

$$A'\Gamma A = I_r, \qquad B(\frac{1}{T}\boldsymbol{X}\boldsymbol{X}')B' = \Lambda^2, \tag{4.9}$$

where Λ^2 is diagonal. If the autocorrelation structure implied by (4.3) is ignored by setting $\Phi = 0$, then one is led to the least squares estimator considered by Robinson (1973). He has shown that the resulting least squares estimator is consistent but not efficient. To arrive at the asymptotic theory for the estimator that is obtained by minimizing (4.8) subject to (4.9), we shall again use the results of Silvey (1959) and Robinson (1972, 1973). For the normalization conditions (4.9), we shall be particularly interested in the choice of $\Gamma = \hat{\Sigma}_{\epsilon\epsilon}^{-1}$ for the asymptotic results where $\hat{\Sigma}_{\epsilon\epsilon}$ is the sample covariance matrix of the residuals $\hat{\boldsymbol{\epsilon}}$ obtained from the maximum likelihood estimation of the model (4.1) and (4.3) with the rank restriction on C. Note because of this choice of Γ, the normalization conditions vary with sample quantities. As in Chapters 2 and 3, however, to provide more explicit results the asymptotic theory presented here is for the estimator with the choice of $\Gamma = \Sigma_{\epsilon\epsilon}^{-1}$, which is a population quantity. For the sake of brevity we omit the details but the main result follows.

Theorem 4.1 Consider the model (4.1) and (4.3) under the stated assumptions, with $C = AB$, where A and B satisfy the normalization conditions (4.4), and let $h(\theta) = 0$ denote the $r^2 \times 1$ vector of normalization conditions (4.4) obtaining by stacking the conditions (4.4) without duplication. The first $r(r+1)/2$ components of $h(\theta)$ are of the form $\alpha_i'\Sigma_{\epsilon\epsilon}^{-1}\alpha_j - \delta_{ij}$, $i \le j$, where α_i denotes the ith column of the matrix A and $\delta_{ij} = 1$ for $i = j$ and 0 otherwise, and the remaining $r(r-1)/2$ elements are $\beta_i'\Sigma_{xx}\beta_j$, $i < j$, where β_i' denotes the ith row of B. Assume the limits

$$\lim_{T\to\infty} \frac{1}{T}\sum_{t=1}^{T} X_t X_t' = \Sigma_{xx} = \Gamma_x(0), \qquad \lim_{T\to\infty} \frac{1}{T}\sum_{t=2}^{T} X_{t-1} X_t' = \Gamma_x(1)$$

exist almost surely, and define

$$\begin{aligned} G &= (\Sigma_{\epsilon\epsilon}^{-1} \otimes \Gamma_x(0)) - (\Sigma_{\epsilon\epsilon}^{-1}\Phi \otimes \Gamma_x(1)') \\ &\quad -(\Phi'\Sigma_{\epsilon\epsilon}^{-1} \otimes \Gamma_x(1)) + (\Phi'\Sigma_{\epsilon\epsilon}^{-1}\Phi \otimes \Gamma_x(0)) \\ &= \mathrm{Cov}[\mathrm{vec}(X_t\epsilon_t'\Sigma_{\epsilon\epsilon}^{-1} - X_{t-1}\epsilon_t'\Sigma_{\epsilon\epsilon}^{-1}\Phi)]. \end{aligned}$$

Let $\hat{\theta}$ denote the estimator that absolutely minimizes (4.8) subject to (4.9) for the choice $\Gamma = \Sigma_{\epsilon\epsilon}^{-1}$. Then as $T \to \infty$, $\hat{\theta} \to \theta$ almost surely and $\sqrt{T}(\hat{\theta}-\theta)$ has a multivariate normal distribution with zero mean vector and covariance matrix given by the leading $(r(m+n)+m^2)$-order square submatrix of the $(r(m+n+r)+m^2)$-order matrix

$$\begin{bmatrix} B_\theta & -H_\theta \\ -H_\theta' & 0 \end{bmatrix}^{-1} \begin{bmatrix} B_\theta & 0 \\ 0 & 0 \end{bmatrix} \begin{bmatrix} B_\theta & -H_\theta \\ -H_\theta' & 0 \end{bmatrix}^{-1} \tag{4.10}$$

where

$$B_\theta = M' \begin{bmatrix} G & 0 \\ 0 & \Sigma_{\epsilon\epsilon}^{-1} \otimes \Sigma_u \end{bmatrix} M, \qquad M = \begin{bmatrix} I_m \otimes B' & A \otimes I_n & 0 \\ 0 & 0 & I_m \otimes I_m \end{bmatrix},$$

$H_\theta = \{(\partial h_j(\theta)/\partial\theta_i)\}$ is the matrix of first derivatives of the constraints in (4.4) and is of order $r(m+n)+m^2 \times r^2$, and $\Sigma_u = \Phi\Sigma_u\Phi' + \Sigma_{\epsilon\epsilon}$, which can be obtained from $\mathrm{vec}(\Sigma_u) = [I_{m^2} - \Phi\otimes\Phi]^{-1}\mathrm{vec}(\Sigma_{\epsilon\epsilon})$. Specifically, it can be shown as in Theorem 3.2 of Chapter 3 that the covariance matrix of this limiting distribution is given by $W = (B_\theta + H_\theta H_\theta')^{-1}B_\theta(B_\theta + H_\theta H_\theta')^{-1}$. We note that the choice of normalization condition (4.4) guarantees the existence of the inverse matrix in (4.10).

Proof: We shall only briefly indicate the proof here since many of the details follow along similar lines as Silvey (1959) and have been discussed in the proof of Theorem 3.2 in Chapter 3. Initially we consider estimates that minimize $S_{1T}^*(\theta) - h(\theta)'\lambda$, where λ is the Lagrangian multiplier and $S_{1T}^*(\theta)$ denotes the criterion (4.8) with the choice $\Gamma = \Sigma_{\epsilon\epsilon}^{-1}$. Let $\hat{\theta}$ and $\hat{\lambda}$ be solutions of the first-order equations. It can be shown that $\{\partial^2 S_{1T}^*(\theta)/\partial\theta\partial\theta'\} \to B_\theta$ almost surely. So following Silvey (1959), we have the following result [similar to (3.29)]

$$\begin{bmatrix} B_\theta + o(1) & -H_\theta + o(1) \\ -H_\theta' + o(1) & 0 \end{bmatrix} \begin{bmatrix} \sqrt{T}(\hat{\theta}-\theta) \\ \sqrt{T}\hat{\lambda} \end{bmatrix} = \begin{bmatrix} -\sqrt{T}\partial S_{1T}^*(\theta)/\partial\theta \\ 0 \end{bmatrix}. \tag{4.11}$$

Thus the asymptotic distribution of $\sqrt{T}(\hat{\theta}-\theta)$ follows from the asymptotic distribution of the quantity $\sqrt{T}\partial S_{1T}^*(\theta)/\partial\theta$. For the model considered, we can express in vector form as

$$\begin{aligned} y^* &= \mathrm{vec}(\boldsymbol{Y}^{*\prime}) = \boldsymbol{X}^*\mathrm{vec}(C') + \mathrm{vec}(\epsilon') \\ &= \boldsymbol{X}^*(I_m \otimes B')\alpha + e = \boldsymbol{X}^*(A \otimes I_n)\beta + e, \end{aligned}$$

where $e = \text{vec}(\epsilon')$ has covariance matrix $\Sigma_{\epsilon\epsilon} \otimes I_T$. The vector AR(1) error part of the model, $\boldsymbol{U} = \Phi \boldsymbol{U}_{-1} + \boldsymbol{\epsilon}$, can similarly be written as $\text{vec}(\boldsymbol{U}') = \text{vec}(\boldsymbol{U}'_{-1}\Phi') + \text{vec}(\epsilon') = (I_m \otimes \boldsymbol{U}'_{-1})\phi + e$. The criterion in (4.8) with $\Gamma = \Sigma_{\epsilon\epsilon}^{-1}$ can also be rewritten as $S_{1T}^*(\theta) = \frac{1}{2T} e'(\Sigma_{\epsilon\epsilon}^{-1} \otimes I_T)e$, and we therefore find that

$$\frac{\partial S_{1T}^*(\theta)}{\partial \theta} = \frac{1}{T}\frac{\partial e'}{\partial \theta}(\Sigma_{\epsilon\epsilon}^{-1} \otimes I_T)e = -\frac{1}{T}\begin{bmatrix} (I_m \otimes B)\boldsymbol{X}^{*\prime} \\ (A' \otimes I_n)\boldsymbol{X}^{*\prime} \\ (I_m \otimes \boldsymbol{U}_{-1}) \end{bmatrix} (\Sigma_{\epsilon\epsilon}^{-1} \otimes I_T)e$$

$$= -\frac{1}{T}\begin{bmatrix} I_m \otimes B & 0 \\ A' \otimes I_n & 0 \\ 0 & I_m \otimes I_m \end{bmatrix} \begin{bmatrix} (\Sigma_{\epsilon\epsilon}^{-1} \otimes \boldsymbol{X})e - (\Phi'\Sigma_{\epsilon\epsilon}^{-1} \otimes \boldsymbol{X}_{-1})e \\ (\Sigma_{\epsilon\epsilon}^{-1} \otimes \boldsymbol{U}_{-1})e \end{bmatrix},$$

where $\boldsymbol{X}^* = (I_m \otimes \boldsymbol{X}') - (\Phi \otimes \boldsymbol{X}'_{-1})$.

Hence, noting that $(\Phi'\Sigma_{\epsilon\epsilon}^{-1} \otimes \boldsymbol{X}_{-1})e = (\Phi'\Sigma_{\epsilon\epsilon}^{-1} \otimes I_n)\text{vec}(\boldsymbol{X}_{-1}\boldsymbol{\epsilon}')$ and so on, the vector of partial derivatives can be written as $\sqrt{T}\partial S_{1T}^*(\theta)/\partial\theta = M'F'\sqrt{T}Z_T$, where $Z_T = \text{vec}\{\boldsymbol{X}_{-1}\boldsymbol{\epsilon}', \boldsymbol{X}\boldsymbol{\epsilon}', \boldsymbol{U}_{-1}\boldsymbol{\epsilon}'\}/T$ and

$$F' = \begin{bmatrix} \Phi'\Sigma_{\epsilon\epsilon}^{-1} \otimes I_n & -(\Sigma_{\epsilon\epsilon}^{-1} \otimes I_n) & 0 \\ 0 & 0 & -(\Sigma_{\epsilon\epsilon}^{-1} \otimes I_m) \end{bmatrix}.$$

It is known [e.g., Anderson (1971, Chap. 5)] that $\sqrt{T}Z_T \xrightarrow{d} N(0, V)$ as $T \to \infty$, where

$$V = \begin{bmatrix} \Sigma_{\epsilon\epsilon} \otimes \Gamma_x(0) & \Sigma_{\epsilon\epsilon} \otimes \Gamma_x(1) & 0 \\ \Sigma_{\epsilon\epsilon} \otimes \Gamma_x(1)' & \Sigma_{\epsilon\epsilon} \otimes \Gamma_x(0) & 0 \\ 0 & 0 & \Sigma_{\epsilon\epsilon} \otimes \Sigma_u \end{bmatrix}.$$

It thus follows that $\sqrt{T}\partial S_{1T}^*(\theta)/\partial\theta \xrightarrow{d} N(0, B_\theta)$, where $B_\theta = M'F'VFM$ is as defined, and also that $\{\partial^2 S_{1T}^*(\theta)/\partial\theta\partial\theta'\} \to E\{\frac{1}{T}\frac{\partial e'}{\partial \theta}(\Sigma_{\epsilon\epsilon}^{-1} \otimes I_T)\frac{\partial e}{\partial \theta'}\} = B_\theta$. The asymptotic theory of Theorem 4.1 then follows from (4.11).

Note that $\text{rank}(B_\theta) = \text{rank}(M) = r(m+n-r) + m^2 < r(m+n) + m^2$ because of the existence of r^2 linearly independent restrictions between the columns. Hence the information matrix B_θ is singular and this makes the computation of the asymptotic covariances of the estimators in Theorem 4.1 more difficult. Similar to the discussion in Section 3.5, Silvey (1959, p. 401) suggested alternatively that the asymptotic covariance matrix of $\hat{\theta}$ is given by the leading submatrix of

$$\begin{bmatrix} B_\theta + H_\theta H_\theta' & -H_\theta \\ -H_\theta' & 0 \end{bmatrix}^{-1} \begin{bmatrix} B_\theta & 0 \\ 0 & 0 \end{bmatrix} \begin{bmatrix} B_\theta + H_\theta H_\theta' & -H_\theta \\ -H_\theta' & 0 \end{bmatrix}^{-1},$$

which can be shown to be equivalent to the matrix W given in Theorem 4.1.

The asymptotic theory for the estimators $\hat{A}$ and $\hat{B}$ is useful for inferences concerning whether certain components of X_t contribute in the formation of the explanatory indexes BX_t and also whether certain components of Y_t are influenced by certain of these indexes. The overall regression coefficient matrix $C = AB$ is, of course, also of interest, and the asymptotic distribution of the estimator $\hat{C} = \hat{A}\hat{B}$ follows directly from results of Theorem 4.1. Based on the relation $\hat{C} - C = \hat{A}\hat{B} - AB = (\hat{A} - A)B + A(\hat{B} - B) + O_p(T^{-1})$, it follows directly that $\sqrt{T}\text{vec}(\hat{C}' - C') \xrightarrow{d} N(0, M_1 W_1 M_1')$, where $M_1 = [I_m \otimes B', A \otimes I_n]$ and W_1 denotes the asymptotic covariance matrix of $\sqrt{T}\{(\hat{\alpha} - \alpha)', (\hat{\beta} - \beta)'\}'$, that is, the leading $r(m+n)$-order square submatrix of the matrix W in Theorem 4.1.

Remark–Asymptotic Independence of Estimators. As will be discussed in more detail in the next section, we note that the asymptotic covariance matrix W in Theorem 4.1 is block diagonal of the form $W = \text{diag}(W_1, \Sigma_{\epsilon\epsilon}^{-1} \otimes \Sigma_u)$, where $W_1 = (B_\delta + H_1 H_1')^{-1} B_\delta (B_\delta + H_1 H_1')^{-1}$, with $\delta = (\alpha', \beta')'$, $B_\delta = M_1' G M_1$, and $H_1 = \partial h(\theta)'/\partial \delta$. This indicates that the estimates $\hat{A}$ and $\hat{B}$ of the regression component of the model (4.1) and (4.3) are asymptotically independent of the estimate $\hat{\Phi}$ of the AR(1) error component of the model. In addition, it follows that the asymptotic covariance matrix of $\sqrt{T}\text{vec}(\hat{C}' - C')$, where $\hat{C} = \hat{A}\hat{B}$, can be expressed more explicitly as

$$\begin{aligned} M_1 W_1 M_1' &= M_1 (M_1' G M_1 + H_1 H_1')^{-1} M_1' G M_1 (M_1' G M_1 + H_1 H_1')^{-1} M_1' \\ &\equiv M_1 (M_1' G M_1)^{-} M_1'. \end{aligned}$$

4.4 Computational Algorithms for Efficient Estimators

We now briefly comment on the procedure for computing the efficient estimator. An iterative procedure can be formulated along the lines of Aitchison and Silvey's (1958) modification of the Newton–Raphson method to account for constraints as discussed in Section 3.5. It is based on a modification of the first-order equations in (4.11) to

$$\begin{bmatrix} B_\theta + H_\theta H_\theta' & -H_\theta \\ -H_\theta' & 0 \end{bmatrix} \begin{bmatrix} \hat{\theta} - \theta \\ \hat{\lambda} \end{bmatrix} = \begin{bmatrix} -\partial S_{1T}^*(\theta)/\partial \theta \\ 0 \end{bmatrix}, \qquad (4.12)$$

where explicitly

$$\partial S_{1T}^*(\theta)/\partial \theta = \text{vec}(\partial S_{1T}^*/\partial A',\ \partial S_{1T}^*(\theta)/\partial B',\ \partial S_{1T}^*(\theta)/\partial \Phi'),$$

with

$$\partial S_{1T}^*(\theta)/\partial A' = T^{-1}(B\boldsymbol{X}_{-1}\boldsymbol{\epsilon}'\Sigma_{\epsilon\epsilon}^{-1}\Phi - B\boldsymbol{X}\boldsymbol{\epsilon}'\Sigma_{\epsilon\epsilon}^{-1}),$$

$$\partial S_{1T}^*(\theta)/\partial B' = T^{-1}(\boldsymbol{X}_{-1}\boldsymbol{\epsilon}'\Sigma_{\epsilon\epsilon}^{-1}\Phi A - \boldsymbol{X}\boldsymbol{\epsilon}'\Sigma_{\epsilon\epsilon}^{-1}A),$$

$$\partial S_{1T}^*(\theta)/\partial \Phi' = -T^{-1}\boldsymbol{U}_{-1}\boldsymbol{\epsilon}'\Sigma_{\epsilon\epsilon}^{-1},$$

$\boldsymbol{U} = \boldsymbol{Y} - AB\boldsymbol{X}$, and $\boldsymbol{\epsilon} = \boldsymbol{U} - \Phi\boldsymbol{U}_{-1}$. Hence, at the ith step the iterative scheme is given by

$$\begin{bmatrix} \hat{\theta}_i \\ \hat{\lambda}_i \end{bmatrix} = \begin{bmatrix} \hat{\theta}_{i-1} \\ 0 \end{bmatrix} - \begin{bmatrix} W_{i-1} & Q_{i-1} \\ Q'_{i-1} & -I_{r^2} \end{bmatrix} \begin{bmatrix} \partial S_{1T}^*(\theta)/\partial\theta|_{\hat{\theta}_{i-1}} \\ h(\theta)|_{\hat{\theta}_{i-1}} \end{bmatrix}, \tag{4.13}$$

where

$$\begin{bmatrix} W & Q \\ Q' & -I_{r^2} \end{bmatrix} = \begin{bmatrix} B_\theta + H_\theta H_\theta' & -H_\theta \\ -H_\theta' & 0 \end{bmatrix}^{-1},$$

so that

$$W = (B_\theta + H_\theta H_\theta')^{-1} B_\theta (B_\theta + H_\theta H_\theta')^{-1},$$

and

$$Q = -(B_\theta + H_\theta H_\theta')^{-1} H_\theta.$$

The covariance matrix of $\hat{\theta}_i$ is approximated by

$$(1/T)(B_\theta + H_\theta H_\theta')^{-1} B_\theta (B_\theta + H_\theta H_\theta')^{-1} = (1/T)W$$

evaluated at $\hat{\theta}_i$. The iterations are carried out until either the successive differences between $\hat{\theta}_i$ or the magnitudes of the Lagrangian multipliers $\hat{\lambda}_i$ or the magnitudes of first order quantities $\partial S_{1T}^*(\hat{\theta}_{i-1})/\partial\theta$ are reasonably small.

However, as in the full-rank estimation discussed in Section 4.3, the calculations in the iterative scheme for $\hat{\theta}$ can be presented in a somewhat simpler form, in which one alternates between estimation of A and B (α and β), for given estimate of Φ, and estimation of Φ for given estimates of A and B. Let $\delta = (\alpha', \beta')'$ so that $\theta = (\delta', \phi')'$. Then note that the matrix B_θ is block diagonal, having the form $B_\theta = \text{diag}(M_1' G M_1, \Sigma_{\epsilon\epsilon}^{-1} \otimes \Sigma_u) \equiv \text{diag}(B_\delta, B_\phi)$, where $M_1 = [I_m \otimes B', A \otimes I_n]$. In addition, because the normalization constraints $h(\theta) = 0$ ($h(\delta) = 0$) do not involve Φ, the matrix of first derivatives of $h(\theta)$ with respect to θ can be partitioned as $H_\theta = \partial h(\theta)'/\partial\theta = [H_1', 0']'$ where $H_1 = \partial h(\theta)'/\partial\delta$. Therefore, upon rearranging we see that the first-order equations in (4.12) separate as

$$\begin{bmatrix} B_\delta + H_1 H_1' & -H_1 \\ -H_1' & 0 \end{bmatrix} \begin{bmatrix} \hat{\delta} - \delta \\ \hat{\lambda} \end{bmatrix} = \begin{bmatrix} -\partial S_{1T}^*(\theta)/\partial\delta \\ 0 \end{bmatrix}, \tag{4.14}$$

and $B_\phi(\hat{\phi} - \phi) = -\partial S_{1T}^*(\theta)/\partial\phi$. These last equations are equivalent to $(\hat{\Phi} - \Phi)\Sigma_u = T^{-1}\boldsymbol{\epsilon}\boldsymbol{U}_{-1}' \equiv T^{-1}(\boldsymbol{U} - \Phi\boldsymbol{U}_{-1})\boldsymbol{U}_{-1}'$. On substituting the sample estimate $\tilde{\Sigma}_u = T^{-1}U_{-1}U'_{-1}$ for Σ_u we obtain the solution $\hat{\Phi} = \hat{U}\hat{U}'_{-1}(\hat{U}_{-1}\hat{U}'_{-1})^{-1}$ for the estimate of Φ, where $\hat{U} = Y - \hat{C}X$ with $\hat{C} = \hat{A}\hat{B}$. The iteration scheme in (4.13) is then reduced in dimension, based on the uncoupled first-order equations in (4.14), by replacing $\hat{\theta}_i$ by $\hat{\delta}_i$ and making the additional appropriate simplifications. Specifically, since

$$\begin{bmatrix} B_\delta + H_1H_1' & -H_1 \\ -H_1' & 0 \end{bmatrix}^{-1} = \begin{bmatrix} W_1 & Q_1 \\ Q_1' & -I_{r^2} \end{bmatrix},$$

where $Q_1 = -(M_1'GM_1 + H_1H_1')^{-1}H_1$, this leads to the iterative step to obtain the estimates $\hat{A}$ and $\hat{B}$ as follows:

$$\hat{\delta}_i = \hat{\delta}_{i-1} - W_{1,i-1}(\partial S_{1T}^*(\theta)/\partial\delta)|_{\hat{\theta}_{i-1}} - Q_{1,i-1}h(\delta)|_{\hat{\delta}_{i-1}}$$

$$\hat{\lambda}_i = -Q'_{1,i-1}(\partial S_{1T}^*(\theta)/\partial\delta)|_{\hat{\theta}_{i-1}} + h(\delta)|_{\hat{\delta}_{i-1}}$$

The iterations thus alternate between the above equations to update the estimates of A and B and the step given to update the estimate of Φ. The covariance matrices of the final estimates $\hat{\delta}$ and $\hat{\phi}$ are provided by $\frac{1}{T}W_1$ and $\frac{1}{T}W_2$ where $W_2 = \Sigma_{\epsilon\epsilon} \otimes \Sigma_u^{-1} \equiv B_\phi^{-1}$.

4.5 Alternative Estimators and Their Properties

Initial specification of an appropriate rank of the regression coefficient matrix is desirable before the iterative estimation procedure is employed and the initial estimates of the component matrices are important for the convergence of iterative methods. This dictates a need for an alternative preliminary estimator which is computationally simpler and which may be useful at the initial model specification stages to identify an appropriate rank for the model. In practice, the full-rank estimator may be computed first, and if there is a structure to the true parameter matrix, we should be able to detect it by using the full-rank estimator directly. The alternative initial estimator we propose below is essentially based on this simple concept.

To motivate the alternative estimator we first recall the solution to the reduced-rank regression problem for the single regressor case under independent errors, as presented in Chapter 2. The sample version of the criterion which led to the solution in the single regressor case reduces to finding A and B by minimizing [see equation (2.12′)]

$$\begin{aligned} &\mathrm{tr}[\Gamma^{1/2}(\tilde{C} - AB)\hat{\Sigma}_{xx}(\tilde{C} - AB)'\Gamma^{1/2}] \\ &= [\mathrm{vec}\{(\tilde{C} - AB)'\}]'(\Gamma \otimes \hat{\Sigma}_{xx})\mathrm{vec}\{(\tilde{C} - AB)'\}, \end{aligned} \tag{4.15}$$

where $\tilde{C} = \hat{\Sigma}_{yx}\hat{\Sigma}_{xx}^{-1}$ is the full-rank least squares estimator. Using the Householder–Young Theorem, the problem is solved explicitly as in (2.17). This is not the case with an AR(1) error model since the full-rank estimate of C, as displayed in (4.7) of Section 4.3, does not have an analytically simple form. For given Φ and $\Sigma_{\epsilon\epsilon}$ ($\equiv \Gamma^{-1}$), minimization of the efficient criterion (4.8) reduces to the problem of finding A and B such that

$$[\text{vec}\{(\tilde{C} - AB)'\}]' \, \tilde{G} \, \text{vec}\{(\tilde{C} - AB)'\} \tag{4.16}$$

is minimized, where $\tilde{C}$ is the corresponding full-rank GLS estimate as described in Section 4.3, $\text{vec}(\tilde{C}') = \tilde{G}^{-1}\boldsymbol{X}^{*\prime}(\tilde{\Sigma}_{\epsilon\epsilon}^{-1} \otimes I_T)\text{vec}(\boldsymbol{Y}^{*\prime})/T$, and $\tilde{G} = \boldsymbol{X}^{*\prime}(\tilde{\Sigma}_{\epsilon\epsilon}^{-1} \otimes I_T)\boldsymbol{X}^*/T$. Now comparison of this latter expression with (4.15) indicates that an explicit solution via the Householder–Young Theorem is not possible unless the matrix $\tilde{G}$ can be written as the Kronecker product of two matrices of appropriate dimensions. This can happen only under rather special circumstances, for instance, when $\Phi = \rho I$. Hence, the solution is not simple as in the standard independent errors case.

Motivated by a desire to obtain initial estimates for the reduced-rank model (for various choices of the rank r) which are relatively easy to compute in terms of the full-rank estimator $\tilde{C}$ of model (4.1)-(4.3), we consider an alternate preliminary estimator. This estimator essentially is chosen to minimize (4.16), but with $\tilde{G}$ replaced by $(\Gamma \otimes \boldsymbol{X}\boldsymbol{X}'/T)$, the weights matrix that appears in (4.15), the inefficient least squares criterion. Note since this results in a criterion that is essentially the least squares criterion (4.15) with a better estimate $\tilde{C}$ of C in place of the least squares estimate $\hat{\Sigma}_{yx}\hat{\Sigma}_{xx}^{-1}$, the natural choice of Γ is $\Gamma = \tilde{\Sigma}_u^{-1}$, where $\tilde{\Sigma}_u = (\boldsymbol{Y} - \tilde{C}\boldsymbol{X})(\boldsymbol{Y} - \tilde{C}\boldsymbol{X})'/T$. This proposed estimation procedure, it should be kept in mind, does not have any overall optimality properties corresponding to our model, but is motivated by the standard model and computational convenience.

As in the single regressor case, we can obtain the alternative estimators $\tilde{A}^*$ and $\tilde{B}^*$ using the Householder–Young Theorem, which satisfy the normalization conditions $\tilde{A}^{*\prime}\tilde{\Sigma}_u^{-1}\tilde{A}^* = I_r$, $\tilde{B}^*(\boldsymbol{X}\boldsymbol{X}'/T)\tilde{B}^{*\prime} = \tilde{\Lambda}^{*2}$. From the results in Chapter 2, it follows that the columns of $\hat{\Sigma}_{xx}^{1/2}\tilde{B}^{*\prime}$ are chosen to be the eigenvectors corresponding to the first r eigenvalues of the matrix $\hat{\Sigma}_{xx}^{1/2}\tilde{C}'\tilde{\Sigma}_u^{-1}\tilde{C}\hat{\Sigma}_{xx}^{1/2}$, the positive diagonal elements of $\tilde{\Lambda}^*$ are the square roots of the eigenvalues, and $\tilde{A}^* = \tilde{C}\hat{\Sigma}_{xx}\tilde{B}^{*\prime}\tilde{\Lambda}^{*-2}$. Equivalently, we can represent the alternate estimators by taking $\tilde{A}^* = \tilde{\Sigma}_u^{1/2}\tilde{V}_{(r)}$ and $\tilde{B}^* = \tilde{V}_{(r)}'\tilde{\Sigma}_u^{-1/2}\tilde{C}$, where $\tilde{V}_{(r)} = [\tilde{V}_1, \ldots, \tilde{V}_r]$ and the $\tilde{V}_j$ are normalized eigenvectors of the matrix $\tilde{\Sigma}_u^{-1/2}\tilde{C}\hat{\Sigma}_{xx}\tilde{C}'\tilde{\Sigma}_u^{-1/2}$ corresponding to its r largest eigenvalues. Although primarily of interest for preliminary model specification and use as an initial estimator, the asymptotic distribution of the alternative estimator can also be developed along similar lines as in Theorem 4.1. We shall not present details, but if we let $\tilde{\alpha}^* = \text{vec}(\tilde{A}^{*\prime})$, $\tilde{\beta}^* = \text{vec}(\tilde{B}^{*\prime})$, $\tilde{\delta}^* = (\tilde{\alpha}^{*\prime}, \tilde{\beta}^{*\prime})'$ denote the alternative estimator and $\delta^* = (\alpha^{*\prime}, \beta^{*\prime})'$ the parameter values satisfying the corresponding population normalizations, then it can be shown that $\sqrt{T}(\tilde{\delta}^* - \delta^*)$ has a limiting multivariate normal distribution with zero mean vector and covariance matrix given by the leading $r(m+n)$-order square submatrix of the $r(m+n+r)$-order matrix

$$\begin{bmatrix} M_1^{*\prime}(\Sigma_u^{-1} \otimes \Sigma_{xx})M_1^* & -H_1^* \\ -H_1^{*\prime} & 0 \end{bmatrix}^{-1}$$
$$\times \begin{bmatrix} M_1^{*\prime}(\Sigma_u^{-1} \otimes \Sigma_{xx})G^{-1}(\Sigma_u^{-1} \otimes \Sigma_{xx})M_1^* & 0 \\ 0 & 0 \end{bmatrix}$$
$$\times \begin{bmatrix} M_1^{*\prime}(\Sigma_u^{-1} \otimes \Sigma_{xx})M_1^* & -H_1^* \\ -H_1^{*\prime} & 0 \end{bmatrix}^{-1},$$

where $M_1^* = [I_m \otimes B^{*\prime}, A^* \otimes I_n]$ and $H_1^* = \{(\partial h_j(\delta^*)/\partial \delta_i^*)\}$ is $r(m+n) \times r^2$.

4.5.1 A Comparison Between Efficient and Other Estimators

Though efficient estimators are desirable, the alternative estimator discussed above, as well as the least squares estimator, are more directly computable and may be quite useful as initial estimators and for model specification purposes. We will briefly consider their merits in terms of asymptotic efficiencies relative to the efficient estimator.

To obtain relatively explicit results, we shall restrict attention to the rank 1 case only, so that $C = \alpha\beta'$, where α and β are column vectors which satisfy the normalizations $\alpha'\Sigma_{\epsilon\epsilon}^{-1}\alpha = 1$, $\beta'\Gamma_x(0)\beta = \lambda^2$. In addition, we assume for simplicity that Φ is known and has the form $\Phi = \rho I$, where $|\rho| < 1$. In this case we can show that the asymptotic covariance matrix of the efficient estimator $(\hat{\alpha}', \hat{\beta}')'$ is

$$W_1 = \begin{bmatrix} \frac{1}{a}(\Sigma_{\epsilon\epsilon} - \alpha\alpha') & 0 \\ 0 & \Psi^{-1} \end{bmatrix}, \tag{4.17}$$

with

$$\Psi = (1+\rho^2)\Gamma_x(0) - \rho(\Gamma_x(1) + \Gamma_x(1)'),$$

and

$$a = (1+\rho^2)\lambda^2 - 2\rho(\beta'\Gamma_x(1)\beta) = \beta'\Psi\beta.$$

In the classical independent errors case, $\rho = 0$, W_1 reduces to the result given in Robinson (1974), $W_1 = \text{diag}\{(\beta'\Gamma_x(0)\beta)^{-1}(\Sigma_{\epsilon\epsilon} - \alpha\alpha'), \Gamma_x(0)^{-1}\}$ as given in the discussion following the proof of Theorem 2.4 in Section 2.5.

For the alternative estimator note the decomposition is $C = \alpha^*\beta^{*\prime}$, where the column vector α^* satisfies the normalization condition $\alpha^{*\prime}\Sigma_u^{-1}\alpha^* = 1$. Because of this change in the constraint on α^*, it is necessary to make a transformation of the parameter estimator so that the alternative estimator will be comparable to the efficient estimator. This transformation is easy to obtain [since $\Sigma_u = \{1/(1-\rho^2)\}\Sigma_{\epsilon\epsilon}$ for the special case $\Phi = \rho I$] as follows: $\beta = \{1/(1-\rho^2)^{1/2}\}\beta^*$, $\alpha = (1-\rho^2)^{1/2}\alpha^*$ and $\lambda = \lambda^*/(1-\rho^2)^{1/2}$. Hence the asymptotic covariance matrix of the alternative estimator when normalized in the same way as the efficient estimator is

$$R = \begin{bmatrix} \frac{a^*}{\lambda^4}(\Sigma_{\epsilon\epsilon} - \alpha\alpha') & 0 \\ 0 & \Psi^{-1} \end{bmatrix}, \tag{4.18}$$

where

$$a^* = \beta'\Gamma_x(0)\Psi^{-1}\Gamma_x(0)\beta.$$

Comparison of the matrix W_1 in (4.17) with R in (4.18) indicates that, asymptotically, the alternative estimator of β is efficient but the alternative estimator of α is not. In fact, the inefficiency of the alternative estimator of α can be readily established by verifying that $a^*/\lambda^4 \geq 1/a$ through the Cauchy–Schwarz inequality. This asymptotic comparison has an interesting interpretation which indicates that an efficient estimator can be achieved indirectly by a two-step procedure in which the efficient alternative estimator $\tilde{\beta}$ is first obtained, and then an efficient estimate of α is obtained by regression of Y_t on the index $\tilde{\beta}'X_t$, taking into account the AR error structure. We note that one special case in which the alternate estimator $\tilde{\alpha}$ is efficient occurs when the X_t have first-order autocovariance matrix of the form $\Gamma_x(1) = \gamma\Gamma_x(0)$, since then $\Psi = (1+\rho^2 - 2\gamma\rho)\Gamma_x(0)$ and hence $a^*/\lambda^4 = 1/a$, and so in particular $\tilde{\alpha}$ is efficient when the X_t are white noise ($\gamma = 0$).

Now we shall compare the least squares (LS) estimator with the alternative estimator for the rank one case. From the asymptotic results of Robinson (1973) and using a similar parameter transformation as in the case of the alternative estimator, we obtain the covariance matrix for the least squares estimator under the normalization $\alpha'\Sigma_{\epsilon\epsilon}^{-1}\alpha = 1$ as

$$S = \begin{bmatrix} \frac{a_1}{\lambda^4(1-\rho^2)}(\Sigma_{\epsilon\epsilon} - \alpha\alpha') & 0 \\ 0 & \frac{1}{(1-\rho^2)}\Gamma_x(0)^{-1}\Gamma_x^*(0)\Gamma_x(0)^{-1} \end{bmatrix} \tag{4.19}$$

with

$$a_1 = \lambda^2 + 2\sum_{j=1}^{\infty}(\beta'\Gamma_x(j)\beta)\rho^j,$$

and

$$\Gamma_x^*(0) = \Gamma_x(0) + \sum_{j=1}^{\infty}(\Gamma_x(j) + \Gamma_x(j)')\rho^j.$$

It is difficult to analytically compare the above covariance matrix with R in (4.18) in general. However, for illustration, suppose the X_t's are first-order autoregressive with coefficient matrix equal to γI. With these simplifying conditions we have shown before that the alternative estimator is efficient. However, the LS estimator performs poorly. We shall particularly concentrate on the estimators of α and compare the elements in the first diagonal block in S and R. Thus we compare $a_1/\lambda^4(1-\rho^2)$ with a^*/λ^4, where $a_1 = \lambda^2(1+\gamma\rho)/(1-\gamma\rho)$ and $a^* = \lambda^2/(1+\rho^2-2\gamma\rho)$. For $\gamma = \rho = 0.8$, the efficiency of the LS estimator of α relative to the alternative estimator of α is 0.22. This simple comparison suggests that the alternative estimator may perform substantially better than the LS estimator and as mentioned before the alternative estimator is also computationally convenient.

4.6 Identification of Rank of the Regression Coefficient Matrix

It is obviously important to be able to determine the rank of the coefficient matrix C for it is a key element in the structure of the reduced-rank regression problem. From the asymptotic distribution theory presented earlier, we do not learn about the rank of C because the asymptotic results are derived assuming the rank conditions are true. In the single regressor model with independent errors, the relationship between canonical correlation analysis and reduced-rank regression allows one to check on the rank by testing if certain canonical correlations are zero (see Section 2.6). When the rank of C is r, under independence and normality assumptions the statistic suggested by Bartlett (1947), $T\sum_{j=r+1}^{m}\log(1+\hat{\lambda}_j^2) \equiv -T\sum_{j=r+1}^{m}\log(1-\hat{\rho}_j^2)$, which is also the likelihood ratio statistic as discussed in Section 2.6, has a limiting $\chi^2_{(m-r)(n-r)}$ distribution [Hsu (1941)]. However, this test procedure is not valid under the autoregressive error model of (4.3). We need to develop other large sample quantities to test the hypothesis of interest in the framework of a correlated error structure.

Likelihood ratio test procedures have been shown to be valid for rather general autocorrelated error structures by Kohn (1979, Sec. 8). The likelihood ratio test statistic takes the form

$$\chi^2 = T\log(|\hat{\Sigma}_{\epsilon\epsilon}|/|\tilde{\Sigma}_{\epsilon\epsilon}|), \tag{4.20}$$

where $\tilde{\Sigma}_{\epsilon\epsilon}$ denotes the unrestricted maximum likelihood estimate of $\Sigma_{\epsilon\epsilon}$ under the autoregressive errors structure and $\hat{\Sigma}_{\epsilon\epsilon}$ the restricted estimate based on the restricted maximum likelihood estimate $\hat{C} = \hat{A}\hat{B}$ of C subject to the constraint that $\text{rank}(C) = r$. Hence this statistic can be used to test the null hypothesis that $\text{rank}(C) = r$, and the statistic will follow a χ^2 distribution with $mn - \{r(m+n) - r^2\} = (m-r)(n-r)$ degrees of freedom, the number of constrained parameters, under the null hypothesis. This statistic can be used to test the validity of the assumption of a reduced-rank structure of rank r for C, once the efficient estimates $\hat{A}$ and $\hat{B}$ have been obtained as in Section 4.3 with $\hat{C} = \hat{A}\hat{B}$. However, this procedure would not be useful at the preliminary specification stage where we are interested in selecting an appropriate rank for C prior to efficient estimation. This leads to consideration of an 'approximate' likelihood ratio procedure, as a rough guide to preliminary model specification [e.g., see Robinson (1973, p. 159) for a related suggestion], in which the alternative estimator of C derived under the assumption of AR(1) errors is used to form the estimate of $\Sigma_{\epsilon\epsilon}$ in the statistic (4.20) as an approximation to the efficient constrained estimator $\hat{\Sigma}_{\epsilon\epsilon}$, with the resulting statistic treated as having roughly a chi-squared distribution with $(m-r)(n-r)$ degrees of freedom under H_0. Thus, if $\tilde{C}^* = \tilde{A}^*\tilde{B}^*$ denotes the alternative estimator of C from Section 4.5, then we use $\hat{\Sigma}^*_{\epsilon\epsilon} = \hat{\boldsymbol{\epsilon}}^*\hat{\boldsymbol{\epsilon}}^{*\prime}/T$ in (4.20) in place of the restricted maximum likelihood estimate $\hat{\Sigma}_{\epsilon\epsilon}$, where $\hat{\boldsymbol{\epsilon}}^* = \hat{\boldsymbol{U}}^* - \hat{\Phi}^*\hat{\boldsymbol{U}}^*_{-1}$, $\hat{\Phi}^* = \hat{\boldsymbol{U}}^*\hat{\boldsymbol{U}}^{*\prime}_{-1}(\hat{\boldsymbol{U}}^*_{-1}\hat{\boldsymbol{U}}^{*\prime}_{-1})^{-1}$, and $\hat{\boldsymbol{U}}^* = \boldsymbol{Y} - \tilde{C}^*\boldsymbol{X}$ is obtained from the alternative estimate.

It should be emphasized that the main motivation (and advantage) of using this approximate test, with the alternative estimator approximating the exact restricted ML estimate, is that preliminary tests for models of various rank can all be tested simultaneously based on one set of eigenvalue-eigenvector calculations for the alternative estimator. The validity of the model (rank) selected at this preliminary stage would be checked further, of course, following efficient estimation of the parameters A and B by use of the chi-squared likelihood ratio statistic (4.20) with the efficient estimates $\hat{A}$, $\hat{B}$, and $\hat{\Sigma}_{\epsilon\epsilon}$. However, it should be noted that use of the preliminary approximate likelihood ratio test would tend to be 'conservative', in the sense of yielding larger values for the test statistic than the exact procedure, since the value $|\hat{\Sigma}^*_{\epsilon\epsilon}|$ based on the preliminary alternative estimator must be larger than that of $|\hat{\Sigma}_{\epsilon\epsilon}|$ from the final restricted ML estimate. It is possible that a computationally simple preliminary test which has an exact asymptotic chi-squared distribution might be obtainable, and an investigation of this matter may be a worthwhile topic for further research.

4.7 Inference for the Numerical Example

The initial analysis of the macroeconomic data indicated that the serial correlations must be taken into account for proper analysis. Thus, having tentatively identified that the data set may be analyzed in the framework of the models described in this chapter, we first compute the full-rank estimator under an autoregressive structure for the errors. A first-order autoregressive errors model is considered for the purpose of illustration, although one might argue towards a second-order model. The trace of the residual covariance matrix $\tilde{\Sigma}_{\epsilon\epsilon}$ for the AR(1) model is 133.52 and the trace under the ordinary least squares fit, i.e., the zero autocorrelation model, is 215.83, which indicates that significant improvement occurs when an AR(1) model for the errors is assumed. The full-rank estimate of the regression coefficient matrix under this assumption on the errors is given as follows:

$$\tilde{C} = \begin{bmatrix} 5.42^* & -0.90^* & -0.06 & -0.42^* & 1.90^* \\ 1.24^* & -0.10 & 0.13 & -0.32^* & 0.56^* \\ -5.92 & 0.29 & -0.37 & 1.21 & -1.05 \\ 3.13^* & -0.51 & -0.11 & -0.13 & 1.38^* \\ 9.46^* & -0.37 & 0.71^* & -0.86^* & 0.39 \end{bmatrix}$$

The starred entries indicate that the t-ratios of these elements are greater than 1.65. Although the elements of $\tilde{C}$ are not too different from the elements of the least squares estimate, there are some differences in the significance of individual elements when autocorrelation in the errors is accounted for. The corresponding estimate of the coefficient matrix Φ in the AR(1) errors model (4.3) is given by

$$\tilde{\Phi} = \begin{bmatrix} -0.186 & -0.323 & -0.121 & -0.147 & 0.170 \\ -0.121 & -0.450 & -0.031 & -0.106 & 0.082 \\ -0.235 & -0.324 & 0.853 & 0.596 & 0.123 \\ 0.151 & 0.393 & -0.266 & -0.081 & -0.096 \\ -0.946 & 0.801 & -0.116 & -0.275 & 0.266 \end{bmatrix}$$

with eigenvalues of $\tilde{\Phi}$ equal to 0.649, −0.508, 0.094, and $0.083 \pm 0.240i$. This provides an additional indication that autocorrelation in the error term U_t in the regression model (4.1) is substantial.

In order to initially specify the rank of the regression coefficient matrix C, the approximate chi-squared likelihood ratio statistic (4.20), which is based on the full-rank estimator $\tilde{C}$ and the alternative estimators under the AR(1) errors structure, was computed for various values of the rank r and the results are presented in Table 4.1. It is seen from this table that the rank of the matrix C may be taken to be equal to 2 when serial correlations are accounted for. For the purpose of illustration, we take the rank to be equal to 2 and the efficient estimates of the component matrices and their standard errors can be computed using the iterative algorithm given in Section 4.4. These are presented below:

$$\hat{A}' = \begin{bmatrix} \begin{matrix} 2.70 \\ (0.13) \end{matrix} & \begin{matrix} 1.09 \\ (0.07) \end{matrix} & \begin{matrix} -1.69 \\ (1.16) \end{matrix} & \begin{matrix} 2.54 \\ (0.34) \end{matrix} & \begin{matrix} 2.84 \\ (0.43) \end{matrix} \\ \begin{matrix} 0.21 \\ (0.43) \end{matrix} & \begin{matrix} 0.31 \\ (0.21) \end{matrix} & \begin{matrix} 0.15 \\ (1.28) \end{matrix} & \begin{matrix} -0.86 \\ (0.49) \end{matrix} & \begin{matrix} 3.80 \\ (0.50) \end{matrix} \end{bmatrix},$$

$$\hat{B} = \begin{bmatrix} \begin{matrix} 0.81 \\ (0.41) \end{matrix} & \begin{matrix} -0.13 \\ (0.09) \end{matrix} & \begin{matrix} 0.08 \\ (0.07) \end{matrix} & \begin{matrix} -0.24 \\ (0.08) \end{matrix} & \begin{matrix} 0.61 \\ (0.16) \end{matrix} \\ \begin{matrix} 1.57 \\ (0.47) \end{matrix} & \begin{matrix} 0.19 \\ (0.12) \end{matrix} & \begin{matrix} 0.28 \\ (0.08) \end{matrix} & \begin{matrix} -0.24 \\ (0.10) \end{matrix} & \begin{matrix} -0.49 \\ (0.19) \end{matrix} \end{bmatrix}.$$

We compute the matrix $\hat{C} = \hat{A}\hat{B}$ to see how well the full-rank estimate $\tilde{C}$ is reproduced by the product of the lower-rank matrices,

$$\hat{C} = \begin{bmatrix} 2.51 & -0.30 & 0.27 & -0.70 & 1.54 \\ 1.38 & -0.08 & 0.17 & -0.34 & 0.51 \\ -1.14 & 0.24 & -0.09 & 0.37 & -1.10 \\ 0.72 & -0.49 & -0.05 & -0.41 & 1.97 \\ 8.26 & 0.37 & 1.30 & -1.61 & -0.12 \end{bmatrix}.$$

Table 4.1 Summary of computations of approximate chi-squared statistics used for specification of rank of the model.

r = rank of the matrix C under H_0	Chi-squared statistic	d.f.	Critical value (5%)
	78.42	25	37.70
1	31.09	16	26.30
2	13.48	9	16.90
3	3.35	4	9.49
4	0.20	1	3.84

It can be seen that the entries of the full-rank estimate $\tilde{C}$ are fairly reproduced in $\hat{C}$ and this could certainly be improved by entertaining the possibility that the rank of the matrix is 3. Alternatively, we can compare the trace of the residual covariance matrix $\hat{\Sigma}_{\epsilon\epsilon}$, 145.42, obtained from estimation of the rank 2 model, with the trace of $\tilde{\Sigma}_{\epsilon\epsilon}$, which we had earlier mentioned as 133.52. The 9 percent increase in the trace value has to be viewed in conjunction with the reduction in the number of (functionally) independent regression parameters from 25 under the full-rank model to 16 under the rank 2 model. Under the null hypothesis that $\text{rank}(C) = 2$, we also compute the exact likelihood ratio statistic (4.20) using the restricted ML estimate $\hat{C}$ to check the validity of the reduced-rank specification. The computed value is $\chi^2 = 6.92$ with 9 degrees of freedom, which does not suggest any serious inadequacy in the reduced-rank specification. Subsequent calculation of the restricted ML estimate for a rank 1 model specification and of the associated likelihood ratio statistic (4.20) led to rejection of a regression model of rank 1.

It is not our intention to give a detailed interpretation of the estimates of the component matrices but some observations are worth noting. The two linear combination of X_t, $X_t^* = \hat{B}X_t$, are the most useful linear combinations of the exogenous variables for predicting the endogenous variables Y_t and these are plotted in Figure 4.2. Based on the standard errors, note that the index of weekly wage rates does not appear to play any significant role in explaining the endogenous variables considered here. The first linear

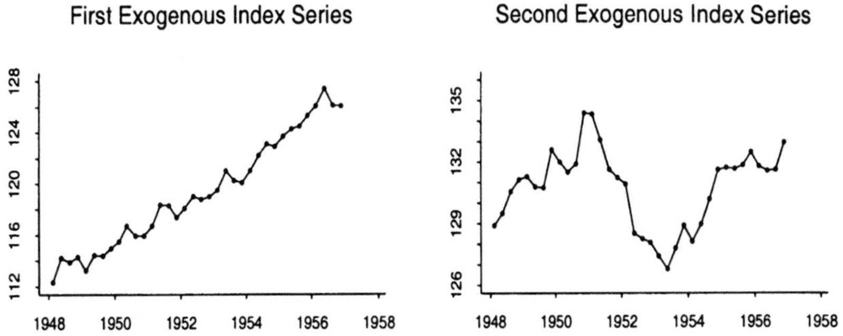

Figure 4.2. First two exogenous 'index' series for U.K. economic data

combination is (approximately) $x_{1t}^* = 0.61x_{5t} - 0.24x_{4t} + 0.81x_{1t}$, an index based on only the price index of total consumption, the price index of total exports and the labor force. This combination is essentially a trend variable and affects all the endogenous variables except unemployment. The second linear combination $x_{2t}^* = 1.57x_{1t} + 0.28x_{3t} - 0.24x_{4t} - 0.49x_{5t}$ seems to be especially useful only to explain y_{5t}, the index of volume of total exports. Also, it may be observed that neither combination significantly influences the unemployment series. To come up with a serious model of the U.K. economy, we may have to expand the set of exogenous variables and the nature of that investigation would be much more detailed than that considered here.

5

Multiple Time Series Modeling With Reduced Ranks

5.1 Introduction and Time Series Models

There has been growing interest in multiple time series modeling, particularly through use of vector autoregressive moving average models. The subject has found appeal and has applications in various disciplines, including engineering, physical sciences, business and economics, and the social sciences. In general, multiple time series analysis is concerned with modeling and estimation of dynamic relationships among m related time series $y_{1t}, \ldots, y_{mt}$, based on observations on these series over T equally spaced time points $t = 1, \ldots, T$, and also between these series and potential input or exogenous time series variables $x_{1t}, \ldots, x_{nt}$, observed over the same time period. In this chapter, we shall explore the use of certain reduced-rank modeling techniques for analysis of multiple time series in practice. We first introduce a general model for multiple time series modeling, but will specialize to multivariate autoregressive (AR) models for more detailed investigation.

Let $Y_t = (y_{1t}, \ldots, y_{mt})'$ be an $m \times 1$ multiple time series vector of response variables and let $X_t = (x_{1t}, \ldots, x_{nt})'$ be an $n \times 1$ vector of input or predictor time series variables. Let ϵ_t denote an $m \times 1$ white noise vector of errors, independently distributed over time with $E(\epsilon_t) = 0$ and $\text{Cov}(\epsilon_t) = \Sigma_{\epsilon\epsilon}$, a positive-definite matrix. We consider the multivariate time series model

$$Y_t = \sum_{s=0}^{p^*} C_s X_{t-s} + \epsilon_t, \tag{5.1}$$

where the C_s are $m \times n$ matrices of unknown parameters. In the general setting, the 'input' vectors X_t could include past (lagged) values of the response series Y_t. Even for moderate values of the dimensions m and n, the number of parameters that must be estimated in model (5.1), when no constraints are imposed on the matrices C_s, can become quite large. Because of the potential complexity of these models, it is often useful to consider canonical analyses or other dimension reduction procedures such as reduced-rank regression methods described in earlier chapters. This may also lead to more efficient models due to the reduction in the number of unknown parameters.

The multivariate time series model in (5.2) below, which places some structure on the coefficient matrices C_s in (5.1), has some useful properties and is in the spirit of reduced-rank regression models. We consider the model

$$Y_t = \sum_{s=0}^{p_2} \sum_{u=0}^{p_1} A_s B_u X_{t-s-u} + \epsilon_t, \tag{5.2}$$

where the A_s are $m \times r$ and the B_u are $r \times n$ matrices, respectively, with $1 \leq r \leq \min(m, n)$, and we shall assume $m \leq n$ for convention. Let L denote the lag operator such that $LY_t = Y_{t-1}$, and define the $m \times r$ and $r \times n$ matrix polynomial operators, of degrees p_2 and p_1, respectively, by

$$A(L) = \sum_{s=0}^{p_2} A_s L^s, \qquad B(L) = \sum_{u=0}^{p_1} B_u L^u.$$

Model (5.2) can then be written as

$$Y_t = A(L)B(L)X_t + \epsilon_t. \tag{5.3}$$

Brillinger (1969) considered a generalization of (5.3) where $A(L)$ and $B(L)$ are two-sided infinite order operators. He gives the interpretation already mentioned in Section 2.1, namely, that the $r \times 1$ series $X_t^* = B(L)X_t$ represents a reduction of the $n \times 1$ input series X_t. The reduction might be chosen, in terms of the operator $B(L)$ and, eventually, the operator $A(L)$, so that the information about Y_t contained in the series X_t is recovered as closely as possible by the factor $A(L)X_t^* = A(L)B(L)X_t$. Observe that in the standard reduced-rank regression model considered in Chapter 2, we have simply $A(L) = A$ and $B(L) = B$.

The model (5.3) can accommodate many situations that may be useful in practice. Suppose we partition X_t to explicitly allow for lagged values of the series Y_t, as $X_t = (Z_t', Y_{t-1}')'$. Partitioning the matrices B_u in (5.2) conformably, with $B(L) = [B_0(L), B_1(L)]$, allows us to write (5.3) as

$$Y_t - A(L)B_1(L)Y_{t-1} = A(L)B_0(L)Z_t + \epsilon_t. \tag{5.4}$$

The quantities $Z_t^* = B_0(L)Z_t$ have, in certain situations, interpretations as dynamic indices for the Z_t and consequently, a model such as (5.4) is sometimes referred to as an index model (e.g., Sims, 1981).

We now focus on the class of multivariate autoregressive (AR) models that results as a special case of (5.2). When $m = n$ and X_t contains only lagged values of Y_t so that $X_t = Y_{t-1}$, model (5.3) becomes

$$Y_t = A(L)B(L)Y_{t-1} + \epsilon_t, \tag{5.5}$$

and this model for Y_t represents a $p^* = p_1 + p_2 + 1$ order multivariate AR process. We mention two particular cases of special interest. First, with $p_2 = 0$ and $p_1 = p - 1$ we have

$$\begin{aligned} Y_t = A_0B(L)Y_{t-1} + \epsilon_t &= A_0B_0Y_{t-1} + \cdots + A_0B_{p-1}Y_{t-p} + \epsilon_t \\ &= A_0BY_{t-1}^* + \epsilon_t \equiv CY_{t-1}^* + \epsilon_t, \end{aligned} \tag{5.6}$$

where $B = [B_0, B_1, \ldots, B_{p-1}]$, $C = A_0B$, and $Y_{t-1}^* = (Y_{t-1}', \ldots, Y_{t-p}')'$. Since rank$(C) = r \leq m$, model (5.6) has the same structure as the multivariate reduced-rank regression model (2.5). Various aspects of this particular reduced-rank AR model form were studied by Velu, Reinsel, and Wichern (1986), and are also presented by Lütkepohl (1993, Chap. 5). Second, with $p_1 = 0$ and $p_2 = p - 1$ we obtain

$$\begin{aligned} Y_t &= A_0B_0Y_{t-1} + \cdots + A_{p-1}B_0Y_{t-p} + \epsilon_t \\ &= A_0Y_{t-1}^* + \cdots + A_{p-1}Y_{t-p}^* + \epsilon_t, \end{aligned} \tag{5.7}$$

where $Y_t^* = B_0Y_t$ is an $r \times 1$ vector of linear combinations of Y_t, $r \leq m$. In model (5.7), the Y_t^* may be regarded as a lower-dimensional set of indices for the Y_t, whose p lagged values are sufficient to represent the past behavior of Y_t. So this model may be called an index model. Note that on multiplying (5.7) on the left by B_0, we find that the series Y_t^* also has the structure of an AR process of order p. Estimation of parameters and other aspects of model (5.7) have been considered by Reinsel (1983).

The class of models stated in (5.2) is rather flexible and provides many options for modeling multivariate time series data. The structure of coefficient matrices in (5.2) in its various forms is within a reduced-rank framework. There are other forms and extensions of reduced-rank model for multiple time series, which may not directly follow from (5.2), that are found to also be of practical importance. For example, the extended reduced-rank regression model (3.3) considered by Anderson (1951) and the model (3.23) may have utility in multivariate time series modeling. Specifically, consider the vector AR(p) model

$$Y_t = \sum_{j=1}^{p} \Phi_j Y_{t-j} + \epsilon_t. \tag{5.8}$$

Suppose that for some p_1, $1 \leq p_1 < p$, the matrix $[\Phi_{p_1+1}, \ldots, \Phi_p]$ is of reduced rank while $[\Phi_1, \ldots, \Phi_{p_1}]$ is of full rank. (For instance, a particular case occurs with $p = 2$ where Φ_2 is reduced rank but Φ_1 is full rank.) Such models with this structure fit into the framework of model (3.3), with the correspondences $Z_t = (Y'_{t-1}, \ldots, Y'_{t-p_1})'$ and $X_t = (Y'_{t-p_1-1}, \ldots, Y'_{t-p})'$. This type of model might be useful if dimension reduction is possible in representing the influence on Y_t of more distant past values whereas the recent lagged values must be retained in tact. In the case where both matrices $[\Phi_1, \ldots, \Phi_{p_1}]$ and $[\Phi_{p_1+1}, \ldots, \Phi_p]$ are of reduced ranks, we have an example of the model (3.23).

A more general reduced-rank setup for vector AR models was proposed by Ahn and Reinsel (1988). In the vector AR(p) model (5.8), suppose it is assumed that the matrices Φ_j have a particular reduced-rank structure, such that $\text{rank}(\Phi_j) = r_j \geq \text{rank}(\Phi_{j+1}) = r_{j+1}$, $j = 1, \ldots, p-1$. The Φ_j then have the decomposition $\Phi_j = A_j B_j$, where A_j is $m \times r_j$ and B_j is $r_j \times m$, and it is further assumed that $\text{range}(A_j) \supset \text{range}(A_{j+1})$. These models are referred to as nested reduced-rank AR models. We will discuss this more general case further in Section 5.3 [also see Reinsel (1997, Section 6.1) for more details on this topic].

5.2 Reduced-Rank Autoregressive Models

The dynamic relationships among components of the vector time series Y_t can be modeled in various ways. For example, see Brillinger (1969), Parzen (1969), Akaike (1976), Sims (1981), Jenkins and Alavi (1981), Tiao and Box (1981), Hannan and Kavalieris (1984), Tiao and Tsay (1989), and Lütkepohl (1993) for various approaches. Doan, Litterman, and Sims (1984) considered Bayesian methods applied to time-varying coefficients vector autoregressive models, with particular emphasis on forecasting. Some considerations to the dimension reduction aspects in modeling have been given by the authors above, as well as by Anderson (1963), Brillinger (1981), Box and Tiao (1977), Priestley, Rao, and Tong (1974a, b), and others. In this section we consider the vector AR model (5.8) where the coefficient matrices Φ_j are of reduced-rank structure similar to the standard reduced-rank regression model. Because the vector AR model (5.8) is in the same framework as the multivariate regression model (1.1), we can readily make use of earlier results from Chapters 1 and 2. Therefore, we will avoid unnecessary repetition of details and shall be brief in our presentation.

5.2.1 *Estimation and Inference*

We consider the vector autoregressive model (5.8) of order p, given by $Y_t = \sum_{j=1}^{p} \Phi_j Y_{t-j} + \epsilon_t$. Defining $\Phi(L) = I - \Phi_1 L - \cdots - \Phi_p L^p$, where L denotes the lag operator, we assume that $\det\{\Phi(z)\} \neq 0$ for all complex numbers $|z| \leq 1$. The series Y_t is thus assumed to be stationary. Set $C = [\Phi_1, \ldots, \Phi_p]$ and assume that $\text{rank}(C) = r < m$, which yields the decomposition $C = AB$ as in (2.4). From model (5.6), the correspondence between the reduced-rank regression model in Chapter 2 and this particular reduced-rank AR model follows by setting $A = A_0$, $B = [B_0, B_1, \ldots, B_{p-1}]$, and $X_t = Y^*_{t-1} = (Y'_{t-1}, \ldots, Y'_{t-p})'$ in (2.5). Estimation of the component matrices A and B follows easily from Theorem 2.2 and subsequent results presented in Section 2.3.

For the stationary vector AR(p) process $\{Y_t\}$ above, define the covariance matrix at lag j as $\Gamma(j) = E(Y_{t-j}Y'_t)$, and set $\Gamma_* = [\Gamma(1)', \ldots, \Gamma(p)']' = E(Y^*_{t-1}Y'_t)$, and $\Gamma_p = E(Y^*_{t-1}Y^{*\prime}_{t-1})$ as the $mp \times mp$ matrix which has $\Gamma(i-j)$ in the (i,j)th block. As in Chapter 2, under the assumption of reduced rank, for any choice of positive-definite matrix Γ, the population coefficient matrices A and B in the reduced-rank AR model can be expressed as

$$A = \Gamma^{-1/2}[V_1, \ldots, V_r] = \Gamma^{-1/2}V = [\alpha_1, \ldots, \alpha_r]\,, \tag{5.9a}$$

$$B = V'\Gamma^{1/2}\Gamma'_*\Gamma_p^{-1} = [\beta_1, \ldots, \beta_r]', \tag{5.9b}$$

where the V_j are (normalized) eigenvectors of the matrix $\Gamma^{1/2}\Gamma'_*\Gamma_p^{-1}\Gamma_*\Gamma^{1/2}$ associated with the r largest eigenvalues λ_j^2. This also implies that A and B in (5.9) satisfy the normalizations $B\Gamma_pB' = \Lambda^2$ and $A'\Gamma A = I_r$, where $\Lambda^2 = \text{diag}(\lambda_1^2, \ldots, \lambda_r^2)$.

Based on a sample $Y_1, \ldots, Y_T$, let $\hat{\Gamma}_p = (1/T)\sum_t Y^*_{t-1}Y^{*\prime}_{t-1}$ and $\hat{\Gamma}_* = (1/T)\sum_t Y^*_{t-1}Y'_t$. Then the (conditional) ML estimates of A and B corresponding to the population quantities in (5.9) are given by

$$\hat{A} = \tilde{\Sigma}_{\epsilon\epsilon}^{1/2}[\hat{V}_1, \ldots, \hat{V}_r] = \tilde{\Sigma}_{\epsilon\epsilon}^{1/2}\hat{V} = [\hat{\alpha}_1, \ldots, \hat{\alpha}_r]\,, \tag{5.10a}$$

$$\hat{B} = \hat{V}'\tilde{\Sigma}_{\epsilon\epsilon}^{-1/2}\hat{\Gamma}'_*\hat{\Gamma}_p^{-1} = [\hat{\beta}_1, \ldots, \hat{\beta}_r]', \tag{5.10b}$$

where $\hat{V} = [\hat{V}_1, \ldots, \hat{V}_r]$ and $\hat{V}_j$ is the (normalized) eigenvector of the matrix $\tilde{\Sigma}_{\epsilon\epsilon}^{-1/2}\hat{\Gamma}'_*\hat{\Gamma}_p^{-1}\hat{\Gamma}_*\tilde{\Sigma}_{\epsilon\epsilon}^{-1/2}$ corresponding to the jth largest eigenvalue $\hat{\lambda}_j^2$, and

$$\tilde{\Sigma}_{\epsilon\epsilon} = (1/T)[\boldsymbol{YY}' - \boldsymbol{YY}^{*\prime}_{-1}(\boldsymbol{Y}^*_{-1}\boldsymbol{Y}^{*\prime}_{-1})^{-1}\boldsymbol{Y}^*_{-1}\boldsymbol{Y}'] \,.$$

Here, $\boldsymbol{Y} = [Y_1, \ldots, Y_T]$ and $\boldsymbol{Y}^*_{-1} = [Y^*_0, \ldots, Y^*_{T-1}]$ are the $m \times T$ and $mp \times T$ data matrices containing the observations on Y_t and Y^*_{t-1}, respectively, for $t = 1, \ldots, T$. It follows that the ML estimators $\hat{A} = [\hat{\alpha}_1, \ldots, \hat{\alpha}_r]$ and $\hat{B} = [\hat{\beta}_1, \ldots, \hat{\beta}_r]'$ above, with $\Gamma = \Sigma_{\epsilon\epsilon}^{-1}$ taken as fixed, possess the same form of asymptotic distribution as given in Theorem 2.4 for the multivariate reduced-rank regression model. The proof of this result is essentially the same as given for Theorem 2.4, and so to avoid repetition, we omit any further details. These asymptotic results are useful in the calculation of approximate standard errors for the ML estimates in (5.10), and will be used for the numerical example on the analysis of U.S. hog data that will be presented later in the chapter. It also follows that the same asymptotic theory holds for the LR testing for the rank of the coefficient matrix C in the reduced-rank autoregressive model as was presented in Section 2.6 for the reduced-rank regression model (e.g., Kohn, 1979). That is, under the null hypothesis H_0: rank$(C) \leq r$, the LR test statistic

$$\mathcal{M} = [T - mp + (mp - m - 1)/2] \sum_{j=r+1}^{m} \log(1 + \hat{\lambda}_j^2)$$

has an asymptotic $\chi^2_{(m-r)(mp-r)}$ distribution.

5.2.2 *Relationship to Canonical Analysis of Box and Tiao*

In Section 2.4, the connection between reduced-rank regression and canonical correlation analysis was demonstrated. In the multiple time series context, Box and Tiao (1977) considered a canonical analysis of a vector time series from a prediction point of view. Consider the m-variate autoregressive process $\{Y_t\}$ defined by (5.8), with $\text{Cov}(Y_t) = \Gamma(0)$ and $\text{Cov}(\epsilon_t) = \Sigma_{\epsilon\epsilon}$, both positive-definite. For an arbitrary linear combination of Y_t, $\omega_t = \ell' Y_t$, we have

$$\omega_t = \ell' Y_t = \ell' \hat{Y}_t + \ell' \epsilon_t \equiv \hat{\omega}_t + e_t,$$

where $\hat{Y}_t = \Phi_1 Y_{t-1} + \cdots + \Phi_p Y_{t-p}$, $\hat{\omega}_t = \ell' \hat{Y}_t$, and $e_t = \ell' \epsilon_t$. Thus we find that the variance of ω_t can be decomposed into the sum of a 'predictor' variance and an 'error' variance as $\sigma^2_\omega = \sigma^2_{\hat{\omega}} + \sigma^2_e$. A quantity that measures the predictability of the linear combination ω_t, based on the past values of the vector series $\{Y_t\}$, is the ratio

$$\rho^2 = \frac{\text{Var}(\hat{\omega}_t)}{\text{Var}(\omega_t)} = \frac{\sigma^2_{\hat{\omega}}}{\sigma^2_\omega} = \frac{\ell' \Gamma_*(0) \ell}{\ell' \Gamma(0) \ell}, \tag{5.11}$$

where $\Gamma_*(0) = \text{Cov}(\hat{Y}_t) = \sum_{j=1}^{p} \Phi_j \Gamma(j) = C\,\Gamma_*$. It follows that, for maximum predictability, the quantity ρ^2 must be the largest eigenvalue of the matrix $\Gamma(0)^{-1}\Gamma_*(0)$ and ℓ is the corresponding eigenvector. Similarly, the eigenvector corresponding to the smallest eigenvalue of $\Gamma(0)^{-1}\Gamma_*(0)$ gives the least predictable linear combination of Y_t. In general, there will be m eigenvalues $1 \geq \rho_1^2 \geq \rho_2^2 \geq \cdots \geq \rho_m^2 \geq 0$, and let $\ell_1, \ldots, \ell_m$ denote the corresponding eigenvectors. Then define the $m \times m$ matrix L' that has ℓ_i' as its ith row, and consider the transformed process $\mathcal{W}_t = L' Y_t = (\omega_{1t}, \ldots, \omega_{mt})'$. It follows that the transformation $\mathcal{W}_t = L' Y_t$ produces m new component series that are ordered from 'most predictable' to 'least predictable'.

The developments of this canonical analysis approach are due to Box and Tiao (1977), and we will now indicate that they also correspond to the classical canonical correlation analysis originally developed by Hotelling (1935, 1936). Therefore, we see that this approach is directly related to the reduced-rank AR modeling and the ML estimates $\hat{A}$ and $\hat{B}$ given in (5.10) because of the previously established connection of reduced-rank regression with classical canonical correlation analysis.

For the stationary vector AR(p) model (5.8), we know that

$$C = [\Phi_1, \ldots, \Phi_p] = \Gamma_*' \Gamma_p^{-1}$$

and that $\Gamma_*(0) = \text{Cov}(\hat{Y}_t) = \sum_{j=1}^{p} \Phi_j \Gamma(j) = C\Gamma_* = \Gamma_*' \Gamma_p^{-1} \Gamma_*$ [e.g., see Reinsel (1997, Sections 2.2 and 6.4)]. From (5.11) it follows that the choice of ℓ that gives the maximum predictability is then the eigenvector ℓ_1 satisfying

$$\Gamma(0)^{-1} \Gamma_*' \Gamma_p^{-1} \Gamma_* \, \ell_1 = \rho_1^2 \, \ell_1, \tag{5.12}$$

with ρ_1^2 being the largest eigenvalue of $\Gamma(0)^{-1}\Gamma_*'\Gamma_p^{-1}\Gamma_*$. Note that this is the same as the largest eigenvalue of the matrix $\Gamma(0)^{-1/2}\Gamma_*'\Gamma_p^{-1}\Gamma_*\Gamma(0)^{-1/2}$, which is of the same form as the matrix in (2.28) of Section 2.4. Thus we see that this is equivalent to a canonical correlation analysis between the vectors Y_t and $X_t \equiv Y_{t-1}^* = (Y_{t-1}', \ldots, Y_{t-p}')'$. If $h_i'Y_t$ and $g_i'X_t$ denote the ith canonical variables, then it is easy to see that $h_i = \ell_i$ and $g_i = C'\ell_i$, for $i = 1, \ldots, m$. The ρ_i are the canonical correlations between Y_t and X_t. Hence, the canonical analysis based on the predictability measure of (5.11) is the same as the classical canonical correlation analysis.

It is to be noted that the number of nonzero eigenvalues associated with (5.12) is the same as the rank of Γ_*, which in turn is equal to rank(C). So the preceding canonical analysis techniques for vector AR processes considered by Box and Tiao are related to the reduced-rank regression model (2.5). Specifically, from the results in Section 2.4 and from (5.9), $\ell_i = \Sigma_{\epsilon\epsilon}^{-1/2} V_i$, $i = 1, \ldots, m$, where V_i denotes the (normalized) eigenvector in (5.9) for the choice of $\Gamma = \Sigma_{\epsilon\epsilon}^{-1}$ and ℓ_i is the eigenvector corresponding to the ith largest root ρ_i^2 in (5.12), normalized such that $\ell_i'\Gamma(0)\ell_i = (1 - \rho_i^2)^{-1}$. Hence, under a reduced-rank structure for the coefficient matrix C in the vector AR(p) model, the component matrices A and B in (5.9) can equivalently be expressed in terms of $L_{(r)} = [\ell_1, \ldots, \ell_r]$ as $A = \Sigma_{\epsilon\epsilon} L_{(r)}$ and $B = L_{(r)}' \Gamma_*' \Gamma_p^{-1}$. With $m-r$ zero eigenvalues in (5.12), we see that $\ell_i'C = 0$ for $i = r+1, \ldots, m$. Thus, there are $m-r$ corresponding linear combinations $\omega_{it} = \ell_i'Y_t \equiv \ell_i'\epsilon_t$ which form white noise series. It is useful to identify such white noise series in the multiple time series modeling, since that will simplify the structure of the model, and this can be accomplished through the canonical correlation analysis and associated reduced-rank regression modeling.

5.3 Extended and Nested Reduced-Rank Autoregressive Models

5.3.1 An Extended Reduced-Rank Autoregressive Model

We discuss briefly the use of the extended reduced-rank regression model (3.3) considered by Anderson (1951), as presented in Section 3.1, within the time series framework of the vector autoregressive model (5.8). As noted in the discussion following (5.8), for the vector AR(p) model in (5.8), if for some p_1, $1 \leq p_1 < p$, the matrix $[\Phi_{p_1+1}, \dots, \Phi_p]$ of AR coefficients for more distant lags is of reduced rank $r_1 < m$ but the matrix $[\Phi_1, \dots, \Phi_{p_1}]$ of AR coefficients for the recent past is of full rank, then the resulting time series model is in the form of (3.3). Writing $C = [\Phi_{p_1+1}, \dots, \Phi_p]$ and $D = [\Phi_1, \dots, \Phi_{p_1}]$, with $Z_t = (Y'_{t-1}, \dots, Y'_{t-p_1})'$ and $X_t = (Y'_{t-p_1-1}, \dots, Y'_{t-p})'$, the AR($p$) model can be restated as

$$Y_t = CX_t + DZ_t + \epsilon_t, \qquad t = 1, \dots, T, \tag{5.13}$$

where $\text{rank}(C) = r_1 < m$. Hence, we have $C = AB$ where A is an $m \times r_1$ matrix and B is an $r_1 \times m(p-p_1)$ matrix.

The ML estimation of the component parameter matrices A and B, and D in (5.13) follows from the results developed in Section 3.1. Likewise, the corresponding population results, which we discuss first, follow from Theorem 3.1. As in Section 5.2, denote $\Gamma_*(p) = [\Gamma(1)', \dots, \Gamma(p)']' = E(Y^*_{t-1}Y'_t)$ and $\Gamma_p = E(Y^*_{t-1}Y^{*\prime}_{t-1})$. Then note that $\Sigma_{yz} \equiv E(Y_t Z'_t) = \Gamma_*(p_1)'$, $\Sigma_{zz} \equiv E(Z_t Z'_t) = \Gamma_{p_1}$, $\Sigma_{xx} \equiv E(X_t X'_t) = \Gamma_{p-p_1}$, and denote $\Sigma_{yx} \equiv E(Y_t X'_t) = [\Gamma(p_1+1)', \dots, \Gamma(p)'] \equiv \Gamma_{**}(p-p_1)'$ and $\Sigma_{xz} \equiv E(X_t Z'_t) = \Gamma_{(p-p_1,p_1)}$. From the results of Theorem 3.1, and relation (3.5) in particular, it follows that the population coefficient matrices in the extended reduced-rank vector AR model (5.13) can be written as

$$A^{(r_1)} = \Sigma_{\epsilon\epsilon}^{1/2}[V_1, \dots, V_{r_1}] = \Sigma_{\epsilon\epsilon}^{1/2} V_{(r_1)}, \tag{5.14a}$$

$$B^{(r_1)} = V'_{(r_1)} \Sigma_{\epsilon\epsilon}^{-1/2} \Sigma_{yx.z} \Sigma_{xx.z}^{-1}, \tag{5.14b}$$

and $D = (\Sigma_{yz} - A^{(r_1)}B^{(r_1)}\Sigma_{xz})\Sigma_{zz}^{-1} = (\Gamma_*(p_1)' - A^{(r_1)}B^{(r_1)}\Gamma_{(p-p_1,p_1)})\Gamma_{p_1}^{-1}$, where the V_j are (normalized) eigenvectors of $\Sigma_{\epsilon\epsilon}^{-1/2}\Sigma_{yx.z}\Sigma_{xx.z}^{-1}\Sigma_{xy.z}\Sigma_{\epsilon\epsilon}^{-1/2}$ associated with the r_1 largest eigenvalues λ^2_{1j}, and

$$\begin{aligned}
\Sigma_{yx.z} &= \Sigma_{yx} - \Sigma_{yz}\Sigma_{zz}^{-1}\Sigma_{zx} = \Gamma_{**}(p-p_1)' - \Gamma_*(p_1)'\Gamma_{p_1}^{-1}\Gamma'_{(p-p_1,p_1)}, \\
\Sigma_{xx.z} &= \Sigma_{xx} - \Sigma_{xz}\Sigma_{zz}^{-1}\Sigma_{zx} = \Gamma_{p-p_1} - \Gamma_{(p-p_1,p_1)}\Gamma_{p_1}^{-1}\Gamma'_{(p-p_1,p_1)}
\end{aligned}$$

are partial covariance matrices eliminating the linear effects of Z_t from both Y_t and X_t. This implies that $A^{(r_1)}$ and $B^{(r_1)}$ in (5.14) satisfy the normalizations $B^{(r_1)}\Sigma_{xx.z}B^{(r_1)\prime} = \Lambda_1^2$ and $A^{(r_1)\prime}\Sigma_{\epsilon\epsilon}^{-1}A^{(r_1)} = I_{r_1}$, where $\Lambda_1^2 = \mathrm{diag}(\lambda_{11}^2, \dots, \lambda_{1r_1}^2)$. Also recall that the rank of C in (5.13) is equal to the number of nonzero partial canonical correlations $[\rho_{1j} = \lambda_{1j}/(1+\lambda_{1j}^2)^{1/2}]$ between Y_t and the more distant lagged values $X_t = (Y'_{t-p_1-1}, \dots, Y'_{t-p})'$, eliminating the influence of the closer lagged values $Z_t = (Y'_{t-1}, \dots, Y'_{t-p_1})'$.

For the sample $Y_1, \dots, Y_T$, let

$$\hat{\Gamma}_*(p_1)' = \frac{1}{T}\sum_t Y_t Z_t', \quad \hat{\Gamma}_{p_1} = \frac{1}{T}\sum_t Z_t Z_t', \quad \hat{\Gamma}_{**}(p-p_1)' = \frac{1}{T}\sum_t Y_t X_t'$$

and so on, denote the sample covariance matrices corresponding to the population quantities above, with $\hat{\Sigma}_{yx.z} = \hat{\Gamma}_{**}(p-p_1)' - \hat{\Gamma}_*(p_1)'\hat{\Gamma}_{p_1}^{-1}\hat{\Gamma}'_{(p-p_1,p_1)}$ and $\hat{\Sigma}_{xx.z} = \hat{\Gamma}_{p-p_1} - \hat{\Gamma}_{(p-p_1,p_1)}\hat{\Gamma}_{p_1}^{-1}\hat{\Gamma}'_{(p-p_1,p_1)}$ the associated sample partial covariance matrices. Then it follows directly from the derivation in Section 3.1, leading to (3.11), that the (conditional) ML estimates are

$$\hat{A}^{(r_1)} = \tilde{\Sigma}_{\epsilon\epsilon}^{1/2}[\hat{V}_1, \dots, \hat{V}_{r_1}] = \tilde{\Sigma}_{\epsilon\epsilon}^{1/2}\hat{V}_{(r_1)}, \tag{5.15a}$$

$$\hat{B}^{(r_1)} = \hat{V}'_{(r_1)}\tilde{\Sigma}_{\epsilon\epsilon}^{-1/2}\hat{\Sigma}_{yx.z}\hat{\Sigma}_{xx.z}^{-1}, \tag{5.15b}$$

and $\hat{D} = (\hat{\Sigma}_{yz} - \hat{A}^{(r_1)}\hat{B}^{(r_1)}\hat{\Sigma}_{xz})\hat{\Sigma}_{zz}^{-1} = (\hat{\Gamma}_*(p_1)' - \hat{A}^{(r_1)}\hat{B}^{(r_1)}\hat{\Gamma}_{(p-p_1,p_1)})\hat{\Gamma}_{p_1}^{-1}$, where the $\hat{V}_j$ are (normalized) eigenvectors of $\tilde{\Sigma}_{\epsilon\epsilon}^{-1/2}\hat{\Sigma}_{yx.z}\hat{\Sigma}_{xx.z}^{-1}\hat{\Sigma}_{xy.z}\tilde{\Sigma}_{\epsilon\epsilon}^{-1/2}$ associated with the r_1 largest eigenvalues $\hat{\lambda}_{1j}^2$. As with the ML estimators for the regression model (3.3) in Section 3.1, the asymptotic distribution theory for the ML estimators $\hat{A}^{(r_1)}$, $\hat{B}^{(r_1)}$, and $\hat{D}$ in (5.15) above follows easily from the results in Theorem 2.4. It also follows that LR testing for the appropriate rank of the coefficient matrix C in the extended reduced-rank AR model (5.13) can be carried out in the same manner as described in Section 3.1. In particular, for testing the null hypothesis $H_0\!:\mathrm{rank}(C) \le r_1$, we can use the LR test statistic

$$\begin{aligned}\mathcal{M} &= [T - mp + (m(p-p_1) - m - 1)/2] \sum_{j=r_1+1}^{m} \log(1 + \hat{\lambda}_{1j}^2) \\ &= -[T - mp + (m(p-p_1) - m - 1)/2] \sum_{j=r_1+1}^{m} \log(1 - \hat{\rho}_{1j}^2),\end{aligned}$$

which has an asymptotic $\chi^2_{(m-r_1)(m(p-p_1)-r_1)}$ null distribution, where $\hat{\rho}_{1j}^2 = \hat{\lambda}_{1j}^2/(1+\hat{\lambda}_{1j}^2)$ are the squared sample partial canonical correlations between Y_t and X_t eliminating Z_t.

To conclude, we observe that the extended reduced-rank vector autoregressive model (5.13) has the feature that $m - r_1$ linear combinations of Y_t exist such that their dependence on the past can be represented in terms of only the recent past lags $Z_t = (Y'_{t-1}, \dots, Y'_{t-p_1})'$. This type of feature may be useful in modeling and interpreting time series data with several component variables. The specification of the appropriate reduced rank r_1 and corresponding ML estimation of the model parameters can be performed using the sample partial canonical correlation analysis and associated methods as highlighted above and as described in more detail in Section 3.1.

5.3.2 Nested Reduced-Rank Autoregressive Models

We now briefly discuss analysis of vector AR(p) models that have simplifying nested reduced-rank structures in their coefficient matrices Φ_j. Specifically, we consider the vector AR(p) model in (5.8), in which it is assumed that the matrices Φ_j have a particular reduced-rank structure, such that $\text{rank}(\Phi_j) = r_j \geq \text{rank}(\Phi_{j+1}) = r_{j+1}$, $j = 1, 2, \ldots, p-1$. Then, the Φ_j can be represented in the form $\Phi_j = A_j B_j$, where A_j is $m \times r_j$ and B_j is $r_j \times m$, and we further assume that $\text{range}(A_j) \supset \text{range}(A_{j+1})$. Thus, we can write (5.8) as

$$Y_t = \sum_{j=1}^{p} A_j B_j Y_{t-j} + \epsilon_t. \tag{5.16}$$

We refer to such models as *nested reduced-rank* AR models, and these models, which generalize the earlier work by Velu, Reinsel, and Wichern (1986), were studied by Ahn and Reinsel (1988). These models can result in more parsimonious parameterization, provide more details about structure possibly with simplifying features, and offer more interesting and useful interpretations concerning the interrelations among the m time series.

Specification of Ranks Through Partial Canonical Analysis

To obtain information on the ranks of the matrices Φ_j in the nested reduced-rank model, a partial canonical correlation analysis approach can be used. One fundamental consequence of the model above is that there will exist at least $m - r_j$ zero partial canonical correlations between Y_t and Y_{t-j}, given $Y_{t-1}, \ldots, Y_{t-j+1}$. This follows because we can find a $(m - r_j) \times m$ matrix F_j', whose rows are linearly independent, such that $F_j' A_j = 0$ and, hence, $F_j' A_i = 0$ for all $i \geq j$ because of the nested structure of the A_j. Therefore,

$$F_j'(Y_t - \sum_{i=1}^{j-1} \Phi_i Y_{t-i}) = F_j'(Y_t - \sum_{i=1}^{p} \Phi_i Y_{t-i}) \equiv F_j' \epsilon_t \tag{5.17}$$

is independent of $(Y_{t-1}', \ldots, Y_{t-j}')'$ and consists of $m - r_j$ linear combinations of $(Y_t', \ldots, Y_{t-j+1}')'$. Thus, $m - r_j$ zero partial canonical correlations between Y_t and Y_{t-j} occur. Hence, performing a (partial) canonical correlation analysis for various values of $j = 1, 2, \ldots$ will identify the rank structure of the nested reduced-rank model, as well as the overall order p of the AR model.

The sample test statistic that can be used to (tentatively) specify the ranks is

$$\mathcal{M}(j, s) = -(T - j - jm - 1 - 1/2) \sum_{i=(m-s)+1}^{m} \log(1 - \hat{\rho}_i^2(j)) \tag{5.18}$$

for $s = 1, 2, \ldots, m$, where $1 \geq \hat{\rho}_1(j) \geq \cdots \geq \hat{\rho}_m(j) \geq 0$ are the *sample partial canonical correlations* between Y_t and Y_{t-j}, given $Y_{t-1}, \ldots, Y_{t-j+1}$. Under the null hypothesis that $\text{rank}(\Phi_j) \leq m - s$ within the nested reduced-rank model framework, the statistic $\mathcal{M}(j, s)$ is asymptotically distributed as chi-squared with s^2 degrees of freedom (e.g., Tiao and Tsay, 1989). Hence, if the value of the test statistic is not 'unusually large', we would not reject the null hypothesis and might conclude that Φ_j has reduced rank equal to the smallest value r_j ($\equiv m - s_j$) for which the test does not reject the null hypothesis. Note, in particular, that when $s = m$ the statistic in (5.18) is the same as the LR test statistic for testing $H_0 : \Phi_j = 0$ in an AR(j) model, since $\log(U_j) = \log\{|S_j|/|S_{j-1}|\} = \sum_{i=1}^{m} \log(1 - \hat{\rho}_i^2(j))$, where S_j denotes the residual sum of squares matrix after LS estimation of the AR model of order j.

Canonical Form for the Reduced-Rank Model

It follows from the structure of the matrices $\Phi_j = A_j B_j$ that one can construct a nonsingular $m \times m$ matrix P such that the transformed series $\mathcal{Z}_t = PY_t$ will have the following simplified model structure. The model for $\mathcal{Z}_t$ is AR(p),

$$\begin{aligned} \mathcal{Z}_t = PY_t &= \sum_{i=1}^{p} P(A_i B_i) P^{-1} P Y_{t-i} + P\epsilon_t \\ &= \sum_{i=1}^{p} A_i^* B_i^* \mathcal{Z}_{t-i} + e_t \equiv \sum_{i=1}^{p} \Phi_i^* \mathcal{Z}_{t-i} + e_t, \end{aligned} \tag{5.19}$$

where $A_i^* = PA_i$, $B_i^* = B_i P^{-1}$, $\Phi_i^* = A_i^* B_i^*$, and $e_t = P\epsilon_t$. The rows of P will consist of a basis of the rows of the various matrices F_j' indicated in (5.17). Thus, the matrix P can be chosen so that its last $m - r_j$ rows are orthogonal to the columns of A_j, and, hence, $A_i^* = [A_{i1}^{*\prime}, 0']'$ has its last $m - r_i$ rows equal to zero. This implies that $\Phi_i^* = A_i^* B_i^*$ also has its last $m - r_i$ rows being zero. Hence, the "canonical form" of the model has

$$\mathcal{Z}_t = \sum_{i=1}^{p} \begin{bmatrix} \Phi_{i1}^* \\ 0 \end{bmatrix} \mathcal{Z}_{t-i} + e_t,$$

with the number $m - r_i$ of zero rows for $\Phi_i^* = [\Phi_{i1}^{*\prime}, 0']'$ increasing as i increases. Therefore, the components z_{jt} of $\mathcal{Z}_t = (z_{1t}, z_{2t}, \ldots, z_{mt})'$ are represented in an AR(p) model by fewer and fewer past lagged variables $\mathcal{Z}_{t-i}$ as j increases from 1 to m.

For the nested reduced-rank AR model with parameters $\Phi_j = A_j B_j$, for any $r_j \times r_j$ nonsingular matrix Q we can write $\Phi_j = A_j Q^{-1} Q B_j = \bar{A}_j \bar{B}_j$, where $\bar{A}_j = A_j Q^{-1}$ and $\bar{B}_j = QB_j$. Therefore, some normalization conditions on the A_j and B_j are required to ensure a unique set of parameters. Assuming the components of Y_t have been arranged so that the upper $r_j \times r_j$ block matrix of each A_j is full rank, this parameterization can be obtained by expressing the Φ_j as $\Phi_j = A_j B_j = A_1 D_j B_j$, where A_1 is $m \times r_1$ and has certain elements "normalized" to fixed values of ones and zeros, and $D_j = [I_{r_j}, 0]'$ is $r_1 \times r_j$. More specifically, the matrix A_1 can always be formed to be lower triangular with ones on the main diagonal, and when augmented with the last $m - r_1$ columns of the identity matrix, the inverse of the resulting matrix A can form the necessary matrix P of the canonical transformation in (5.19). [See Ahn and Reinsel (1988) for further details concerning the normalization.] Thus, with these conventions we can write the model as $Y_t = A_1 \sum_{i=1}^{p} D_i B_i Y_{t-i} + \epsilon_t$. In addition, if we set $\Phi_0^{\#} = A^{-1} = P$ and define the $m \times m$ coefficient matrices $\Phi_i^{\#} = [B_i', 0]'$, $i = 1, \ldots, p$, the nested reduced-rank model can be written in an equivalent form as $\Phi_0^{\#} Y_t = \sum_{i=1}^{p} \Phi_i^{\#} Y_{t-i} + \Phi_0^{\#} \epsilon_t$, which is sometimes referred to as an "echelon canonical form" for the vector AR(p) model [e.g., see Reinsel (1997, Section 3.1)].

For example, consider the special case where the ranks of all Φ_i are equal, $r_1 = \cdots = r_p \equiv r$, and notice that this special nested model then reduces to the reduced-rank AR model discussed in Section 5.2 where $\text{rank}[\Phi_1, \ldots, \Phi_p] = r$. The normalization for the matrix A_1 discussed above in this case becomes $A_1 = [I_r, A_{10}']'$, where A_{10} is an $(m-r) \times r$ matrix. The augmented matrix A and the matrix $P = A^{-1}$ in the canonical form (5.19) can be taken as

$$A = \begin{bmatrix} I_r & 0 \\ A_{10} & I_{m-r} \end{bmatrix} \quad \text{and} \quad P = A^{-1} = \begin{bmatrix} I_r & 0 \\ -A_{10} & I_{m-r} \end{bmatrix},$$

where the last $m-r$ rows of P consist of the (linearly independent) vectors F_j' from (5.17) such that $F_j' \Phi_i = 0$ for all $i = 1, \ldots, p$. Thus, for this special model, the last $m-r$ components $z_{jt} = F_j' Y_t$ in the transformed canonical form (5.19) will be white noise series.

Maximum Likelihood Estimation of Parameters in the Model

We let β_0 denote the $a \times 1$ vector of unknown parameters in the matrix A_1, where it follows that $a = \sum_{j=1}^{p} (m - r_j)(r_j - r_{j+1})$. Also, let $\beta_j = \text{vec}(B_j')$, a $mr_j \times 1$ vector, for $j = 1, \ldots, p$, and set $\theta = (\beta_0', \beta_1', \ldots, \beta_p')'$. The unknown parameters θ are estimated by (conditional) maximum likelihood, using a Newton–Raphson iterative procedure. The (conditional) log-likelihood for

T observations $Y_1, \ldots, Y_T$ (given $Y_0, \ldots, Y_{1-p}$) is

$$l(\theta, \Sigma_{\epsilon\epsilon}) = -(T/2)\log|\Sigma_{\epsilon\epsilon}| - (1/2)\sum_{t=1}^{T} \epsilon_t' \Sigma_{\epsilon\epsilon}^{-1} \epsilon_t.$$

It can be shown that an approximate Newton–Raphson iterative procedure to obtain the ML estimate of θ is given by

$$\begin{aligned} \hat{\theta}^{(i+1)} &= \hat{\theta}^{(i)} - \{\partial^2 l/\partial\theta\partial\theta'\}^{-1}_{\hat{\theta}^{(i)}} \{\partial l/\partial\theta\}_{\hat{\theta}^{(i)}} \\ &= \hat{\theta}^{(i)} + \left\{ M' \left(\sum_{t=1}^{T} \mathcal{U}_t \Sigma_{\epsilon\epsilon}^{-1} \mathcal{U}_t' \right) M \right\}^{-1}_{\hat{\theta}^{(i)}} \left\{ M' \sum_{t=1}^{T} \mathcal{U}_t \Sigma_{\epsilon\epsilon}^{-1} \epsilon_t \right\}_{\hat{\theta}^{(i)}}, \end{aligned}$$

where $\hat{\theta}^{(i)}$ denotes the estimate at the ith iteration, $\partial\epsilon_t'/\partial\theta = M'\mathcal{U}_t$, $\mathcal{U}_t = [I_m \otimes Y_{t-1}', \ldots, I_m \otimes Y_{t-p}']'$, and M' is a specific matrix of dimension $(a + m\sum_{j=1}^{p} r_j) \times m^2 p$ whose elements are given functions of the parameters θ (but not the observations).

It can be shown that the ML estimate $\hat{\theta}$ has an asymptotic distribution such that $T^{1/2}(\hat{\theta} - \theta)$ converges in distribution to $N(0, V^{-1})$ as $T \to \infty$, where $V = M'E(\mathcal{U}_t \Sigma_{\epsilon\epsilon}^{-1} \mathcal{U}_t')M \equiv M'GM$ and $G = E(\mathcal{U}_t \Sigma_{\epsilon\epsilon}^{-1} \mathcal{U}_t')$ has (i, j)th block element equal to $\Sigma_{\epsilon\epsilon}^{-1} \otimes \Gamma(i - j)$. Thus for large T, the ML estimate $\hat{\theta}$ is approximately distributed as $N(\theta, T^{-1}\hat{V}_T^{-1})$, where $V_T = M'(T^{-1}\sum_{t=1}^{T} \mathcal{U}_t \Sigma_{\epsilon\epsilon}^{-1} \mathcal{U}_t')M$. In addition, the asymptotic distribution of the (reduced-rank) ML estimates $\hat{\Phi}_j = \hat{A}_j \hat{B}_j$ of the AR parameters Φ_j follows directly from the result above. It is determined from the relations

$$\begin{aligned} \hat{\Phi}_j - \Phi_j &= \hat{A}_1 D_j \hat{B}_j - A_1 D_j B_j \\ &= (\hat{A}_1 - A_1) D_j B_j + A_1 D_j (\hat{B}_j - B_j) + (\hat{A}_1 - A_1) D_j (\hat{B}_j - B_j) \\ &= (\hat{A}_1 - A_1) D_j B_j + A_1 D_j (\hat{B}_j - B_j) + O_p(T^{-1}). \end{aligned}$$

From this, we find that

$$\hat{\phi} - \phi = M(\hat{\theta} - \theta) + O_p(T^{-1}),$$

where $\hat{\phi} = \text{vec}\{\hat{\Phi}_1', \ldots, \hat{\Phi}_p'\}$, $\phi = \text{vec}\{\Phi_1', \ldots, \Phi_p'\}$, and $M = \partial\phi/\partial\theta'$, in fact, so that

$$T^{1/2}(\hat{\phi} - \phi) \xrightarrow{d} N(0, MV^{-1}M') \quad \text{as } T \to \infty.$$

In particular, it is noted that the collection of resulting reduced-rank estimates $\hat{\Phi}_j = \hat{A}_j \hat{B}_j$ of the Φ_j has a smaller asymptotic covariance matrix than the corresponding full-rank least squares estimates, since

$$MV^{-1}M' = M\{M'GM\}^{-1}M' < G^{-1}.$$

Note that the preceding developments and estimation results have similarities to the results presented in Section 3.5 concerning the estimation of multivariate reduced-rank regression models with two sets of regressors, except that the reduced-rank normalizations imposed on the vector of autoregressive parameters ϕ are now in the more explicit form $\Phi_j = A_1 D_j B_j$ with constraints that certain elements of A_1 be fixed at zero or one. Also, it follows that LR tests of various hypotheses concerning the ranks and other restrictions on the matrices Φ_j can be performed in the usual manner based on the ratio of determinants of the residual covariance matrix estimators, $\hat{\Sigma}_{\epsilon\epsilon} = (1/T)\sum_{t=1}^{T} \hat{\epsilon}_t \hat{\epsilon}_t'$, in the "full" and "restricted" models, respectively. More details concerning the nested reduced-rank AR models are given by Ahn and Reinsel (1988) and Reinsel (1997, Sec. 6.1).

5.4 Numerical Example on U.S. Hog Data

To illustrate the reduced-rank modeling of the previous sections, we consider U.S. hog, corn, and wage time series previously analyzed by several authors, including Quenouille (1968, Chap. 8), Box and Tiao (1977), Reinsel (1983), Velu, Reinsel, and Wichern (1986), and Tiao and Tsay (1983, 1989). Annual observations from 1867 to 1948 on five variables, farm wage rate (FW), hog supply (HS), hog price (HP), corn price (CP), and corn supply (CS), will be analyzed. These measurements were logarithmically transformed and linearly coded by Quenouille and, following Box and Tiao, the wage rate and hog price time series were shifted backward by one period. We consider this modified set of series as our basic set for analysis. The wage rate was originally included in the set as a feasible trend variable to reflect the general economic level. The other series are almost exclusively related to the hog industry. After corrections for means and recording the data in hundreds, we denote the basic 5-variate time series as Y_t. Time series plots of the five series, unadjusted for means, are shown in Figure 5.1.

The series Y_t was first subjected to a multivariate stepwise autoregressive analysis to identify an appropriate order of an AR model. The LR chi-squared statistic (Tiao and Box, 1981; Reinsel, 1997, Section 4.3) for testing the significance of the coefficient matrix at lag j (i.e., testing $H_0 : \Phi_j = 0$ in the AR model of order j) is given by (5.18) with $s = m$. For the first three lags (orders) $j = 1, 2, 3$, the LR statistic values are obtained as 448.51, 98.13, and 31.93, respectively. Note that the upper 5% critical value for the χ^2_{25} distribution is 37.7. Thus the LR test results suggest that an AR model of order 2 may be acceptable and that it fits much better than, for example, an AR model of order 1. Using more elaborate initial model specification techniques and criterion, Tiao and Tsay (1989) suggested an ARMA(2,1) model for the series.

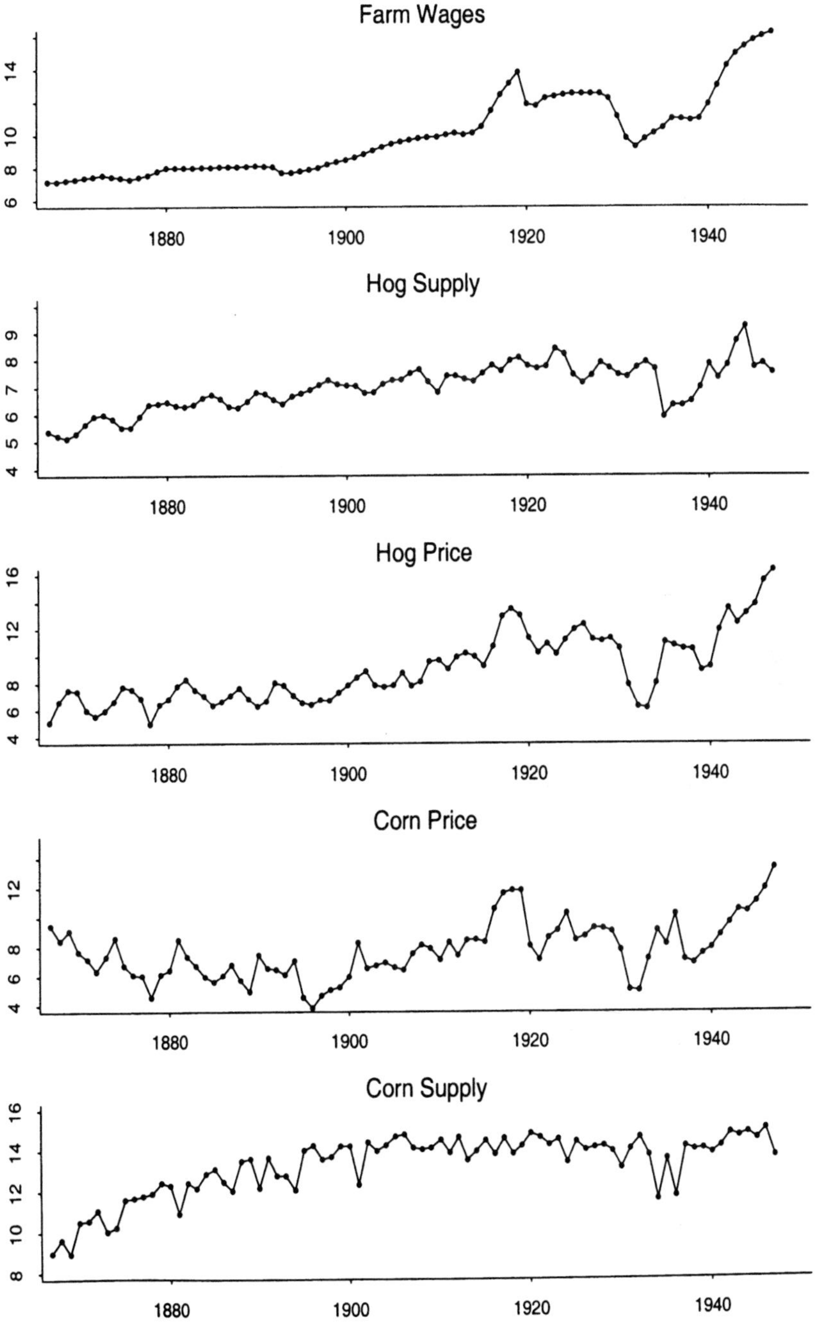

Figure 5.1. Annual U.S. hog industry data, 1867–1947

The AR(2) model $Y_t = \Phi_1 Y_{t-1} + \Phi_2 Y_{t-2} + \epsilon_t$, when fitted without imposing any rank or other constraints, gives (conditional) ML or LS estimates

$$\tilde{\Phi}_1 = \begin{bmatrix} 1.49^* & -0.09 & -0.03 & 0.01 & 0.12 \\ 0.12 & 0.43^* & -0.08 & 0.11^* & 0.29^* \\ 1.21^* & -0.42 & 0.34^* & 0.20 & 0.02 \\ 1.05^* & -0.83 & -0.27 & 0.75^* & 0.49 \\ 0.00 & 0.28 & 0.12 & -0.02 & 0.23 \end{bmatrix}$$

$$\tilde{\Phi}_2 = \begin{bmatrix} -0.35^* & -0.17 & -0.10 & -0.02 & 0.02 \\ -0.16 & 0.04 & 0.28^* & -0.22^* & -0.16^* \\ -0.47 & 0.14 & -0.53^* & 0.31^* & 0.39^* \\ -0.25 & -0.12 & -0.42^* & 0.07 & 0.07 \\ -0.39 & 0.34 & 0.37^* & -0.11 & 0.24 \end{bmatrix}$$

with residual ML covariance matrix estimate $\tilde{\Sigma}_{\epsilon\epsilon}$ such that $|\tilde{\Sigma}_{\epsilon\epsilon}| = 0.1444 \times 10^{-3}$. The asterisks indicate that the individual coefficient estimates have t-ratios greater than two in absolute value.

Two interesting points emerge from these estimation results. First, the wage rate (y_{1t}) is clearly nonstationary and it is not influenced by any other variables; hence, it may be regarded as exogenous. This is consistent with our initial remark that the wage rate was included to reflect the general economic level. Second, the corn supply (y_{5t}) is influenced mainly by hog price (y_{3t}) in the previous year. Historically, the ratios between total feed consumption and hog production have not changed much over the years. This is true regardless of improvements in feeding efficiency; much of the corn produced is intended for hog feed and so the price of hogs influences the corn supply.

One can attempt to simplify the estimated full-rank AR(2) model by sequentially setting to zero those coefficients whose estimates are quite small compared to their standard errors and reestimating the remaining coefficients. However, in general, this can be time-consuming and may not lead to any simplifications or improved interpretation in the structure of the model for the data, while such a simplification in structure may be uncovered by a reduced-rank analysis. Hence, we now investigate the possibility of reduced-rank autoregression models for these data. As discussed previously, the relationship between canonical correlation analysis and reduced-rank regression allows one to check on the rank of the regression coefficient matrix by testing if certain canonical correlations are zero.

The squared sample canonical correlations $\hat{\rho}_i^2$ between Y_t and $X_t = (Y'_{t-1}, Y'_{t-2})'$, which are the same as the (estimated) predictability ratios from (5.11), are equal to 0.981, 0.866, 0.694, 0.193, and 0.084. We test the null hypothesis $H_0 : \text{rank}(C) = r$ against the alternative $H_1 : \text{rank}(C) = 5$ for $r = 1, 2, 3, 4$, where $C = [\Phi_1, \Phi_2]$, using the LR test statistic $\mathcal{M}$ given in Section 5.2. The LR test statistic values corresponding to the four null hypotheses are obtained as $\mathcal{M} = 244.91, 104.07, 21.16$, and 6.14, with 36, 24, 14, and 6 degrees of freedom, and the associated tail probabilities are found to be 0.00, 0.00, 0.10, and 0.41, respectively. That is, $\text{rank}(C) = 3$ is consistent with the data. (Strictly, the asymptotic chi-squared distribution theory for the LR statistics in this setting is developed for the stationary vector AR case. There are indications that the series Y_t in this example is at least 'partially nonstationary'; nevertheless, it is felt that the use of the LR test procedures here is informative for rank determination.) The (conditional) ML estimates of the component matrices obtained from (5.10) for the AR(2) model of reduced rank $r = 3$ are

$$\hat{A} = \begin{bmatrix} 0.33 & 0.06 & 0.04 \\ 0.09 & -0.15 & -0.08 \\ 0.32 & 0.15 & 0.47 \\ 0.20 & 0.17 & 0.42 \\ 0.12 & -0.28 & -0.02 \end{bmatrix}$$

$$\hat{B} = \begin{bmatrix} 4.07 & 0.60 & 0.13 & -0.15 & 0.35 & -1.13 & -0.33 & -0.03 & -0.19 & -0.09 \\ 1.88 & -1.54 & 0.21 & -0.98 & -1.58 & 0.73 & -1.03 & -1.43 & 0.84 & 0.13 \\ -0.60 & -1.14 & 0.32 & 1.05 & 0.48 & -0.47 & 0.76 & -0.63 & 0.46 & 0.83 \end{bmatrix}$$

with ML error covariance matrix estimate $\hat{\Sigma}_{\epsilon\epsilon}$ such that $|\hat{\Sigma}_{\epsilon\epsilon}| = 0.1954 \times 10^{-3}$.

The standard errors of the ML estimates above can be computed using the previously mentioned asymptotic results similar to those in Theorem 2.4. All the elements in $\hat{A}$, except for the coefficient in the (5,3) position, are found to be significant when 'significance' is taken to be greater than twice the asymptotic standard error. The estimated linear combinations

$$\tilde{Y}^*_{t-1} = \hat{B}(Y'_{t-1}, Y'_{t-2})'$$

can be regarded as indices constructed from the past data values. For illustrative purposes, we present these linear combinations with significant terms only, together with standard errors. These are

$$\begin{aligned}
\tilde{y}^*_{1,t-1} &= 4.07(\pm 0.42)y_{1,t-1} - 1.13(\pm 0.43)y_{1,t-2}, \\
\tilde{y}^*_{2,t-1} &= 1.88(\pm 0.42)y_{1,t-1} - 1.54(\pm 0.51)y_{2,t-1} - 0.98(\pm 0.25)y_{4,t-1} \\
&\quad -1.58(\pm 0.26)y_{5,t-1} - 1.03(\pm 0.44)y_{2,t-2} \\
&\quad -1.43(\pm 0.22)y_{3,t-2} + 0.84(\pm 0.23)y_{4,t-2}, \\
\tilde{y}^*_{3,t-1} &= -1.14(\pm 0.50)y_{2,t-1} + 1.05(\pm 0.22)y_{4,t-1} - 0.63(\pm 0.21)y_{3,t-2} \\
&\quad +0.46(\pm 0.22)y_{4,t-2} + 0.83(\pm 0.28)y_{5,t-2}.
\end{aligned}$$

The first linear combination mainly consists of farm wages at lag 1 and lag 2, whereas the third linear combination does not involve farm wages.

The total number of functionally independent parameter estimates in $\hat{A}$ and $\hat{B}$ is only 36, because of the normalization conditions, compared with 50 in the full-rank model. One could also consider a further simplification of the reduced-rank model by use of the procedure which sets to zero those coefficients whose estimates are quite small compared to their standard errors and reestimates the remaining coefficients. However, in this case the explicit eigenvalue-eigenvector solution given by (5.10) would no longer be applicable, and nonlinear iterative estimation procedures must be employed, such as those described by Reinsel (1997, Sections 5.1 and 6.1). Such a procedure was, in fact, applied to the above AR(2) reduced-rank model example and resulted in a quite adequate model containing only 22 parameters, but we omit further details.

In view of the discussion in Section 5.2, the above reduced-rank regression analysis is easy to relate to the canonical analysis. As noted, the ML estimates $\hat{A}$ and $\hat{B}$, specifically the eigenvectors $\hat{V}_i$ in (5.10), are related to the eigenvectors $\hat{\ell}_i$ that are obtained using the sample version of (5.12) by

$$\hat{\ell}_i = \tilde{\Sigma}_{\epsilon\epsilon}^{-1/2}\hat{V}_i, \qquad i = 1, \ldots, m.$$

Since $\hat{\ell}_i = \tilde{\Sigma}_{\epsilon\epsilon}^{-1}\hat{\alpha}_i$, the asymptotic distribution theory for the $\hat{\ell}_i$ follows easily from that for $\hat{A}$, for $i = 1, \ldots, r$, thus providing a useful inferential tool for the canonical analysis of Box and Tiao (1977). The $\hat{\ell}_i$ can be used for the canonical analysis that transforms the series Y_t into 5 new series

$$\mathcal{W}_t = \hat{L}'Y_t,$$

where $\hat{L} = [\hat{\ell}_1, \ldots, \hat{\ell}_m]$, and the accompanying standard errors of the $\hat{\ell}_i$ can be helpful for interpreting the extent of contribution of each original component y_{it} in the canonical variates. The matrix $\hat{L}'$ is obtained as

$$\hat{L}' = \begin{bmatrix} 2.561 & 1.548 & 0.610 & -0.627 & -0.388 \\ -2.286 & 3.747 & 0.180 & 0.805 & 1.627 \\ 2.219 & 0.372 & -1.235 & -1.176 & -1.104 \\ -0.480 & 0.813 & 1.366 & -1.326 & -0.646 \\ 0.869 & -2.978 & -1.115 & 0.634 & 1.647 \end{bmatrix},$$

with nearly all of the estimated coefficients being 'significantly' nonzero. The components of $\mathcal{W}_t$ are contemporaneously uncorrelated and are ordered from most predictable to least predictable. Also recall from Section 5.2 that the components $w_{it} = \hat{\ell}_i'Y_t$ are normalized such that they have sample variance equal to $(1-\hat{\rho}_i^2)^{-1}$ (i.e., $\hat{\ell}_i'\hat{\Gamma}(0)\hat{\ell}_i = (1-\hat{\rho}_i^2)^{-1}$). The transformed series $\mathcal{W}_t$ are plotted in Figure 5.2. This figure indicates that the first two series are nearly nonstationary and the last two fluctuate randomly around constants and hence (on further examination) tend to behave like white noise series, consistent with the reduced rank 3 structure.

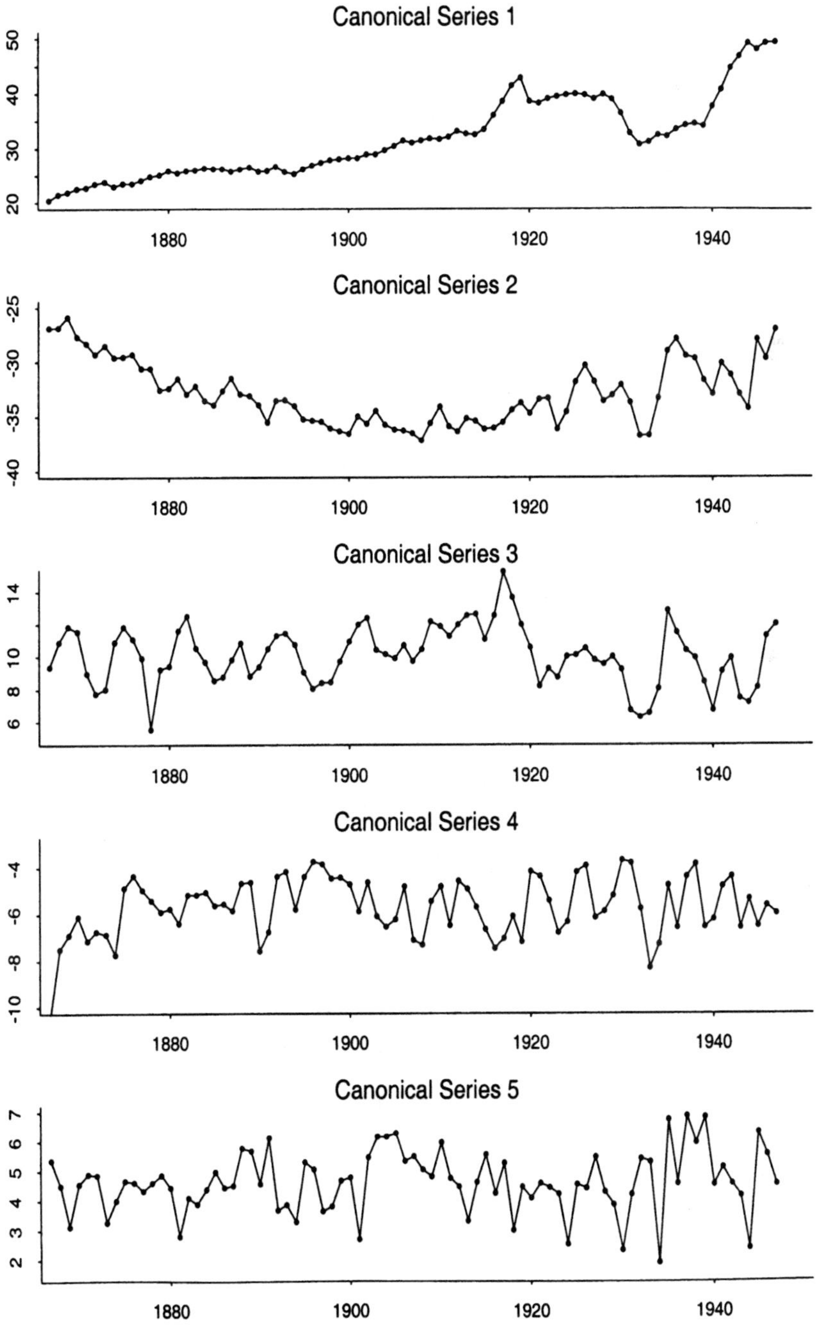

Figure 5.2. Transformed series in canonical analysis for annual U.S. hog industry data, 1867–1947

The first linear combination $\omega_{1t} = \hat{\ell}_1' Y_t$ associated with the largest canonical correlation, that is, the largest predictability ratio, appears to have contributions from all the series, but is dominated by farm wages. The nearly random (white noise) components associated with the two smallest canonical correlations may be of interest due to their relative stability over time. For instance, the sample variance of the final white noise canonical component, $\omega_{5t}/3.739$, when normalized so that the vector of coefficients of the y_{it} has unit norm, is equal to only 0.078. It is instructive to observe that this variance is very small compared to the average variance (equal to 3.58) of the original component series y_{it}, $i = 1, \ldots, 5$.

To arrive at some additional meaningful interpretations involving the last two (white noise) components, we look at certain linear combinations of ω_{4t} and ω_{5t}. By fixing the coefficients of HS_t (y_{2t}) and HP_t (y_{3t}) to be equal to 1, we obtain the linear combination

$$0.589\omega_{4t} - 0.175\omega_{5t} = -0.435y_{1t} + y_{2t} + y_{3t} - 0.892y_{4t} - 0.668y_{5t}.$$

On taking antilogs, this implies, approximately, that the series can be related to a Cobb–Douglas production function,

$$\mathrm{I}_t = \mathrm{HP}_t\mathrm{HS}_t / \{(\mathrm{CP}_t\mathrm{CS}_t)^{0.78}(\mathrm{FW}_t)^{0.43}\},$$

and is a white noise series. [See Box and Tiao (1977) for similar interpretations.] The sample variance of the (essentially) white noise index above, $\omega_t^* = (0.589\omega_{4t} - 0.175\omega_{5t})/1.853$, normalized in a similar way as $\omega_{5t}/3.739$, is 0.135. Hence, we have obtained a quite stable contemporaneous linear relationship among the original 5 variables.

A nested reduced-rank AR model (5.16) can also be considered for these data. The testing procedure (5.18) for investigation of the rank structure of a possible nested reduced-rank model was used. The results of a partial canonical correlation analysis of the data, in terms of the (squared) sample partial canonical correlations $\hat{\rho}_i^2(j)$ for lags $j = 1, 2, 3$, and the associated test statistic values $\mathcal{M}(j, s)$ from (5.18) are displayed in Table 5.1. From the statistics in this table, we reconfirm that an AR model of order 2 may be appropriate, but we also find that the hypothesis that $\mathrm{rank}(\Phi_2) = 3$ is clearly not rejected and that the hypothesis that $\mathrm{rank}(\Phi_1) = 4$ is clearly not rejected. Hence, these results suggest an AR(2) model with a nested reduced-rank structure such that $\mathrm{rank}(\Phi_1) = 4$ and $\mathrm{rank}(\Phi_2) = 3$, that is, $Y_t = A_1(B_1Y_{t-1} + D_2B_2Y_{t-2}) + \epsilon_t$ with A_1 a 5×4 matrix, B_1 a 4×5 matrix, B_2 a 3×5 matrix, and $D_2 = [I_3, 0]'$. Subsequent estimation of this model, by (conditional) ML methods, as outlined in a subsection of Section 5.3, resulted in a fitted model with the following estimates:

$$\hat{A}_1 = \begin{bmatrix} 1.000 & 0.000 & 0.000 & 0.000 \\ 0.000 & 1.000 & 0.000 & 0.000 \\ 0.000 & 0.000 & 1.000 & 0.000 \\ 0.437 & 0.103 & 0.647 & 1.000 \\ -0.701 & 1.765 & 0.430 & -0.394 \end{bmatrix}$$

$$\hat{B}_1 = \begin{bmatrix} 1.482 & -0.032 & -0.022 & 0.012 & 0.116 \\ 0.152 & 0.332 & -0.098 & 0.096 & 0.278 \\ 1.195 & -0.449 & 0.337 & 0.209 & 0.020 \\ -0.152 & -0.919 & -0.551 & 0.526 & 0.331 \end{bmatrix}$$

$$\hat{B}_2 = \begin{bmatrix} -0.362 & -0.160 & -0.097 & -0.020 & -0.007 \\ -0.180 & 0.054 & 0.281 & -0.200 & -0.106 \\ -0.442 & 0.120 & -0.542 & 0.307 & 0.401 \end{bmatrix}$$

with ML error covariance matrix estimate $\hat{\Sigma}_{\epsilon\epsilon}$ such that $|\hat{\Sigma}_{\epsilon\epsilon}| = 0.1584 \times 10^{-3}$. In the results above, the elements in $\hat{A}_1$ equal to 0 and 1 are fixed for normalization purposes.

Notice that when $\hat{A}_1$ is augmented with the last column of the identity matrix, then P equal to the inverse of this matrix provides the canonical transformation $\mathcal{Z}_t = PY_t$ in (5.19) for this nested reduced-rank model. From this, for example, we find that the series

$$z_{5t} = 0.529y_{1t} - 1.805y_{2t} - 0.685y_{3t} + 0.394y_{4t} + y_{5t}$$

forms a white noise process, while the series

$$z_{4t} = -0.437y_{1t} - 0.103y_{2t} - 0.647y_{3t} + y_{4t}$$

has dependence on the past of $\{Y_t\}$ that can be represented by only one lag [an AR(1) scalar component in the terminology of Tiao and Tsay (1989)]. We note that the estimated white noise series z_{5t} under this nested model is almost the same (except for a scale factor) as the estimated 'least predictable' canonical series $\omega_{5t} = \hat{\ell}_5' Y_t$ obtained from the earlier reduced-rank model and associated canonical correlation analysis. Again, further simplification of the nested reduced-rank model above could be considered by setting to zero those coefficients whose estimates are quite small compared to their standard errors and reestimating the remaining coefficients, but we will not go into any further details here.

Table 5.1 Summary of partial canonical correlation analysis between Y_t and Y_{t-j} and associated chi-squared test statistics $\mathcal{M}(j,s)$ from (5.18) for the U.S. hog data. (Test statistics are used to determine the appropriate AR order and ranks $r = m - s$ of coefficient matrices Φ_j.)

		$\mathcal{M}(j,s)$				
j	Partial Correlations	$s=5$	$s=4$	$s=3$	$s=2$	$s=1$
1	0.987, 0.876, 0.725, 0.413, 0.153	448.5	177.5	70.4	15.5	1.75
2	0.741, 0.561, 0.473, 0.140, 0.068	98.1	44.3	18.8	1.7	0.31
3	0.471, 0.386, 0.310, 0.073, 0.023	31.9	16.5	6.6	0.4	0.03
	critical value = $\chi^2_{s^2}(.05)$	37.7	26.3	16.9	9.5	3.84

The empirical results of several analyses of the hog data are summarized in Table 5.2 in terms of the variances of the individual residual series, as well as the determinant of the ML residual covariance matrix $\hat{\Sigma}_{\epsilon\epsilon}$ and the AIC criterion,

$$\text{AIC} = \log(|\hat{\Sigma}_{\epsilon\epsilon}|) + 2n^*/T,$$

where n^* denotes the number of estimated parameters in the model. Included in Table 5.2 are results for both the full-rank and reduced-rank second-order autoregressive models discussed earlier, as well as results from the nested reduced-rank AR(2) model and a second-order index model of rank 3 considered by Reinsel (1983). As mentioned before, various authors have analyzed the U.S. hog data, and in particular Tiao and Tsay (1983) used the hog data to illustrate specification procedures for multivariate models. For comparative purposes they also fit univariate autoregressive moving average models. They observed that the multivariate models clearly outperform the univariate models in terms of residual variance, mainly for the hog supply, hog price, and corn price series. Table 5.2 shows that the results for the reduced-rank models (especially the nested model) are fairly close to those of the full-rank model when compared in terms of residual variances, but the reduced-rank models involve far fewer estimated parameters, and results for both these models are slightly better than those for the index model. Between the two reduced-rank models, although the nested model involves a few more parameters, the overall measures such as the AIC criterion indicate that it represents a somewhat better model fit. The results clearly indicate that little useful information is lost by reducing the dimension of the lagged series.

Table 5.2 Summary of various analyses of the U.S. hog data.

	ML Residual variances				
Variable	Full rank model	Reduced rank model ($r = 3$)	Nested reduced rank ($r_1, r_2 = 4{,}3$)	Index model ($r = 3$)	Uni-variate model
Wage rate	0.1332	0.1374	0.1336	0.1449	0.1456
Hog supply	0.0544	0.0566	0.0554	0.0710	0.1379
Hog price	0.3665	0.3729	0.3668	0.5632	0.8165
Corn price	0.9448	1.0916	0.9580	1.0776	1.4270
Corn supply	0.4623	0.5130	0.4853	0.4845	0.4594
$\lvert\hat{\Sigma}_{\epsilon\epsilon}\rvert(\times 10^{-3})$	0.1444	0.1954	0.1584		
n^* = number of parameters	50	36	42		
AIC	−7.577	−7.629	−7.687		

5.5 Relationship Between Nonstationarity and Canonical Correlations

The reduced-rank models for vector autoregressive time series discussed in the previous sections can be viewed generally in the framework of models related to multivariate linear regression. The associated canonical analysis of Box and Tiao (1977) has been seen to be the same as the classical canonical correlation analysis originally introduced by Hotelling (1935, 1936). Only a few concepts specific to time series analysis, such as the notion of the degree of predictability of linear combinations based on the past of the series $\{Y_t\}$, have thus far entered into the presentation of the reduced-rank AR models and canonical analysis. In this section, we briefly address one additional issue unique to time series analysis. This is the aspect of nonstationarity of the vector time series process $\{Y_t\}$ and the relation of nonstationarity to the canonical analysis.

Consider again the vector AR(p) model (5.8), which we express as $\Phi(L)Y_t = \epsilon_t$, where $\Phi(L) = I - \Phi_1 L - \cdots - \Phi_p L^p$ is the AR(p) operator. Recall that we have assumed the condition that $\det\{\Phi(z)\} \neq 0$ for all complex numbers $|z| \leq 1$ (i.e., all roots of $\det\{\Phi(z)\} = 0$ must lie outside the unit circle), and that this condition ensures the stationarity of the process $\{Y_t\}$. In particular, for the AR(1) process $Y_t = \Phi Y_{t-1} + \epsilon_t$ [$p = 1$ in (5.8)], we have $\Phi(L) = I - \Phi L$ and the condition for stationarity above is equivalent to the condition that all eigenvalues of the matrix Φ lie inside the unit circle.

One interesting result in the work of Box and Tiao (1977) is the development of the correspondence between nonstationarity in the vector AR(1) process, as reflected by the roots of $\det\{I - \Phi z\} = 0$ (equivalently, the eigenvalues of Φ) and the (squared) canonical correlations between Y_t and Y_{t-1} in the canonical analysis, as reflected in the eigenvalues of the matrix

$$R_* = \Gamma(0)^{-1/2}\Gamma(1)'\Gamma(0)^{-1}\Gamma(1)\Gamma(0)^{-1/2},$$

for example, see (5.12) and the discussion following it. To develop the correspondence more easily, note that we can write the matrix R_* as

$$R_* = \Gamma(0)^{-1/2}\Phi\Gamma(0)\Phi'\Gamma(0)^{-1/2},$$

since $\Phi = \Gamma(1)'\Gamma(0)^{-1}$. The specific result to be established is that the number of eigenvalues of Φ approaching the unit circle (the nonstationary boundary) is the same as the number of canonical correlations between Y_t and Y_{t-1} that approach unity. This correspondence was presented as a result in the appendix of the paper by Box and Tiao (1977). Velu, Wichern, and Reinsel (1987) gave an alternative more direct proof of the result, which we now discuss.

As a preliminary, we begin with a statement of the following useful lemma.

Lemma 5.1 For any $m \times m$ complex-valued matrix Ψ, with ordered eigenvalues $|\alpha_1| \geq \cdots \geq |\alpha_m|$,

$$\prod_{j=1}^{k} |\alpha_j| \leq \prod_{j=1}^{k} \rho_j, \qquad k = 1, \ldots, m-1,$$

and

$$\prod_{j=1}^{m} |\alpha_j| = \prod_{j=1}^{m} \rho_j,$$

where ρ_j^2 is the jth largest eigenvalue of $\Psi\Psi^*$ and Ψ^* denotes the conjugate transpose of Ψ.

Proof: See Marshall and Olkin (1979, p. 231).

We now give the main result as follows.

Theorem 5.1 Suppose Y_t follows the stationary vector AR(1) model $Y_t = \Phi Y_{t-1} + \epsilon_t$, where $\text{Cov}(\epsilon_t) = \Sigma_{\epsilon\epsilon}$ is assumed to be positive-definite and fixed. With respect to variation of the matrix Φ, a necessary and sufficient condition for $d \leq m$ of the eigenvalues of $\Gamma(0)^{-1/2}\Phi\Gamma(0)\Phi'\Gamma(0)^{-1/2}$ to tend to unity is that d of the eigenvalues of Φ approach values on the unit circle.

Proof: (Sufficiency) Note that the eigenvalues of Φ are the same as the eigenvalues of $\Psi = \Gamma(0)^{-1/2}\Phi\Gamma(0)^{1/2}$. Denote these eigenvalues by α_j, $j = 1, \ldots, m$. The squared canonical correlations, denoted by ρ_j^2, between Y_t and Y_{t-1} ($\equiv X_t$) are the eigenvalues of $\Gamma(0)^{-1/2}\Phi\Gamma(0)\Phi'\Gamma(0)^{-1/2} = \Psi\Psi^*$ [see Section 2.4 and also (5.12)]. Because $\det\{I - \Phi z\} = \prod_{j=1}^{m}(1 - \alpha_j z)$, from the stationarity assumption it follows that $|\alpha_j| < 1$, for $j = 1, \ldots, m$. The canonical correlations ρ_j all lie between zero and one. From Lemma 5.1 with $k = d$, we have

$$\prod_{j=1}^{d} |\alpha_j| \leq \prod_{j=1}^{d} \rho_j,$$

and when $|\alpha_j| \to 1$, $j = 1, \ldots, d$, then $\prod_{j=1}^{d} \rho_j \to 1$ and hence at least the first d of the canonical correlations $\rho_j \to 1$.

(Necessity) See Velu, Wichern, and Reinsel (1987).

Box and Tiao (1977) developed the sufficiency part of the proof with considerable detail, but in the presentation above a simple proof follows with the use of Lemma 5.1. Some extensions of the results in Theorem 5.1 to higher order AR(p) processes were presented by Velu, Wichern, and Reinsel

(1987), involving partial canonical correlations. However, a complete correspondence between the number of roots α_j of $\det\{\Phi(z)\} = 0$ that approach the unit circle and the number of various canonical correlations and partial canonical correlations of the AR(p) process $\{Y_t\}$ has not been established.

The main implication of Theorem 5.1 is that the existence of d canonical correlations ρ_j close to one in the vector AR(1) process implies that the associated canonical components will be (close to) nonstationary series, and these may reflect the nonstationary trending or dynamic growth behavior in the multivariate system, while the remaining $r = m - d$ canonical variates possess stationary behavior. Hence, this canonical analysis may be useful for exploring the nature of nonstationarity of the multiple autoregressive time series. Specifically, the nature of nonstationarity can be examined by considering the canonical correlations between Y_t and Y_{t-1} and the associated eigenvectors (canonical variates). This approach to investigation of the nonstationary aspects of the vector AR process has more recently been explored by Bossaerts (1988) and Bewley and Yang (1995).

A particularly important instance of nonstationarity in the vector AR process is the case where the only roots of $\det\{\Phi(z)\} = 0$ on the unit circle are unity. This corresponds typically to the situation where the nonstationary vector process Y_t becomes stationary upon first differencing, that is, $W_t = (1 - L)Y_t$ is stationary. In this case, for the vector AR(1) model an alternate canonical analysis is also extremely useful to examine the nature of nonstationarity, in particular, the number of unit roots in the system. The associated methodology is the basis for ML estimation and LR testing procedures concerning processes with unit roots, and it will be discussed and developed in much more detail in the next section. For the present, we will establish a connection between the number of unit eigenvalues of Φ in the vector AR(1) process and the number of zero canonical correlations in a certain canonical analysis.

For this, we first note that the vector AR(1) model, $Y_t = \Phi Y_{t-1} + \epsilon_t$, can be expressed in an equivalent form as

$$W_t \equiv Y_t - Y_{t-1} = -(I - \Phi)Y_{t-1} + \epsilon_t, \tag{5.20}$$

which is referred to as an error-correction form. Notice that if Φ has eigenvalues approaching unity then $I - \Phi$ will have eigenvalues approaching zero (and will become of reduced rank). The methodology of interest for exploring the number of unit roots in the AR(1) model is motivated by the model form (5.20), and consists of a canonical correlation analysis between $W_t = Y_t - Y_{t-1}$ and Y_{t-1} ($\equiv X_t$). Under stationarity of the process Y_t, from (5.20) note that the covariance matrix between W_t and $Y_{t-1} \equiv X_t$ is equal to $\Sigma_{wx} = -(I - \Phi)\Gamma(0)$. Thus, the squared canonical correlations between W_t and Y_{t-1} are the eigenvalues of

$$\Sigma_{ww}^{-1}\Sigma_{wx}\Sigma_{xx}^{-1}\Sigma_{xw} = \Sigma_{ww}^{-1}(I - \Phi)\Gamma(0)(I - \Phi)', \tag{5.21}$$

where

$$\Sigma_{ww} = \text{Cov}(W_t) = 2\Gamma(0) - \Phi\Gamma(0) - \Gamma(0)\Phi' = (I-\Phi)\Gamma(0) + \Gamma(0)(I-\Phi)'.$$

We now present the basic result.

Theorem 5.2 Suppose Y_t follows the stationary vector AR(1) model, with error-correction form given in (5.20), where $\text{Cov}(\epsilon_t) = \Sigma_{\epsilon\epsilon}$ is assumed to be positive-definite and fixed. With respect to variation of the matrix Φ, the condition that $d \leq m$ of the eigenvalues of $\Sigma_{ww}^{-1}(I - \Phi)\Gamma(0)(I - \Phi)'$ (i.e., squared canonical correlations between W_t and Y_{t-1}) tend to zero implies that at least d of the eigenvalues of Φ approach unity.

Proof: Denote the squared canonical correlations between W_t and Y_{t-1}, that is, the eigenvalues of (5.21), by σ_j^2, $j = 1, \ldots, m$. Now notice that

$$\begin{aligned}\Gamma(0)^{-1/2}\Sigma_{ww}\Gamma(0)^{-1/2} &= \Gamma(0)^{-1/2}(I-\Phi)\Gamma(0)^{1/2} + \Gamma(0)^{1/2}(I-\Phi)'\Gamma(0)^{-1/2} \\ &\equiv U + U',\end{aligned}$$

where $U = \Gamma(0)^{-1/2}(I - \Phi)\Gamma(0)^{1/2}$. In addition, then, the matrix in (5.21) can be reexpressed as

$$\begin{aligned}&\Gamma(0)^{-1/2}\{\Gamma(0)^{-1/2}\Sigma_{ww}\Gamma(0)^{-1/2}\}^{-1} \\ &\qquad \times \Gamma(0)^{-1/2}(I - \Phi)\Gamma(0)(I - \Phi)'\Gamma(0)^{-1/2}\Gamma(0)^{1/2} \\ &\quad \equiv \Gamma(0)^{-1/2}\{U + U'\}^{-1}UU'\Gamma(0)^{1/2},\end{aligned}$$

and this matrix has the same eigenvalues as the matrix

$$\{U + U'\}^{-1/2}UU'\{U + U'\}^{-1/2}.$$

Let $|\alpha_1^*| \geq \cdots \geq |\alpha_m^*|$ denote the ordered eigenvalues of the matrix $\{U + U'\}^{-1/2}U$. Now suppose that $d = m - r$ canonical correlations σ_j approach 0, so that $\sigma_{r+1} \to 0$ in particular. Then, by Lemma 5.1, $\prod_{j=1}^{r+1} |\alpha_j^*| \leq \prod_{j=1}^{r+1} \sigma_j \to 0$. This implies that $|\alpha_{r+1}^*| \to 0$ (so also $|\alpha_j^*| \to 0$ for $j = r+1, \ldots, m$), and hence that at least $d = m-r$ eigenvalues of $\{U+U'\}^{-1/2}U$ approach 0. This, in turn, implies that at least d eigenvalues of U approach 0. Since the eigenvalues of $U = \Gamma(0)^{-1/2}(I - \Phi)\Gamma(0)^{1/2}$ are the same as the eigenvalues of $I - \Phi$, and the latter eigenvalues are equal to $1 - \alpha_j$, $j = 1, \ldots, m$, where α_j are the eigenvalues of Φ, it follows that at least d values $(1 - \alpha_j) \to 0$, equivalently, at least d eigenvalues $\alpha_j \to 1$.

To consider the (partial) converse result, suppose that $d = m - r$ eigenvalues α_j of Φ approach unity, and all other $|\alpha_j|$ remain bounded in the limit by a constant strictly less than one. Hence, d eigenvalues $1 - \alpha_j$ of

$I - \Phi$ approach zero, and, in addition, we will assume that the *rank* of $I - \Phi$ approaches $r = m - d$ in the limit. Then it will follow that at least d squared canonical correlations σ_j^2 [i.e., eigenvalues of (5.21)] will tend to zero. Finally, then, we notice that the results of Theorems 5.1 and 5.2 combine to establish the following relation. Namely, if $d \le m$ canonical correlations between $W_t = Y_t - Y_{t-1}$ and Y_{t-1} tend to zero then at least d of the canonical correlations between Y_t and Y_{t-1} approach unity.

To discuss certain relationships concerning nonstationarity in more specific detail, we consider further the particular case of a vector AR(1) model $Y_t = \Phi Y_{t-1} + \epsilon_t$ in which $d = m - r$ eigenvalues α_i of Φ approach unity, and all other $|\alpha_i|$ remain bounded in the limit by a constant strictly less than one. We will also assume that Φ approaches its limit in such a way that the rank of $I - \Phi$ is $r = m - d$ in the limit, that is, there are d linearly independent left eigenvectors $Q_1', \ldots, Q_d'$ associated with the unit eigenvalues of Φ. Informally, then, in a canonical correlation analysis between $W_t = Y_t - Y_{t-1}$ and Y_{t-1}, the d linear combinations $w_{1t}^* = Q_1' W_t, \ldots,$ $w_{dt}^* = Q_d' W_t$ tend to provide the canonical variates having associated canonical correlations that approach zero. The corresponding canonical variates for Y_{t-1} are $-Q_i'(I - \Phi)Y_{t-1}$, with the squared canonical correlations

$$\begin{aligned}
\sigma_i^2 &= \operatorname{Corr}^2(Q_i' W_t, -Q_i'(I - \Phi)Y_{t-1}) \\
&= \frac{(Q_i'(I - \Phi)\Gamma(0)(I - \Phi)' Q_i)^2}{(Q_i' \Sigma_{ww} Q_i)\,(Q_i'(I - \Phi)\Gamma(0)(I - \Phi)' Q_i)} \\
&= \frac{Q_i'(I - \Phi)\Gamma(0)(I - \Phi)' Q_i}{Q_i'(I - \Phi)\Gamma(0) Q_i + Q_i'\Gamma(0)(I - \Phi)' Q_i} \to 0,
\end{aligned}$$

since $Q_i'(I - \Phi)\Gamma(0)(I - \Phi)'Q_i/[Q_i'(I - \Phi)\Gamma(0)Q_i + Q_i'\Gamma(0)(I - \Phi)'Q_i] \sim (1 - \alpha_i)^2 Q_i'\Gamma(0)Q_i/[2(1 - \alpha_i)Q_i'\Gamma(0)Q_i] = (1 - \alpha_i)/2 \to 0$ in the limit.

5.6 Cointegration for Nonstationary Series – Reduced Rank in Long Term

In Sections 5.2 and 5.3 we considered certain reduced-rank models for vector autoregressive time series, and in Section 5.5 we presented a relationship between nonstationarity of the process and canonical correlation analyses. In this section, reduced-rank regression methods will be shown to be useful in recognizing and explicitly modeling nonstationary aspects of a vector process. While the computational procedures of the reduced-rank methods used in modeling relationships in this nonstationary setting are similar to those presented for other reduced-rank models in previous chapters and sections, the asymptotic theory of the estimation and testing procedures differs from the usual asymptotic normal theory because of the presence of nonstationary characteristics of the series involved in the analysis.

Many time series in practice exhibit nonstationary behavior, often of a homogeneous nature, that is, drifting or trending behavior such that apart from a shifting local level or local trend the series has homogeneous patterns with respect to shifts over time. Often, in univariate integrated ARMA time series models, a nonstationary time series can be reduced to stationarity through differencing of the series. For the vector AR model (5.8), $\Phi(L)Y_t = \epsilon_t$ where $\Phi(L) = I - \Phi_1 L - \cdots - \Phi_p L^p$, the condition for stationarity of the vector process Y_t is that all roots of $\det\{\Phi(z)\} = 0$ be greater than one in absolute value. To generalize this to nonstationary but nonexplosive processes, one can consider a general form of the vector AR model where some of the roots of $\det\{\Phi(z)\} = 0$ are allowed to have absolute value equal to one. More specifically, because of the prominent role of the differencing operator $(1-L)$ in univariate models, we might only allow some roots to equal one (unit roots) while the remaining roots are all greater than one in absolute value.

The nonstationary (unit-root) aspects of a vector process Y_t become more complicated in multivariate time series modeling compared to the univariate case, due in part to the possibility of cointegration among the component series y_{it} of a nonstationary vector process Y_t. For instance, the possibility exists for each component series y_{it} to be nonstationary with its first difference $(1-L)y_{it}$ stationary (in which case y_{it} is said to be integrated of order one), but such that certain linear combinations $z_{it} = b_i' Y_t$ of Y_t will be stationary. In such circumstances, the process Y_t is said to be *cointegrated* with cointegrating vectors b_i (e.g., Engle and Granger, 1987). An interpretation of cointegrated vector series Y_t, particularly related to economics, is that the individual components y_{it} share some common nonstationary components or "common trends" and, hence, they tend to have certain similar movements in their long-term behavior. A related interpretation is that the component series y_{it}, although they may exhibit nonstationary behavior, satisfy (approximately) a long-run equilibrium relation $b_i' Y_t \approx 0$ such that the process $z_{it} = b_i' Y_t$, which represents the deviations from the equilibrium, exhibits stable behavior and so is a stationary process. A specific nonstationary AR structure for which cointegration occurs is the model where $\det\{\Phi(z)\} = 0$ has $d < m$ roots equal to one and all other roots are greater than one in absolute value, and also the matrix $\Phi(1) = I - \Phi_1 - \cdots - \Phi_p$ has rank equal to $r = m - d$. Then, for such a process, it can be established that r linearly independent vectors b_i exist such that $b_i' Y_t$ is stationary, and Y_t is said to have cointegrating rank r. Properties of nonstationary cointegrated systems have been investigated by Engle and Granger (1987), among others, and basic properties and estimation of cointegrated vector AR models will be considered in some detail in the remainder of this section. In particular, we will illustrate the usefulness of reduced-rank estimation methods in the analysis of such models.

Thus, we consider multivariate AR models for nonstationary processes $\{Y_t\}$. We focus on situations where the only "nonstationary roots" in the AR operator $\Phi(L)$ are roots equal to one (unit roots), and we assume there are $d \leq m$ such unit roots, with all other roots of $\det\{\Phi(z)\} = 0$ outside the unit circle. We note immediately that this implies that $\det\{\Phi(1)\} = 0$ so that the matrix $\Phi(1) = I - \sum_{j=1}^{p} \Phi_j$ does not have full rank. It is also assumed that $\text{rank}\{\Phi(1)\} = r$, with $r = m - d$, and that $\Phi(1)$ has (exactly) d zero eigenvalues. It is further noted that this last condition implies that each component of the first differences $W_t = Y_t - Y_{t-1}$ will be stationary (rather than any component of Y_t being integrated of order higher than one). The AR(p) model, $\Phi(L)Y_t = Y_t - \sum_{j=1}^{p} \Phi_j Y_{t-j} = \epsilon_t$, can also be represented in the *error-correction form* (Engle and Granger, 1987) as $\Phi^*(L)(1-L)Y_t = -\Phi(1)Y_{t-1} + \epsilon_t$, that is,

$$W_t = CY_{t-1} + \sum_{j=1}^{p-1} \Phi_j^* W_{t-j} + \epsilon_t, \tag{5.22}$$

where $W_t = (1-L)Y_t = Y_t - Y_{t-1}$,

$$\Phi^*(L) = I - \sum_{j=1}^{p-1} \Phi_j^* L^j,$$

with $\Phi_j^* = -\sum_{i=j+1}^{p} \Phi_i$, and $C = -\Phi(1) = -(I - \sum_{j=1}^{p} \Phi_j)$.

The form (5.22) is obtained from direct matrix algebra, since $\Phi(L)$ can always be reexpressed as $\Phi(L) = \Phi^*(L)(1-L) + \Phi(1)L$ with $\Phi^*(L)$ as defined. For example, in the AR(2) model we have

$$\begin{aligned} \Phi(L) &= I - \Phi_1 L - \Phi_2 L^2 = (I - IL) + (I - \Phi_1 - \Phi_2)L + \Phi_2(I - IL)L \\ &= (I + \Phi_2 L)(I - IL) + (I - \Phi_1 - \Phi_2)L \equiv \Phi^*(L)(1-L) + \Phi(1)L, \end{aligned}$$

where $\Phi^*(L) = I + \Phi_2 L \equiv I - \Phi_1^* L$ with $\Phi_1^* = -\Phi_2$. Under the assumptions on $\Phi(L)$, it can also be shown that $\Phi^*(L)$ is a stationary AR operator having all roots of $\det\{\Phi^*(z)\} = 0$ outside the unit circle. The error-correction form (5.22) is particularly useful because the number of unit roots in the AR operator $\Phi(L)$ can conveniently be incorporated through the "error-correction" term CY_{t-1}, so that the nature of nonstationarity of the model is conveniently concentrated in the behavior of the single coefficient matrix C in this form.

From the assumptions, the matrix $I - \Phi(1) = \sum_{j=1}^{p} \Phi_j$ has d linearly independent eigenvectors associated with its d unit eigenvalues, while its remaining eigenvalues are less than one in absolute value. Let P and $Q = P^{-1}$ be $m \times m$ matrices such that $Q(\sum_{j=1}^{p} \Phi_j)P = \text{Diag}[I_d, \Lambda_r] = J$, where J is the Jordan canonical form of $\sum_{j=1}^{p} \Phi_j$, so that $QCP = J - I = \text{Diag}[0, \Lambda_r - I_r]$. Hence, $C = P(J-I)Q = P_2(\Lambda_r - I_r)Q_2'$, where $P = [P_1, P_2]$

and $Q' = [Q_1, Q_2]$, with P_1 and Q_1 being $m \times d$ matrices, so that C is of *reduced rank* $r < m$. Therefore, the error-correction form (5.22) can be written as

$$W_t = AQ_2'Y_{t-1} + \sum_{j=1}^{p-1} \Phi_j^* W_{t-j} + \epsilon_t \equiv AZ_{2t-1} + \sum_{j=1}^{p-1} \Phi_j^* W_{t-j} + \epsilon_t, \qquad (5.22')$$

where $A = P_2(\Lambda_r - I_r)$ is $m \times r$ of rank r and $Z_{2t} = Q_2'Y_t$. We also define $Z_{1t} = Q_1'Y_t$, and let $Z_t = (Z_{1t}', Z_{2t}')' = QY_t$.

It follows from the above that although Y_t is nonstationary (with first differences that are stationary), the r linear combinations $Z_{2t} = Q_2'Y_t$ are stationary [e.g., since through (5.22) the variables Z_{2t} are such that $Z_{2t} - \Lambda_r Z_{2t-1}$ is linearly expressible in terms of the stationary series $\{W_t\}$ and $\{\epsilon_t\}$ only and Λ_r is stable]. In this situation, Y_t is *cointegrated* of rank r, and the rows of Q_2' are *cointegrating vectors.* The stationary linear combinations $Z_{2t} = Q_2'Y_t$ may be interpreted as long-term stable equilibrium relations among the variables Y_t, and the error-correction model (5.22), as written in (5.22′), formulates that the changes $W_t = Y_t - Y_{t-1}$ in the variables Y_t depend on the deviations from the equilibrium relations in the previous time period (as well as on previous changes W_{t-j}). These deviations, Z_{2t-1}, are thus viewed as being useful explanatory variables for the next change in the process Y_t, and in the context of (5.22′) they are referred to as an *error-correction mechanism.*

Conversely, the d-dimensional series $Z_{1t} = Q_1'Y_t$ is purely nonstationary, with the unit-root nonstationary behavior of the series Y_t generated by Z_{1t}. That is, from $Z_t = (Z_{1t}', Z_{2t}')' = QY_t$, we obtain the relation $Y_t = PZ_t = P_1Z_{1t} + P_2Z_{2t}$. We then see that Y_t is a linear combination of the d-dimensional purely nonstationary component Z_{1t} and the r-dimensional stationary component Z_{2t}. The purely nonstationary component Z_{1t} may be viewed as the common nonstationary component or the d "common trends" among the Y_t, with the interpretation that the nonstationary behavior in each of the m component series Y_{it} of Y_t is actually driven by d $(< m)$ common underlying stochastic trends (Z_{1t}).

5.6.1 LS and ML Estimation and Inference

Least squares estimates $\tilde{C}$, $\tilde{\Phi}_1^*, \ldots, \tilde{\Phi}_{p-1}^*$ for the model (5.22) in error-correction form can be obtained, and the limiting distribution theory for these estimators has been derived by Ahn and Reinsel (1990). To describe the asymptotic results, note that the model (5.22) can be expressed as

$$W_t = CPZ_{t-1} + \sum_{j=1}^{p-1} \Phi_j^* W_{t-j} + \epsilon_t$$

$$= CP_1 Z_{1t-1} + CP_2 Z_{2t-1} + \sum_{j=1}^{p-1} \Phi_j^* W_{t-j} + \epsilon_t,$$

where $Z_t = QY_t = (Z'_{1t}, Z'_{2t})'$, with $Z_{1t} = Q'_1 Y_t$ purely nonstationary, whereas $Z_{2t} = Q'_2 Y_t$ is stationary, and note also that $CP_1 = 0$. Because of the presence of the nonstationary "regressor" variables (Z_{1t}) in the model above, the least squares and maximum likelihood estimation theory is nonstandard compared to the situation with stationary regressors. In addition, from (5.22), the process $u_t = Q(W_t - CY_{t-1}) = Z_t - JZ_{t-1}$, where $J = \text{Diag}[I_d, \Lambda_r]$, is equal to $u_t = Q(\sum_{j=1}^{p-1} \Phi_j^* W_{t-j} + \epsilon_t)$ and, hence, u_t is stationary. Therefore, u_t has the (stationary) infinite MA representation of the form $u_t = \Psi(L)a_t = \sum_{j=0}^{\infty} \Psi_j a_{t-j}$, with $a_t = Q\epsilon_t$, and we let $\Psi = \Psi(1) = \sum_{j=0}^{\infty} \Psi_j$.

Now let $F = [C, \Phi_1^*, \ldots, \Phi_{p-1}^*]$, $X_{t-1} = (Y'_{t-1}, W'_{t-1}, \ldots, W'_{t-p+1})'$, and $X_{t-1}^* = (Z'_{t-1}, W'_{t-1}, \ldots, W'_{t-p+1})' \equiv Q^* X_{t-1}$, where $Q^* = \text{Diag}[Q, I_{k(p-1)}]$, and assume that observations $Y_{1-p}, \ldots, Y_0, Y_1, \ldots, Y_T$ are available (for convenience of notation), with $W_t = Y_t - Y_{t-1}$. Then the error-correction model (5.22) can be written as

$$W_t = FX_{t-1} + \epsilon_t = FP^* X_{t-1}^* + \epsilon_t,$$

where $P^* = \text{Diag}[P, I_{k(p-1)}] = Q^{*-1}$. Thus, the least squares (LS) estimator of F can be represented by

$$\tilde{F} = \left(\sum_{t=1}^{T} W_t X'_{t-1} \right) \left(\sum_{t=1}^{T} X_{t-1} X'_{t-1} \right)^{-1}$$

$$= \left(\sum_{t=1}^{T} W_t X_{t-1}^{*\prime} \right) \left(\sum_{t=1}^{T} X_{t-1}^* X_{t-1}^{*\prime} \right)^{-1} Q^*,$$

so that

$$(\tilde{F} - F)P^* = \left(\sum_{t=1}^{T} \epsilon_t X_{t-1}^{*\prime} \right) \left(\sum_{t=1}^{T} X_{t-1}^* X_{t-1}^{*\prime} \right)^{-1}. \qquad (5.23)$$

With $U_{t-1}^* = (Z'_{2t-1}, W'_{t-1}, \ldots, W'_{t-p+1})'$, which is stationary, and so $X_{t-1}^* = (Z'_{1t-1}, U_{t-1}^{*\prime})'$, it has been shown that $T^{-3/2} \sum_{t=1}^{T} U_{t-1}^* Z'_{1t-1} \xrightarrow{P} 0$ as $T \to \infty$. Hence it follows that the LS estimator for the model (5.22) has the representation

$$[T\tilde{C}P_1,\ T^{1/2}(\tilde{C}P_2 - CP_2, \tilde{\Phi}_1^* - \Phi_1^*, \ldots, \tilde{\Phi}_{p-1}^* - \Phi_{p-1}^*)]$$

$$= \left[P \left(T^{-1} \sum a_t Z'_{1t-1} \right) \left(T^{-2} \sum Z_{1t-1} Z'_{1t-1} \right)^{-1}, \right.$$

$$\left. \left(T^{-1/2} \sum \epsilon_t U_{t-1}^{*\prime} \right) \left(T^{-1} \sum U_{t-1}^* U_{t-1}^{*\prime} \right)^{-1} \right] + o_p(1).$$

Then, the distribution theory [see Lemma 1 and Theorem 1 in Ahn and Reinsel (1990)] for the least squares estimator is such that $T\tilde{C}P_1 \xrightarrow{d} PM$, where

$$M = \Sigma_a^{1/2}\left\{\int_0^1 B_d(u)dB_m(u)'\right\}'\left\{\int_0^1 B_d(u)B_d(u)'du\right\}^{-1}\Sigma_{a_1}^{-1/2}\Psi_{11}^{-1}, \quad (5.24)$$

where $\Sigma_a = Q\Sigma_{\epsilon\epsilon}Q' = \text{Cov}(a_t)$ with $\Sigma_{\epsilon\epsilon} = \text{Cov}(\epsilon_t)$, $\Sigma_{a_1} = [I_d, 0]\Sigma_a[I_d, 0]'$ is the upper-left $d \times d$ block of the matrix Σ_a, and Ψ_{11} is the upper-left $d \times d$ block of the matrix Ψ, $B_m(u)$ is a m-dimensional standard Brownian motion process, and $B_d(u) = \Sigma_{a_1}^{-1/2}[I_d, 0]\Sigma_a^{1/2}B_m(u)$ is a d-dimensional standard Brownian motion. It can be shown (Reinsel, 1997, Section 6.3.8) that Ψ_{11} is expressible more explicitly as $\Psi_{11} = \{Q_1'\Phi^*(1)P_1\}^{-1}$. It also holds that $T^{1/2}(\tilde{C}P_2 - CP_2)'$ and $T^{1/2}(\tilde{\Phi}_j^* - \Phi_j^*)'$, $j = 1, \ldots, p{-}1$, have a joint limiting multivariate normal distribution as in stationary model situations, such that the "vec" of these terms has limiting distribution $N(0, \Sigma_{\epsilon\epsilon}\otimes\Gamma_{U^*}^{-1})$, where $\Gamma_{U^*} = \text{Cov}(U_t^*)$. Also, the LS estimates $\tilde{\Phi}_1^*, \ldots, \tilde{\Phi}_{p-1}^*$ have the same asymptotic distribution as in the "stationary case," where Q_2 is known and one regresses W_t on the stationary variables $Z_{2t-1} = Q_2'Y_{t-1}$ and $W_{t-1}, \ldots, W_{t-p+1}$.

When maximum likelihood estimation of the parameter matrix C in (5.22) is considered subject to the reduced-rank restriction that rank$(C) = r$, it is convenient to express C as $C = AB$, where A and B are $m \times r$ and $r \times m$ full-rank matrices, respectively. As usual, some normalization conditions on A and B are required for uniqueness. One possibility is to have B normalized so that $B = [I_r, B_0]$, where B_0 is an $r \times (m-r)$ matrix of unknown parameters. The elements of the vector Y_t can always be arranged so that this normalization is possible. Implicit in this arrangement of Y_t is that if we write $Y_t = (Y_{1t}', Y_{2t}')'$, where Y_{1t} is $r \times 1$ and Y_{2t} is $d \times 1$, then there is no cointegration among the components of Y_{2t}. It is emphasized that the estimation of the model with the reduced-rank constraint imposed on C is equivalent to the estimation of the AR model with d unit roots explicitly imposed in the model. Hence, this is an alternative to (arbitrarily) differencing each component of the series Y_t prior to fitting a model in situations where the individual components tend to exhibit nonstationary behavior. This explicit form of modeling the nonstationarity may lead to a better understanding of the nature of nonstationarity among the different component series and may also improve longer-term forecasting over unconstrained model fits that do not explicitly incorporate unit roots in the model. Hence, there may be many desirable reasons to formulate and estimate the AR model with an appropriate number of unit roots explicitly incorporated in the model, and it is found that a particularly convenient way to do this is through the use of model (5.22) with the constraint rank$(C) = r$ imposed.

Maximum likelihood estimation of $C = AB = P_2(\Lambda_r - I_r)Q_2'$ and the Φ_j^* under the constraint that rank$(C) = r$, using an iterative Newton–Raphson procedure with the normalization $B = [I_r, B_0]$, is presented by Ahn and Reinsel (1990), and the limiting distribution theory for these estimators is derived. Specifically, note that the model can be written as

$$W_t = ABY_{t-1} + \sum_{j=1}^{p-1} \Phi_j^* W_{t-j} + \epsilon_t \equiv ABY_{t-1} + DW_{t-1}^* + \epsilon_t, \qquad (5.25)$$

where $D = [\Phi_1^*, \ldots, \Phi_{p-1}^*]$ and $W_{t-1}^* = (W_{t-1}', \ldots, W_{t-p+1}')'$. It is recognized that this takes the same form as the extended reduced-rank model (3.3) considered in Section 3.1. We define the vector of unknown parameters as $\theta = (\beta_0', \alpha')'$, where $\beta_0 = \text{vec}(B_0')$, and $\alpha = \text{vec}\{[A, D]'\} = \text{vec}\{[A, \Phi_1^*, \ldots, \Phi_{p-1}^*]'\}$. With $b = r(m-r) + rm + m^2(p-1)$ representing the number of unknown parameters in θ, define the $b \times m$ matrices

$$\mathcal{U}_{t-1} = [\,(A' \otimes H'Y_{t-1})', I_m \otimes \tilde{U}_{t-1}']',$$

where $\tilde{U}_{t-1} = [(BY_{t-1})', W_{t-1}', \ldots, W_{t-p+1}']'$ is stationary and the matrix $H' = [0, I_{m-r}]$ is $(m-r) \times m$ such that $Y_{2t} = H'Y_t$ is taken to be purely nonstationary by assumption. Then, based on T observations $Y_1, \ldots, Y_T$, the Gaussian estimator of θ is obtained by the iterative approximate Newton–Raphson relations

$$\hat{\theta}^{(i+1)} = \hat{\theta}^{(i)} + \left\{ \sum_{t=1}^{T} \mathcal{U}_{t-1} \Sigma_{\epsilon\epsilon}^{-1} \mathcal{U}_{t-1}' \right\}_{\hat{\theta}^{(i)}}^{-1} \left\{ \sum_{t=1}^{T} \mathcal{U}_{t-1} \Sigma_{\epsilon\epsilon}^{-1} \epsilon_t \right\}_{\hat{\theta}^{(i)}}, \qquad (5.26)$$

where $\hat{\theta}^{(i)}$ denotes the estimate at the ith iteration.

Concerning the asymptotic distribution theory of the Gaussian estimators under the model where the unit roots are imposed, it is assumed that the iterations in (5.26) are started from an initial consistent estimator $\hat{\theta}^{(0)}$ such that $D^*(\hat{\theta}^{(0)} - \theta) = \{T(\hat{\beta}_0^{(0)} - \beta_0)', T^{1/2}(\hat{\alpha}^{(0)} - \alpha)'\}'$ is $O_p(1)$, where $D^* = \text{Diag}[TI_{rd}, T^{1/2}I_{(b-rd)}]$. Then, it has been established (Ahn and Reinsel, 1990) that

$$T(\hat{B}_0 - B_0) \xrightarrow{d} (A'\Sigma_{\epsilon\epsilon}^{-1}A)^{-1} A'\Sigma_{\epsilon\epsilon}^{-1} P\, M\, P_{21}^{-1},$$

where the distribution of M is specified in (5.24), and P_{21} is the $d \times d$ lower submatrix of P_1. For the remaining parameters, α, in the model, it follows that

$$T^{1/2}(\hat{\alpha} - \alpha) = \left[I_m \otimes \left(\frac{1}{T} \sum_{t=1}^{T} \tilde{U}_{t-1} \tilde{U}_{t-1}' \right)^{-1} \right] \frac{1}{\sqrt{T}} \sum_{t=1}^{T} (I_m \otimes \tilde{U}_{t-1}) \epsilon_t + o_p(1).$$

Hence, it is also shown that $T^{1/2}(\hat{\alpha} - \alpha) \xrightarrow{d} N(0, \Sigma_{\epsilon\epsilon} \otimes \Gamma_{\tilde{U}}^{-1})$, where $\Gamma_{\tilde{U}} = \text{Cov}(\tilde{U}_t)$, similar to results in stationary situations.

The iterative procedure (5.26) to obtain the Gaussian estimator $\hat{\theta}$ can readily be modified to incorporate additional constraints, such as zero constraints or nested reduced-rank constraints on the stationary parameters Φ_j^* in (5.22), beyond the cointegrating constraint that rank$(C) = r$. However, when there are no constraints imposed other than rank$(C) = r$, the Gaussian reduced-rank estimator in model (5.22) can also be obtained explicitly through the *partial canonical correlation analysis*, based on the previous work as discussed in Section 3.1, since the model (5.25) has the same form as (3.3). This approach was presented by Johansen (1988, 1991). To describe the results, let $\tilde{W}_t$ and $\tilde{Y}_{t-1}$ denote the "adjusted" or residual vectors from the least squares regressions of W_t and Y_{t-1}, respectively, on the lagged values $W_{t-1}^* = (W_{t-1}', \ldots, W_{t-p+1}')'$, and let

$$S_{\tilde{w}\tilde{w}} = \frac{1}{T}\sum_{t=1}^{T} \tilde{W}_t \tilde{W}_t', \qquad S_{\tilde{w}\tilde{y}} = \frac{1}{T}\sum_{t=1}^{T} \tilde{W}_t \tilde{Y}_{t-1}', \qquad S_{\tilde{y}\tilde{y}} = \frac{1}{T}\sum_{t=1}^{T} \tilde{Y}_{t-1} \tilde{Y}_{t-1}'.$$

Then, the sample partial canonical correlations $\hat{\rho}_i$ between W_t and Y_{t-1}, given $W_{t-1}, \ldots, W_{t-p+1}$, and the corresponding vectors $\hat{V}_i^*$ are the solutions to

$$(\hat{\rho}_i^2 I_m - S_{\tilde{w}\tilde{w}}^{-1/2} S_{\tilde{w}\tilde{y}} S_{\tilde{y}\tilde{y}}^{-1} S_{\tilde{y}\tilde{w}} S_{\tilde{w}\tilde{w}}^{-1/2}) \hat{V}_i^* = 0, \qquad i = 1, \ldots, m. \qquad (5.27)$$

The *reduced-rank Gaussian estimator* of $C = AB$ can then be obtained from results of Sections 2.3-2.4 and 3.1. It can be expressed explicitly as $\hat{C} = S_{\tilde{w}\tilde{w}}^{1/2} \hat{V}_* \hat{V}_*' S_{\tilde{w}\tilde{w}}^{-1/2} \tilde{C}$, where $\tilde{C} = S_{\tilde{w}\tilde{y}} S_{\tilde{y}\tilde{y}}^{-1}$ is the full-rank LS estimator, and $\hat{V}_* = [\hat{V}_1^*, \ldots, \hat{V}_r^*]$ are the (normalized) vectors corresponding to the r largest partial canonical correlations $\hat{\rho}_i$, $i = 1, \ldots, r$. This form of the estimator provides the reduced-rank factorization as

$$\hat{C} = (S_{\tilde{w}\tilde{w}}^{1/2} \hat{V}_*)(\hat{V}_*' S_{\tilde{w}\tilde{w}}^{-1/2} \tilde{C}) \equiv \hat{A}\hat{B},$$

with $\hat{A} = S_{\tilde{w}\tilde{w}}^{1/2} \hat{V}_*$ satisfying the normalization $\hat{A}' S_{\tilde{w}\tilde{w}}^{-1} \hat{A} = I_r$. The Gaussian estimator for the other parameters $\Phi_1^*, \ldots, \Phi_{p-1}^*$ can be obtained by ordinary least squares regression of $W_t - \hat{C} Y_{t-1}$ on the lagged variables $W_{t-1}, \ldots, W_{t-p+1}$. As an alternate form of the reduced-rank estimator, let $\hat{V}_i^{**} = S_{\tilde{w}\tilde{w}}^{-1/2} \hat{V}_i^* (1 - \hat{\rho}_i^2)^{-1/2}$, $i = 1, \ldots, m$, and define the $m \times r$ matrix $\hat{V}_{**} = S_{\tilde{w}\tilde{w}}^{-1/2} \hat{V}_* (I_r - \hat{\Lambda}_*^2)^{-1/2}$, where $\hat{\Lambda}_* = \text{Diag}(\hat{\rho}_1, \ldots, \hat{\rho}_r)$. Then the reduced-rank estimator can be expressed equivalently as $\hat{C} = \tilde{\Sigma}_{\epsilon\epsilon} \hat{V}_{**} \hat{V}_{**}' \tilde{C}$, where

$$\tilde{\Sigma}_{\epsilon\epsilon} = T^{-1} \sum_{t=1}^{T} \tilde{\epsilon}_t \tilde{\epsilon}_t' = S_{\tilde{w}\tilde{w}} - S_{\tilde{w}\tilde{y}} S_{\tilde{y}\tilde{y}}^{-1} S_{\tilde{y}\tilde{w}}$$

is the error covariance matrix estimate of $\Sigma_{\epsilon\epsilon}$ from the full-rank LS estimation. The vectors $\hat{V}_i^{**}$ in $\hat{V}_{**}$ are normalized by $\hat{V}_i^{**\prime}\tilde{\Sigma}_{\epsilon\epsilon}\hat{V}_i^{**} = 1$ (so that $\hat{V}_{**}^{\prime}\tilde{\Sigma}_{\epsilon\epsilon}\hat{V}_{**} = I_r$).

Note from the discussion in Section 2.4 (e.g., see Theorem 2.3) that the expression $\hat{B} = \hat{V}_*^{\prime}S_{\tilde{w}\tilde{w}}^{-1/2}\tilde{C}$ shows that the estimated cointegrating vectors (rows of $\hat{B}$) are obtained from the vectors that determine the first r canonical variates of Y_{t-1} in the partial canonical correlation analysis. In addition, it is seen from the form of the reduced-rank estimator $\hat{C} = S_{\tilde{w}\tilde{w}}^{1/2}\hat{V}_*\hat{V}_*^{\prime}S_{\tilde{w}\tilde{w}}^{-1/2}\tilde{C}$ given above that the Gaussian estimates of the "common trends" components $Z_{1t} = Q_1^{\prime}Y_t$ can be obtained with

$$\hat{Q}_1 = S_{\tilde{w}\tilde{w}}^{-1/2}[\hat{V}_m^*, \ldots, \hat{V}_{r+1}^*],$$

so that the rows of $\hat{Q}_1^{\prime}$ are the vectors $\hat{V}_i^{*\prime}S_{\tilde{w}\tilde{w}}^{-1/2}$, $i = r+1, \ldots, m$, corresponding to the d smallest partial canonical correlations, with the property that $\hat{Q}_1^{\prime}\hat{C} = 0$ since $\hat{Q}_1^{\prime}\hat{A} = [\hat{V}_m^*, \ldots, \hat{V}_{r+1}^*]^{\prime}\hat{V}_* = 0$. This method of estimation of the common "trend" or "long-memory" components was explored by Gonzalo and Granger (1995).

5.6.2 *Likelihood Ratio Test for the Number of Cointegrating Relations*

Within the context of model (5.22), it is necessary to specify or determine the rank r of cointegration or the number d of unit roots in the model. Thus, it is also of interest to test the hypothesis H_0: rank$(C) \leq r$, which is equivalent to the hypothesis that the number of unit roots in the AR model is greater than or equal to d ($d = m-r$), against the general alternative that rank$(C) = m$. The *likelihood ratio test* for this hypothesis is considered by Johansen (1988) and by Reinsel and Ahn (1992). The likelihood ratio (LR) test statistic is given by

$$-T\log(U) = -T\log(|S|/|S_0|),$$

where $S = \sum_{t=1}^{T}\hat{\epsilon}_t\hat{\epsilon}_t^{\prime}$ denotes the residual sum of squares matrix in the full or unconstrained model, while S_0 is the residual sum of squares matrix obtained under the reduced-rank restriction on C that rank$(C) = r$. It follows from work of Anderson (1951) in the multivariate linear model with reduced-rank structure (see Section 3.1) that the LR statistic can be expressed equivalently as

$$-T\log(U) = -T\sum_{i=r+1}^{m}\log(1-\hat{\rho}_i^2), \tag{5.28}$$

where the $\hat{\rho}_i$ are the $d = m-r$ smallest sample partial canonical correlations between $W_t = Y_t - Y_{t-1}$ and Y_{t-1}, given $W_{t-1}, \ldots, W_{t-p+1}$. The limiting distribution for the likelihood ratio statistic has been derived by Johansen (1988), and by Reinsel and Ahn (1992) using the distribution theory for the full-rank LS and Gaussian reduced-rank estimators. Specifically, it is given by

$$-T\log(U) \xrightarrow{d} \operatorname{tr}\left\{ \left(\int_0^1 B_d(u)dB_d(u)' \right)' \left(\int_0^1 B_d(u)B_d(u)'du \right)^{-1} \right.$$
$$\left. \times \left(\int_0^1 B_d(u)dB_d(u)' \right) \right\}, \qquad (5.29)$$

where $B_d(u)$ is a d-dimensional standard Brownian motion process. Note that the asymptotic distribution of the likelihood ratio test statistic under H_0 depends only on d and not on any parameters or the order p of the AR model.

Critical values of the asymptotic distribution of (5.29) have been obtained by simulation by Johansen (1988) and Reinsel and Ahn (1992) and can be used in the test of H_0. Some other approaches to testing for unit roots in nonstationary multivariate systems have been examined by Engle and Granger (1987), Stock and Watson (1988), Fountis and Dickey (1989), Phillips and Ouliaris (1986), and Bewley and Yang (1995). It is known that inclusion of a constant term in the estimation of the nonstationary AR model affects the limiting distribution of the estimators and test statistics. Modification of the asymptotic distribution theory for the LS and Gaussian estimators and for LR testing, to allow for inclusion of constant terms in the estimation, has been considered by Johansen (1991), Johansen and Juselius (1990), and Reinsel and Ahn (1992), and critical values for the corresponding limiting distribution of the LR test statistic have been obtained by Monte Carlo simulation.

5.7 Unit Root and Cointegration Aspects for U.S. Hog Data Example

We illustrate the modeling and analysis procedures for nonstationary vector AR processes discussed in the preceding section by further considering the annual U.S. hog data from Section 5.4. The previous analyses for these data in Section 5.4 indicate that an AR(2) model may be adequate, and the basic time series plots in Figure 5.1 suggest nonstationarity in the component series. Hence, we investigate the nature of the nonstationarity in terms of (the number of) unit roots in the AR(2) model operator, equivalently, the number of cointegrating relations among the components. As discussed, the AR(2) model can be expressed in the equivalent error-correction model form (5.22),

$$W_t = CY_{t-1} + \Phi_1^* W_{t-1} + \epsilon_t,$$

where $W_t = Y_t - Y_{t-1}$, $C = -\Phi(1) = -(I - \Phi_1 - \Phi_2)$, and $\Phi_1^* = -\Phi_2$.

Within the context of a vector AR model, we consider LR tests for the number of unit roots d or the rank of cointegration $r = m - d$. Thus, we perform the partial canonical correlation analysis between W_t and Y_{t-1}, given W_{t-j}, $j = 1, \dots, k-1$. For greater generality and to illustrate the effect of the choice of AR order k on the LR test results, we provide results for various values of $k = 1, \dots, 5$. The sample partial canonical correlations $\hat{\rho}_i$ are shown in Table 5.3, as well as the LR test statistic values for H_0 : rank$(C) = r$ for $r = 0, \dots, 4$. Since an AR(p) model of order $p = 2$ has been chosen as appropriate, we concentrate on the LR test statistics for $k = 2$ in Table 5.3. By referring to the upper percentiles of the limiting distribution of the LR test statistic [e.g., see Reinsel (1997, p. 205)], we find that the hypotheses of $d \geq 1$ ($r \leq 4$) or $d \geq 2$ ($r \leq 3$) are clearly not rejected, while $d \geq 4$ ($r \leq 1$) or $d \geq 5$ ($r \leq 0$) are strongly rejected. There is also moderate evidence for rejection of $d \geq 3$ ($r \leq 2$), and so the LR testing procedures suggest that there are $d = 2$ unit roots and $r = 3$ cointegrating vectors. (Note that, with the exception of the value for $r = 0$ at $k = 2$, the LR test statistic values in Table 5.3 remain relatively stable for values of $k \geq 2$, which is consistent with an AR model of order 2.)

Table 5.3 Summary of partial canonical correlation analysis between $W_t = Y_t - Y_{t-1}$ and Y_{t-1} and associated LR test statistics from (5.29) for the U.S. hog data. (Test statistics are used to determine the appropriate rank $r = m - d$, the number of cointegrating relations, of matrix C.)

		LR Statistic				
k	Partial Correlations	$r=0$	$r=1$	$r=2$	$r=3$	$r=4$
1	0.718, 0.708, 0.473, 0.340, 0.063	142.2	84.9	30.0	10.0	0.32
2	0.804, 0.576, 0.473, 0.354, 0.022	142.9	61.7	30.2	10.5	0.04
3	0.593, 0.530, 0.442, 0.369, 0.003	86.7	53.4	28.0	11.3	0.00
4	0.560, 0.490, 0.386, 0.344, 0.052	71.5	42.9	22.0	9.8	0.21
5	0.677, 0.569, 0.393, 0.372, 0.026	99.2	53.1	23.8	11.2	0.05
	critical values at 0.05%	71.1	49.4	31.7	18.0	8.16

The full-rank LS estimates $\tilde{\Phi}_1$ and $\tilde{\Phi}_2$ of the AR(2) model were given previously in Section 5.4, and from these the LS estimates $\tilde{C} = -(I - \tilde{\Phi}_1 - \tilde{\Phi}_2)$ and $\tilde{\Phi}_1^* = -\tilde{\Phi}_2$ in the error-correction form can easily be deduced. The Gaussian reduced-rank estimates with $r = \text{rank}(C) = 3$ imposed, which incorporates the $d = 2$ unit roots in the model, can be obtained from the partial canonical correlation results for AR order $k = 2 \equiv p$. As described in the previous section, we obtain $\hat{C} = \hat{A}\hat{B}$, with $\hat{A} = S_{\tilde{w}\tilde{w}}^{1/2}\hat{V}_*$ and $\hat{B} = \hat{V}_*' S_{\tilde{w}\tilde{w}}^{-1/2}\tilde{C}$, as

$$\hat{C} = \hat{A}\hat{B} = \begin{bmatrix} 0.027 & -0.180 & 0.023 \\ -0.286 & -0.162 & -0.100 \\ 0.781 & -0.252 & -0.126 \\ 0.288 & -1.013 & 0.463 \\ -0.214 & 0.464 & -0.109 \end{bmatrix}$$

$$\times \begin{bmatrix} 0.639 & 0.367 & -1.240 & 0.617 & 0.262 \\ -0.747 & 1.671 & 0.609 & 0.073 & -0.735 \\ -0.214 & 1.465 & 0.613 & -0.741 & -0.812 \end{bmatrix}$$

$$= \begin{bmatrix} 0.147 & -0.258 & -0.130 & -0.013 & 0.121 \\ -0.041 & -0.522 & 0.195 & -0.115 & 0.125 \\ 0.715 & -0.319 & -1.199 & 0.557 & 0.492 \\ 0.843 & -0.910 & -0.691 & -0.240 & 0.444 \\ -0.460 & 0.537 & 0.482 & -0.018 & -0.309 \end{bmatrix}$$

with ML error covariance matrix estimate $\hat{\Sigma}_{\epsilon\epsilon}$ such that $|\hat{\Sigma}_{\epsilon\epsilon}| = 0.1653 \times 10^{-3}$. There is close agreement between the reduced-rank estimate $\hat{C}$ and the (implied) LS estimate $\tilde{C}$, and the corresponding ML estimate (not shown) $\hat{\Phi}_1^*$ of Φ_1^* is found to be very close to the LS estimate $\tilde{\Phi}_1^* = -\tilde{\Phi}_2$.

From the estimation results above, estimates of three cointegrating linear combinations are given by

$$Z_{2t} = (z_{3t}, z_{4t}, z_{5t})' = \hat{\Lambda}_*^{-1}\hat{B}Y_t \equiv \hat{Q}_2'Y_t,$$

which are normalized such that the variances of the corresponding adjusted series $\tilde{Z}_{2t} = \hat{\Lambda}_*^{-1}\hat{B}\tilde{Y}_t$ have unit variances (i.e., $\hat{\Lambda}_*^{-1}\hat{B}S_{\tilde{y}\tilde{y}}\hat{B}'\hat{\Lambda}_*^{-1} = I_r$). These three series are displayed in Figure 5.3, and give the rather clear appearance of being stationary. Also shown in Figure 5.3 are the two linear combinations

$$Z_{1t} = (z_{1t}, z_{2t})' = \hat{Q}_1'Y_t,$$

where $\hat{Q}_1 = S_{\tilde{w}\tilde{w}}^{-1/2}[\hat{V}_m^*, \ldots, \hat{V}_{r+1}^*]$ and is given by

$$\hat{Q}_1' = \begin{bmatrix} 3.237 & -0.744 & -0.228 & -0.362 & 0.085 \\ 2.346 & -1.096 & -0.673 & -1.017 & -2.059 \end{bmatrix}.$$

The series $Z_{1t} = \hat{Q}_1'Y_t$ are normalized such that the variances of the adjusted first differences of these series, $\hat{Q}_1'\tilde{W}_t$, have unit variances (i.e., $\hat{Q}_1'S_{\tilde{w}\tilde{w}}\hat{Q}_1 = I_d$). These linear combinations can be interpreted as the "common trend" series among the five original series. Notice that the first of the common trend series consists predominantly of the farm wage rates series (y_{1t}).

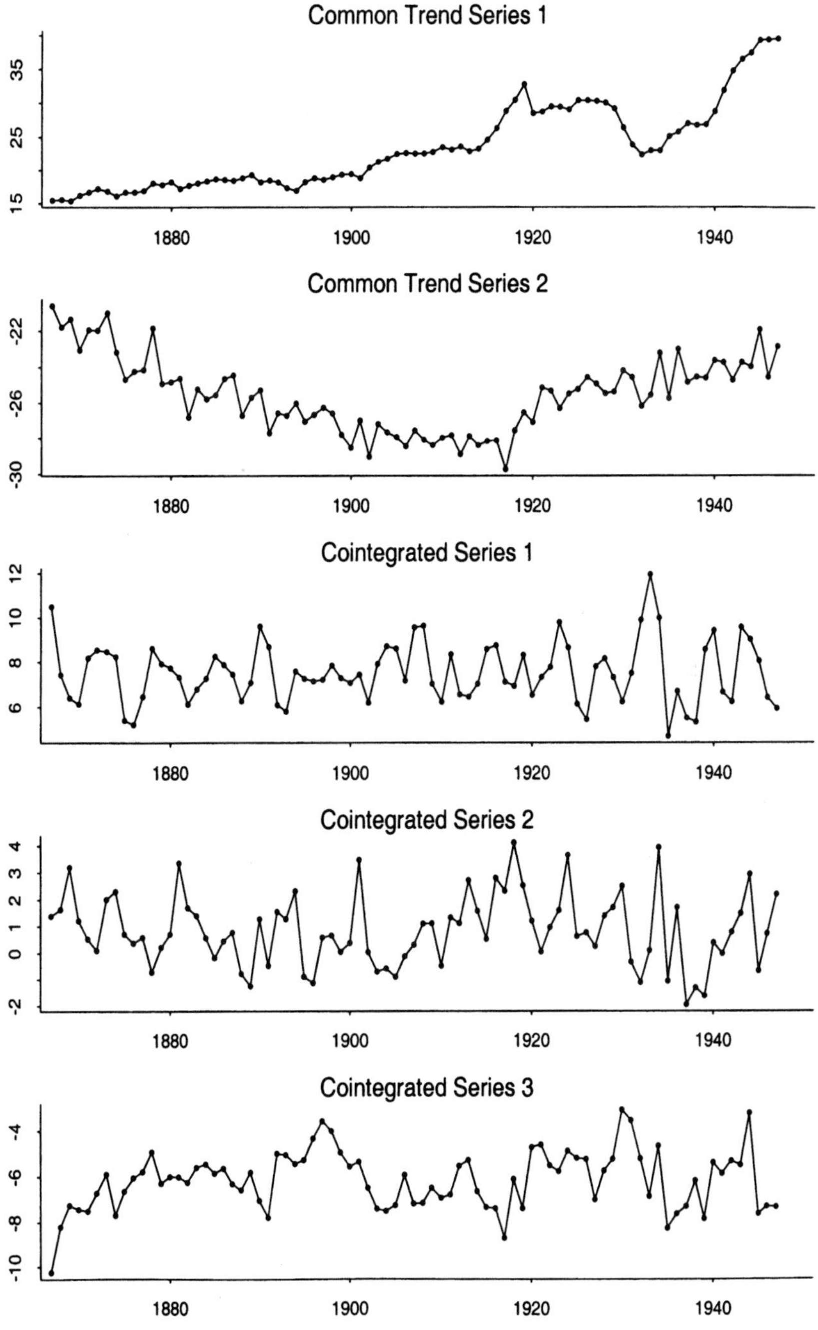

Figure 5.3. Common trends and cointegrating series for annual U.S. hog industry data, 1867–1947

From a model building point of view, we recall from Section 5.4 that a reduced rank of 3 was also indicated for the lag 2 coefficient matrix Φ_2, and so this would also be true for the matrix Φ_1^* in the error-correction form. Thus, for the final model estimation results, we fit a combined reduced-rank AR(2) model in error-correction form, $W_t = CY_{t-1} + \Phi_1^* W_{t-1} + \epsilon_t$, with the constraints that $\text{rank}(C) = 3$ and $\text{rank}(\Phi_1^*) = 3$, the latter constraint as suggested from the LR test statistic results of Table 5.1 in Section 5.4. Notice that there is no "nesting" of the two coefficient matrices in this form of reduced-rank model, and hence, the model has the form of the model (3.23) in Section 3.4. We write $C = AB$ and $\Phi_1^* = EF$, and use the normalizations $B = [I_3, B_0]$ and $E = [I_3, E_0']'$. The final Gaussian reduced-rank estimates are obtained by Newton–Raphson iterative procedures, similar to those described in Sections 5.3 and 5.6, and are given by

$$\hat{C} = \hat{A}\hat{B} = \begin{bmatrix} 0.136 & -0.220 & -0.118 \\ -0.036 & -0.555 & 0.184 \\ 0.714 & -0.333 & -1.196 \\ 0.849 & -1.003 & -0.734 \\ -0.493 & 0.711 & 0.548 \end{bmatrix} \begin{bmatrix} 1 & 0 & 0 & -1.6129 & -0.3114 \\ 0 & 1 & 0 & -0.1694 & -0.4095 \\ 0 & 0 & 1 & -1.3807 & -0.4947 \end{bmatrix}$$

$$= \begin{bmatrix} 0.136 & -0.220 & -0.118 & -0.020 & 0.106 \\ -0.036 & -0.555 & 0.184 & -0.103 & 0.147 \\ 0.714 & -0.333 & -1.196 & 0.557 & 0.506 \\ 0.849 & -1.003 & -0.734 & -0.187 & 0.509 \\ -0.493 & 0.711 & 0.548 & -0.082 & -0.409 \end{bmatrix}$$

$$\hat{\Phi}_1^* = \hat{E}\hat{F} = \begin{bmatrix} 1 & 0 & 0 \\ 0 & 1 & 0 \\ 0 & 0 & 1 \\ 1.511 & -0.986 & 0.072 \\ -2.728 & 3.922 & 1.572 \end{bmatrix}$$

$$\times \begin{bmatrix} 0.364 & 0.117 & 0.084 & 0.033 & 0.008 \\ 0.161 & -0.005 & -0.265 & 0.199 & 0.131 \\ 0.392 & -0.060 & 0.557 & -0.325 & -0.434 \end{bmatrix}$$

$$= \begin{bmatrix} 0.364 & 0.117 & 0.084 & 0.033 & 0.008 \\ 0.161 & -0.005 & -0.265 & 0.199 & 0.131 \\ 0.392 & -0.060 & 0.557 & -0.325 & -0.434 \\ 0.419 & 0.178 & 0.428 & -0.169 & -0.148 \\ 0.256 & -0.433 & -0.392 & 0.179 & -0.193 \end{bmatrix}$$

with ML error covariance matrix estimate $\hat{\Sigma}_{\epsilon\epsilon}$ such that $|\hat{\Sigma}_{\epsilon\epsilon}| = 0.1696 \times 10^{-3}$. Since the ML estimates for Φ_1^* follow the usual normal distribution theory asymptotics, a LR test statistic for the validity of the constraint that $\text{rank}(\Phi_1^*) = 3$ has the usual asymptotic χ_4^2 distribution. The value of this LR statistic is found to be $-(79-2*5-3/2)\log(0.1653/0.1696) = 1.708$, and so the reduced-rank restriction on Φ_1^* is quite reasonable, very similar to the LR test finding from Table 5.1 in Section 5.4. Also notice that the final reduced-rank estimate $\hat{C} = \hat{A}\hat{B}$ is quite similar to the reduced-rank estimate obtained when the rank of Φ_1^* is not constrained.

Finally, it may be of some interest to relate and compare the decomposition of Y_t into trend and stationary components, associated with the above analysis in the error-correction model form, to the canonical analysis associated with the reduced-rank modeling previously considered in Section 5.4. The transformed vectors of variates associated with the reduced-rank and error-correction form analyses, respectively, are the canonical variates $\mathcal{W}_t = \hat{L}'Y_t$ and the common trend and cointegrating variates $Z_t = (Z_{1t}', Z_{2t}')' = \hat{Q}Y_t$, where $\hat{Q} = [\hat{Q}_1, \hat{Q}_2]'$. Hence, these two sets of transformed variates are related by $\mathcal{W}_t = \hat{L}'\hat{Q}^{-1}Z_t$, with the matrix $\hat{L}'\hat{Q}^{-1}$ given by

$$\hat{L}'\hat{Q}^{-1} = \begin{bmatrix} 1.1395 & -0.2594 & 0.0794 & 0.3199 & 0.3702 \\ -0.0097 & -1.1913 & 0.9551 & -0.0831 & 0.7225 \\ 0.0087 & 0.4240 & 0.8146 & -0.7074 & 0.8143 \\ 0.0114 & -0.0864 & -0.5534 & -0.3071 & 0.6034 \\ 0.0276 & -0.0792 & -0.0096 & -0.6063 & -0.4135 \end{bmatrix}.$$

In the error-correction form analysis above, the original series Y_t has been decomposed into a "purely nonstationary" component Z_{1t} and a stationary component Z_{2t}. Of course, any other series formed as linear combinations of the elements of Z_{2t} will also be stationary, whereas linear combinations that include elements of Z_{1t} will be nonstationary. The canonical analysis of Box and Tiao (1977) does not have exactly the same purpose as the analysis above, but there can be some similarities as noted previously. In particular, in the canonical analysis, components associated with canonical correlations close to one will tend to be nonstationary while those associated with the smaller canonical correlations will tend to be stationary; canonical correlations very close to zero give rise to series with (approximate) white noise behavior. From examination of the structure of the matrix $\hat{L}'\hat{Q}^{-1}$ in the relation $\mathcal{W}_t = \hat{L}'\hat{Q}^{-1}Z_t$, it is found that the approximate white noise canonical variate components $\omega_{4t} = \hat{\ell}_4'Y_t$ and $\omega_{5t} = \hat{\ell}_5'Y_t$ in $\mathcal{W}_t$ are related to the components of Z_t roughly as

$$\omega_{4t} \approx -0.55z_{3t} - 0.31z_{4t} + 0.60z_{5t} \qquad \text{and} \qquad \omega_{5t} \approx -0.61z_{4t} - 0.41z_{5t}.$$

Thus, not surprisingly, we see that the (stationary) nearly white noise variates in the canonical correlation analysis are essentially represented as certain linear combinations of only the *stationary* components z_{3t}, z_{4t}, z_{5t} resulting from the cointegration-common trends analysis associated with the error-correction model form. By contrast, in the linear relations that connect the nearly nonstationary components, ω_{1t} and ω_{2t}, of the canonical correlation analysis with the transformed variates Z_t, it is seen that ω_{1t} has a large contribution from the first common trend component z_{1t} and ω_{2t} has a large contribution from z_{2t}.

Also notice that the final error-correction form of the model with the additional reduced-rank structure on Φ_1^* imposed, that is, $W_t = ABY_{t-1} + EFW_{t-1} + \epsilon_t$, can be rewritten in the original AR(2) form as $Y_t = (I + AB + EF)Y_{t-1} - EFY_{t-2} + \epsilon_t$. Hence, in this model any possible white noise structure could be revealed as a linear combination $\ell' Y_t \approx \ell' \epsilon_t$ for which the relations $\ell'(I + AB + EF) \approx 0$ and $\ell'(EF) \approx 0$ hold. However, although the analysis based on the error-correction model form is not particularly designed to identify possible white noise component structure, we see from the discussion in the preceding paragraph that such white noise components (if present) would come from linear combinations of the cointegrating stationary component vector $Z_{2t} = Q_2' Y_t$ in the error-correction model analysis.

6
The Growth Curve Model and Reduced-Rank Regression Methods

6.1 Introduction and the Growth Curve Model

One additional general model class which has aspects of reduced-rank regression, especially in its mathematical structure, is that of the growth curve or generalized multivariate ANOVA (GMANOVA) model. This type of model is often applied in the analysis of longitudinal or repeated measures data arising in biomedical and other areas. For these models, the components y_{jk} of the m-dimensional response vector $Y_k = (y_{1k}, \dots, y_{mk})'$ usually correspond to responses on a single characteristic of a subject made over m distinct times or occasions, with interest in studying features of the mean response over time, and a sample of T independent subjects is available. Corresponding to each subject is an n-dimensional set of time-invariant explanatory variables X_k, which may include factors for treatment assignments held fixed over all occasions. In addition, it may be postulated that the pattern of mean response over time for any subject can be represented by some parametric function, e.g., polynomial function, of time whose coefficients are specified to be linear functions of the time-invariant explanatory variables X_k. That is, it is assumed that $E(y_{jk}) = a_j'\beta_k$ where a_j' is a known (specified) $1 \times r$ vector whose elements are functions of the jth time, for $j = 1, \dots, m$, and the coefficients in β_k are unknown linear functions of X_k. Let $A = [a_1, \dots, a_m]'$ denote the known $m \times r$ matrix, $r < \min(m, n)$, whose columns contain values of the functions of time specified in the parametric form of the mean response function over time. Then the model (2.5),

$$Y_k = A(BX_k) + \epsilon_k, \qquad k = 1, \ldots, T, \tag{6.1}$$

with unknown $r \times n$ matrix B, corresponds to this situation with $\beta_k \equiv BX_k$ interpretable as the coefficients in the mean response function over time for the kth subject, represented as linear functions of the X_k. Another interpretation for the model is that in the standard multivariate regression model, $Y_k = CX_k + \epsilon_k$, the rows of elements of the regression coefficient matrix C are specified to be linear functions of a known basis of functions of time, such as (orthogonal) polynomials up to degree $r-1$. That is, the jth row of C is specified as $C'_{(j)} = a'_j B$ where a'_j is a known $1 \times r$ vector whose elements are functions of the jth time, for $j = 1, \ldots, m$, and B is an $r \times n$ matrix of unknown parameters. Thus $C = AB$, with A specified (known), and hence the model is again $Y_k = ABX_k + \epsilon_k$.

To briefly illustrate a situation more specifically, suppose the T individuals are from n different experimental or treatment groups and there are no other explanatory variables. Then the $n \times T$ overall across-individual design matrix $\boldsymbol{X}$ is simply a matrix of indicator variables as in the one-way ANOVA situation. Assume T_i vector observations Y_{il} are available for the ith group, for $l = 1, \ldots, T_i$, with $T = T_1 + \cdots + T_n$, and observations in each group are measured at the same m time points, $t_1 < t_2 < \cdots < t_m$. In addition, we assume a model for the growth curves over time for each group to be of the form of a polynomial of degree $r-1$, so that the mean value at the jth time point for the lth individual in the ith group is given by

$$E(y_{jil}) = \beta_{1i} + \beta_{2i}\, t_j + \beta_{3i}\, t_j^2 + \cdots + \beta_{ri}\, t_j^{r-1} \equiv a'_j \beta_i,$$

with $a'_j = (1, t_j, \ldots, t_j^{r-1})$ and $\beta_i = (\beta_{1i}, \beta_{2i}, \ldots, \beta_{ri})'$. Then the basic (polynomial) growth curve model for all individuals can be written in matrix form as $\boldsymbol{Y} = AB\boldsymbol{X} + \boldsymbol{\epsilon}$, where $\boldsymbol{Y} = [Y_{11}, \ldots, Y_{1T_1}, \ldots, Y_{n1}, \ldots, Y_{nT_n}]$,

$$A = \begin{bmatrix} 1 & t_1 & . & . & t_1^{r-1} \\ 1 & t_2 & . & . & t_2^{r-1} \\ . & . & . & . & . \\ . & . & . & . & . \\ 1 & t_m & . & . & t_m^{r-1} \end{bmatrix}, \qquad B = \begin{bmatrix} \beta_{11} & . & . & . & \beta_{1n} \\ \beta_{21} & . & . & . & \beta_{2n} \\ . & . & . & . & . \\ . & . & . & . & . \\ \beta_{r1} & . & . & . & \beta_{rn} \end{bmatrix},$$

$$\boldsymbol{X} = \begin{bmatrix} \mathbf{1}'_{T_1} & 0' & 0' & . & . & 0' \\ 0' & \mathbf{1}'_{T_2} & 0' & . & . & 0' \\ . & . & . & . & . & . \\ . & . & . & . & . & . \\ 0' & 0' & 0' & . & . & \mathbf{1}'_{T_n} \end{bmatrix},$$

and $\mathbf{1}_T$ denotes a $T \times 1$ vector of ones.

Techniques for analysis of the basic growth curve model (6.1), and for various extensions of this model, will be described in the remainder of this chapter. Analysis for the growth curve model (6.1) is somewhat simpler than that of the general reduced-rank model (2.5) of Chapter 2, in that the matrix A is specified as known in the growth curve model, but the two cases still share some mathematical similarities. Moreover, application of the general model (2.5), with A *unknown* or unspecified, to the growth curve or longitudinal data setting might be useful or desirable in some situations, especially when m and n are relatively large. The interpretation could be that the columns of A represent the (unknown) basis functions for the "parametric" form of the mean response function over time, which are not explicitly specified. The growth curve model was introduced by Potthoff and Roy (1964), and was also studied extensively in its early stages by Rao (1965, 1966, 1967), Khatri (1966), and Grizzle and Allen (1969).

In some uses of the model (6.1), the covariance matrix $\Sigma_{\epsilon\epsilon}$ of the errors might be assumed to possess some special structure, rather than a general $m \times m$ covariance matrix. A particular case results from the random effects model assumption for the kth individual's regression coefficients (e.g., Laird and Ware, 1982; Reinsel, 1985). For this case it is assumed that $Y_k = A\beta_k + u_k$, where β_k is a random vector with $E(\beta_k) = BX_k$ and $\text{Cov}(\beta_k) = \Sigma_\beta$, and the error vector u_k has $\text{Cov}(u_k) = \sigma^2 I_m$. Hence, with $\tau_k = \beta_k - BX_k$, we have $Y_k = ABX_k + A\tau_k + u_k \equiv ABX_k + \epsilon_k$ where $\epsilon_k = A\tau_k + u_k$ has covariance structure $\Sigma_{\epsilon\epsilon} = \text{Cov}(A\tau_k + u_k) = A\Sigma_\beta A' + \sigma^2 I_m$. Sometimes in the above model, only a subset of β_k is assumed to be random. Other types of special covariance structure, including AR and MA time series structures, are discussed by Jennrich and Schluchter (1986), Diggle (1988), Chi and Reinsel (1989), Rochon and Helms (1989), and Jones (1986, 1990).

6.2 Estimation of Parameters in the Growth Curve Model

Some basic estimation results for the growth curve model written in matrix form, $\boldsymbol{Y} = A B \boldsymbol{X} + \epsilon$, will now be presented. Under normality and independence of the errors ϵ_k, we know from Section 2.3 that the ML estimate of B is the value that minimizes $|(\boldsymbol{Y} - A B \boldsymbol{X})(\boldsymbol{Y} - A B \boldsymbol{X})'|$ with respect to variation of B. It will be established shortly that the ML estimate of B is given by

$$\hat{B} = (A'\tilde{\Sigma}_{\epsilon\epsilon}^{-1}A)^{-1}A'\tilde{\Sigma}_{\epsilon\epsilon}^{-1}\boldsymbol{Y}\boldsymbol{X}'(\boldsymbol{X}\boldsymbol{X}')^{-1} \equiv (A'\tilde{\Sigma}_{\epsilon\epsilon}^{-1}A)^{-1}A'\tilde{\Sigma}_{\epsilon\epsilon}^{-1}\tilde{C}, \tag{6.2}$$

where $\tilde{C} = \boldsymbol{Y}\boldsymbol{X}'(\boldsymbol{X}\boldsymbol{X}')^{-1}$ and $\tilde{\Sigma}_{\epsilon\epsilon} = (1/T)(\boldsymbol{Y} - \tilde{C}\boldsymbol{X})(\boldsymbol{Y} - \tilde{C}\boldsymbol{X})' \equiv (1/T)S$. The corresponding ML estimate of the general error covariance matrix $\Sigma_{\epsilon\epsilon}$

is thus given by

$$\begin{aligned}\hat{\Sigma}_{\epsilon\epsilon} &= \frac{1}{T}(\boldsymbol{Y} - A\hat{B}\boldsymbol{X})(\boldsymbol{Y} - A\hat{B}\boldsymbol{X})' = \tilde{\Sigma}_{\epsilon\epsilon} + (\tilde{C} - A\hat{B})\hat{\Sigma}_{xx}(\tilde{C} - A\hat{B})' \\ &= \tilde{\Sigma}_{\epsilon\epsilon} + (I_m - A(A'\tilde{\Sigma}_{\epsilon\epsilon}^{-1}A)^{-1}A'\tilde{\Sigma}_{\epsilon\epsilon}^{-1})\tilde{C}\hat{\Sigma}_{xx}\tilde{C}' \\ &\qquad \times (I_m - A(A'\tilde{\Sigma}_{\epsilon\epsilon}^{-1}A)^{-1}A'\tilde{\Sigma}_{\epsilon\epsilon}^{-1})'. \end{aligned} \tag{6.3}$$

For derivation of the ML estimate of B, obtained by minimizing $|(\boldsymbol{Y} - AB\boldsymbol{X})(\boldsymbol{Y} - AB\boldsymbol{X})'|$, we first have the decomposition

$$\begin{aligned}(\boldsymbol{Y}-AB\boldsymbol{X})(\boldsymbol{Y}-AB\boldsymbol{X})' &= (\boldsymbol{Y}-AB\boldsymbol{X})[I - \boldsymbol{X}'(\boldsymbol{X}\boldsymbol{X}')^{-1}\boldsymbol{X}](\boldsymbol{Y}-AB\boldsymbol{X})' \\ &\quad + (\boldsymbol{Y}-AB\boldsymbol{X})\boldsymbol{X}'(\boldsymbol{X}\boldsymbol{X}')^{-1}\boldsymbol{X}(\boldsymbol{Y}-AB\boldsymbol{X})' \\ &= (\boldsymbol{Y} - \tilde{C}\boldsymbol{X})(\boldsymbol{Y} - \tilde{C}\boldsymbol{X})' \\ &\quad + (\tilde{C} - AB)\boldsymbol{X}\boldsymbol{X}'(\tilde{C} - AB)' \\ &= S + (\tilde{C} - AB)\boldsymbol{X}\boldsymbol{X}'(\tilde{C} - AB)'\end{aligned}$$

where $S = \boldsymbol{Y}[I - \boldsymbol{X}'(\boldsymbol{X}\boldsymbol{X}')^{-1}\boldsymbol{X}]\boldsymbol{Y}' = (\boldsymbol{Y} - \tilde{C}\boldsymbol{X})(\boldsymbol{Y} - \tilde{C}\boldsymbol{X})'$. So we obtain

$$\begin{aligned}|(\boldsymbol{Y} - AB\boldsymbol{X})(\boldsymbol{Y} - AB\boldsymbol{X})'| &= |S|\,|I_m + S^{-1}(\tilde{C} - AB)\boldsymbol{X}\boldsymbol{X}'(\tilde{C} - AB)'| \\ &= |S|\,|I_n + \boldsymbol{X}\boldsymbol{X}'(\tilde{C} - AB)'S^{-1}(\tilde{C} - AB)| \\ &= |S|\,|I_n + \hat{\Sigma}_{xx}(\tilde{C} - AB)'\tilde{\Sigma}_{\epsilon\epsilon}^{-1}(\tilde{C} - AB)|,\end{aligned}$$

where $\hat{\Sigma}_{xx} = (1/T)\boldsymbol{X}\boldsymbol{X}'$. We can also write

$$\begin{aligned}(\tilde{C}-AB)'\tilde{\Sigma}_{\epsilon\epsilon}^{-1}(\tilde{C}-AB) &= \tilde{C}'[\tilde{\Sigma}_{\epsilon\epsilon}^{-1} - \tilde{\Sigma}_{\epsilon\epsilon}^{-1}A(A'\tilde{\Sigma}_{\epsilon\epsilon}^{-1}A)^{-1}A'\tilde{\Sigma}_{\epsilon\epsilon}^{-1}]\tilde{C} \\ &\quad + (\tilde{C}-AB)'\tilde{\Sigma}_{\epsilon\epsilon}^{-1}A(A'\tilde{\Sigma}_{\epsilon\epsilon}^{-1}A)^{-1}A'\tilde{\Sigma}_{\epsilon\epsilon}^{-1}(\tilde{C}-AB) \\ &= (\tilde{C} - A\hat{B})'\tilde{\Sigma}_{\epsilon\epsilon}^{-1}(\tilde{C} - A\hat{B}) \\ &\quad + (\hat{B} - B)'(A'\tilde{\Sigma}_{\epsilon\epsilon}^{-1}A)(\hat{B} - B) \\ &\equiv \Psi + (\hat{B} - B)'(A'\tilde{\Sigma}_{\epsilon\epsilon}^{-1}A)(\hat{B} - B)\end{aligned}$$

where $\hat{B} = (A'\tilde{\Sigma}_{\epsilon\epsilon}^{-1}A)^{-1}A'\tilde{\Sigma}_{\epsilon\epsilon}^{-1}\tilde{C}$, and $\Psi = (\tilde{C} - A\hat{B})'\tilde{\Sigma}_{\epsilon\epsilon}^{-1}(\tilde{C} - A\hat{B})$. Thus we find that

$$\begin{aligned}|(\boldsymbol{Y}-AB\boldsymbol{X})(\boldsymbol{Y}-AB\boldsymbol{X})'| &= |S|\,|I_n + \hat{\Sigma}_{xx}\Psi|\,|I_n + (I_n + \hat{\Sigma}_{xx}\Psi)^{-1}\hat{\Sigma}_{xx} \\ &\qquad \times (\hat{B}-B)'(A'\tilde{\Sigma}_{\epsilon\epsilon}^{-1}A)(\hat{B}-B)| \\ &= |S|\,|I_n + \hat{\Sigma}_{xx}\Psi|\,|I_n + (\hat{\Sigma}_{xx}^{-1} + \Psi)^{-1} \\ &\qquad \times (\hat{B}-B)'(A'\tilde{\Sigma}_{\epsilon\epsilon}^{-1}A)(\hat{B}-B)|.\end{aligned}$$

Let $\delta_i^2 \geq 0$, $i = 1, \ldots, n$, be the eigenvalues of

$$(\hat{\Sigma}_{xx}^{-1} + \Psi)^{-1}(\hat{B} - B)'(A'\tilde{\Sigma}_{\epsilon\epsilon}^{-1}A)(\hat{B} - B),$$

so

$$|(\boldsymbol{Y}-AB\boldsymbol{X})(\boldsymbol{Y}-AB\boldsymbol{X})'| = |S|\,|I_n + \hat{\Sigma}_{xx}\Psi| \prod_{i=1}^{n}(1+\delta_i^2) \geq |S|\,|I_n + \hat{\Sigma}_{xx}\Psi|$$

with equality (under variation of B) if and only if $\delta_1^2 = \cdots = \delta_n^2 = 0$, that is, if and only if $A(\hat{B} - B) = 0$. Since A is a full-rank matrix, this implies that the value of B given by $\hat{B}$ above yields the minimum, so $\hat{B} = (A'\tilde{\Sigma}_{\epsilon\epsilon}^{-1}A)^{-1}A'\tilde{\Sigma}_{\epsilon\epsilon}^{-1}\tilde{C} \equiv (A'S^{-1}A)^{-1}A'S^{-1}\tilde{C}$ is the ML estimator of B.

We briefly indicate a few simplifications that occur in the estimation for the special case mentioned in Section 6.1 where the across-individual design matrix $\boldsymbol{X}$ is a matrix of indicator variables corresponding to n different treatment groups, that is, $\boldsymbol{X} = \text{diag}(\mathbf{1}'_{T_1}, \mathbf{1}'_{T_2}, \ldots, \mathbf{1}'_{T_n})$. Then we have $\boldsymbol{X}\boldsymbol{X}' = \text{diag}(T_1, \ldots, T_n)$ and

$$\tilde{C} = \boldsymbol{Y}\boldsymbol{X}'(\boldsymbol{X}\boldsymbol{X}')^{-1} = [\bar{Y}_1, \ldots, \bar{Y}_n] \equiv \bar{Y},$$

where $\bar{Y}_i = \sum_{j=1}^{T_i} Y_{ij}/T_i$ is the sample mean vector for the ith group, $i = 1, \ldots, n$. Hence, the ML estimator of B is $\hat{B} = (A'\tilde{\Sigma}_{\epsilon\epsilon}^{-1}A)^{-1}A'\tilde{\Sigma}_{\epsilon\epsilon}^{-1}\bar{Y}$, where $\tilde{\Sigma}_{\epsilon\epsilon} = (1/T)\,S$. In addition,

$$S = \sum_{i=1}^{n}\sum_{j=1}^{T_i}(Y_{ij} - \bar{Y}_i)(Y_{ij} - \bar{Y}_i)'$$

is the usual "within-group" error sum of squares matrix, and from (6.3) the ML estimate of the error covariance matrix $\Sigma_{\epsilon\epsilon}$ is $\hat{\Sigma}_{\epsilon\epsilon} = (1/T)\,S + (1/T)\sum_{i=1}^{n} T_i(\bar{Y}_i - A\hat{\beta}_i)(\bar{Y}_i - A\hat{\beta}_i)'$, where $\hat{\beta}_i = (A'\tilde{\Sigma}_{\epsilon\epsilon}^{-1}A)^{-1}A'\tilde{\Sigma}_{\epsilon\epsilon}^{-1}\bar{Y}_i$ is the ith column of $\hat{B}$.

In the original study of the growth curve model (6.1) by Potthoff and Roy (1964), the following reduction of the model was adopted for the analysis. For an arbitrary $m \times m$ positive-definite matrix Γ, the equation in (6.1) is multiplied by $(A'\Gamma A)^{-1}A'\Gamma$ to obtain

$$(A'\Gamma A)^{-1}A'\Gamma Y_k = BX_k + (A'\Gamma A)^{-1}A'\Gamma\epsilon_k.$$

This can be written as

$$Y_k^* = BX_k + \epsilon_k^*, \qquad k = 1, \ldots, T, \tag{6.4}$$

which is in the form of the classical multivariate regression model. Therefore, from the classical regression framework, the LS estimator of B in (6.4) is $\tilde{B} = \boldsymbol{Y}^*\boldsymbol{X}'(\boldsymbol{X}\boldsymbol{X}')^{-1} = (A'\Gamma A)^{-1}A'\Gamma\tilde{C}$ where $\boldsymbol{Y}^* = (A'\Gamma A)^{-1}A'\Gamma\boldsymbol{Y}$. The optimum choice for Γ is $\Gamma = \Sigma_{\epsilon\epsilon}^{-1}$ which will yield the best linear unbiased estimator of B in model (6.1). With $\Sigma_{\epsilon\epsilon}$ unknown, we see from the previous derivations that the ML estimator $\hat{B}$ in (6.2) has the same form as $\tilde{B}$ using $\Gamma = \tilde{\Sigma}_{\epsilon\epsilon}^{-1}$. Notice also that if the known matrix A were "normalized" so that the columns of $V = \Gamma^{1/2}A$ are orthonormal, that is, $A'\Gamma A = V'V = I_r$, then the above estimator $\tilde{B}$ takes the similar form as in (2.17) of Section 2.3 for the reduced-rank model, $\tilde{B} = V'\Gamma^{1/2}\tilde{C}$, but with the known matrix $V = \Gamma^{1/2}A$ in place of an estimated matrix $\hat{V}$. In addition, for both the reduced-rank and the growth curve model, we have found that the ML estimator is of the same form and is obtained with the choice $\Gamma = \tilde{\Sigma}_{\epsilon\epsilon}^{-1}$. Finally, we mention that if the error covariance matrix $\Sigma_{\epsilon\epsilon}$ has a special "random effects" structure such as $\Sigma_{\epsilon\epsilon} = A\Sigma_\beta A' + \sigma^2 I_m$, as discussed in Section 6.1, then it follows (e.g., see Reinsel, 1985) that the ML estimator of B under this model reduces to the LS estimator $\hat{B} = (A'A)^{-1}A'\boldsymbol{Y}\boldsymbol{X}'(\boldsymbol{X}\boldsymbol{X}')^{-1}$.

To further interpret the ML estimator of B and examine its statistical properties, we can also consider a transformation of the model (6.1) to a conditional model. Let $A_1 = A(A'A)^{-1}$ and A_2 be $m \times r$ and $m \times (m-r)$ matrices, respectively, of full ranks such that $A_1'A = I_r$ and $A_2'A = 0$, which also imply that $A_2'A_1 = 0$. The choice of A_2 as the column basis of $I_m - A(A'A)^{-1}A'$ will satisfy the above conditions. Then let $Y_{1k} = A_1'Y_k$ and $Y_{2k} = A_2'Y_k$, so that the original model (6.1) can be transformed to

$$\begin{bmatrix} Y_{1k} \\ Y_{2k} \end{bmatrix} = \begin{bmatrix} BX_k \\ 0 \end{bmatrix} + \begin{bmatrix} A_1'\epsilon_k \\ A_2'\epsilon_k \end{bmatrix} \tag{6.5}$$

The form (6.5) indicates that the information on B is contained within the relationship between the variables Y_{1k} and X_k and the transformed responses Y_{2k}, which are referred to as the set of covariates, have mean values which do not involve B but Y_{2k} may be correlated with Y_{1k}. The conditional model for Y_{1k}, given Y_{2k}, which follows from (6.5) can be written as

$$Y_{1k} = BX_k + \Theta Y_{2k} + \epsilon_{1k}^* \tag{6.6}$$

where $\Theta = \text{Cov}(Y_{1k}, Y_{2k})\{\text{Cov}(Y_{2k})\}^{-1} = (A_1'\Sigma_{\epsilon\epsilon}A_2)(A_2'\Sigma_{\epsilon\epsilon}A_2)^{-1}$ and the r-dimensional error term ϵ_{1k}^* has zero mean vector and covariance matrix $\Sigma_1^* = (A_1'\Sigma_{\epsilon\epsilon}A_1) - (A_1'\Sigma_{\epsilon\epsilon}A_2)(A_2'\Sigma_{\epsilon\epsilon}A_2)^{-1}(A_2'\Sigma_{\epsilon\epsilon}A_1) \equiv (A'\Sigma_{\epsilon\epsilon}^{-1}A)^{-1}$. The last equality follows from a matrix result presented in Lemma 2b of Rao (1967). The (conditional) model (6.6) is again in the form of the classical regression set up with two sets of predictors. From the discussion in Sections 1.3 and 3.1, it follows that the LS and ML estimator of B in (6.6) is $\hat{B} = \hat{\Sigma}_{y_1x.y_2}\hat{\Sigma}_{xx.y_2}^{-1}$, where $\hat{\Sigma}_{y_1x.y_2}$ and $\hat{\Sigma}_{xx.y_2}$ are sample partial covariance matrices adjusted for Y_{2k}. It will be established below that this estimator can be expressed equivalently in the form in (6.2), namely $\hat{B} = (A'\tilde{\Sigma}_{\epsilon\epsilon}^{-1}A)^{-1}A'\tilde{\Sigma}_{\epsilon\epsilon}^{-1}\tilde{C}$ with $\tilde{C} = \hat{\Sigma}_{yx}\hat{\Sigma}_{xx}^{-1}$. Because the marginal distribution of Y_{2k} does not involve B, it follows that the ML estimator of B from the conditional model (6.6) is also the ML estimator for the unconditional model (6.1), as has already been established.

To verify the equivalence in the two forms of the estimator $\hat{B}$, we first note that the LS estimator of Θ in (6.6) is obtained as

$$\begin{aligned} \hat{\Theta} = \hat{\Sigma}_{y_1 y_2 . x} \hat{\Sigma}_{y_2 y_2 . x}^{-1} &= (A_1' \hat{\Sigma}_{yy.x} A_2)(A_2' \hat{\Sigma}_{yy.x} A_2)^{-1} \\ &\equiv (A_1' \tilde{\Sigma}_{\epsilon\epsilon} A_2)(A_2' \tilde{\Sigma}_{\epsilon\epsilon} A_2)^{-1}, \end{aligned} \tag{6.7}$$

where $\hat{\Sigma}_{yy.x} = \hat{\Sigma}_{yy} - \hat{\Sigma}_{yx} \hat{\Sigma}_{xx}^{-1} \hat{\Sigma}_{xy} \equiv \tilde{\Sigma}_{\epsilon\epsilon}$. It then follows, from the normal equations for LS estimation, that the LS estimator of B in (6.6), $\hat{B} = \hat{\Sigma}_{y_1 x . y_2} \hat{\Sigma}_{xx.y_2}^{-1}$, can be equivalently expressed as

$$\begin{aligned} \hat{B} &= (\hat{\Sigma}_{y_1 x} - \hat{\Theta} \hat{\Sigma}_{y_2 x}) \hat{\Sigma}_{xx}^{-1} = [A_1' \hat{\Sigma}_{yx} - (A_1' \tilde{\Sigma}_{\epsilon\epsilon} A_2)(A_2' \tilde{\Sigma}_{\epsilon\epsilon} A_2)^{-1} A_2' \hat{\Sigma}_{yx}] \hat{\Sigma}_{xx}^{-1} \\ &= [A_1' \tilde{\Sigma}_{\epsilon\epsilon} - (A_1' \tilde{\Sigma}_{\epsilon\epsilon} A_2)(A_2' \tilde{\Sigma}_{\epsilon\epsilon} A_2)^{-1} A_2' \tilde{\Sigma}_{\epsilon\epsilon}] \tilde{\Sigma}_{\epsilon\epsilon}^{-1} \hat{\Sigma}_{yx} \hat{\Sigma}_{xx}^{-1} \\ &= (A_1' A_1)(A_1' \tilde{\Sigma}_{\epsilon\epsilon}^{-1} A_1)^{-1} A_1' \tilde{\Sigma}_{\epsilon\epsilon}^{-1} \hat{\Sigma}_{yx} \hat{\Sigma}_{xx}^{-1} \equiv (A' \tilde{\Sigma}_{\epsilon\epsilon}^{-1} A)^{-1} A' \tilde{\Sigma}_{\epsilon\epsilon}^{-1} \tilde{C} \end{aligned} \tag{6.8}$$

where $\tilde{C} = \hat{\Sigma}_{yx} \hat{\Sigma}_{xx}^{-1}$, and the result in Lemma 2b of Rao (1967) has again been used. Thus the LS estimator of B from the conditional model approach of (6.6) is the ML estimator of B in model (6.1) and can be expressed in the equivalent form as in (6.2).

We note from (6.8) that the ML estimator of B in model (6.1) can be expressed as

$$\begin{aligned} \hat{B} &= A_1' \hat{\Sigma}_{yx} \hat{\Sigma}_{xx}^{-1} - \hat{\Theta} A_2' \hat{\Sigma}_{yx} \hat{\Sigma}_{xx}^{-1} \\ &= (A'A)^{-1} A' \boldsymbol{Y} \boldsymbol{X}' (\boldsymbol{X}\boldsymbol{X}')^{-1} - \hat{\Theta} A_2' \boldsymbol{Y} \boldsymbol{X}' (\boldsymbol{X}\boldsymbol{X}')^{-1}, \end{aligned} \tag{6.9}$$

with $\hat{\Theta} = (A_1' \tilde{\Sigma}_{\epsilon\epsilon} A_2)(A_2' \tilde{\Sigma}_{\epsilon\epsilon} A_2)^{-1}$. In this form, $\hat{B}$ is interpretable as a *covariance adjusted* estimator of the simple LS estimator for model (6.1), $\tilde{B} = (A'A)^{-1} A' \boldsymbol{Y} \boldsymbol{X}' (\boldsymbol{X}\boldsymbol{X}')^{-1}$, where the adjustment involves use of the (mean-zero) covariates $\boldsymbol{Y}_2 = A_2' \boldsymbol{Y}$ through their (estimated) regression relationship with $\boldsymbol{Y}_1 = A_1' \boldsymbol{Y} = (A'A)^{-1} A' \boldsymbol{Y}$. The interpretation and development of properties of the estimator $\hat{B}$ through the covariance adjustment point of view is examined in more detail by Rao (1967) and Grizzle and Allen (1969). This includes the notion of constructing estimators of B through covariance adjustment as in (6.9), but using only a subset of the covariables in $\boldsymbol{Y}_2 = A_2' \boldsymbol{Y}$, whose proper choice would depend on knowledge of special structure for the covariance matrix $\Sigma_{\epsilon\epsilon}$. Specifically, covariables having zero (or very small) covariance with the response variables $\boldsymbol{Y}_1 = A_1' \boldsymbol{Y}$ and with the remaining covariables in $\boldsymbol{Y}_2 = A_2' \boldsymbol{Y}$ would be omitted in (6.9), the idea being that the use of "irrelevant" covariables in (6.9) would result in an increase in variability of the estimator $\hat{B}$ and hence such covariables should be eliminated in the estimation of B.

Notice that an alternate way to view the growth curve model (6.1) is as having the form of the standard multivariate regression model $Y_k =$

$CX_k + \epsilon_k$, with the regression coefficient matrix $C \equiv AB$ constrained or specified to satisfy the linear restriction $A_2'C = 0$ (since $A_2'A = 0$). From the result in (1.18) of Section 1.3 (with $F_1 \equiv A_2'$ and $G_2 \equiv I_n$), it follows that the constrained ML estimator of C under $A_2'C = 0$ can be written as

$$\hat{C} = \tilde{C} - SA_2(A_2'SA_2)^{-1}A_2'\tilde{C},$$

where $\tilde{C} = \boldsymbol{Y}\boldsymbol{X}'(\boldsymbol{X}\boldsymbol{X}')^{-1}$ is the usual LS estimator of C. Hence, since in the growth curve model C is represented in the form $C = AB$ with A known, this implies that the ML estimator of B can be obtained from $\hat{C}$ ($\equiv A\hat{B}$) as

$$\hat{B} = (A'A)^{-1}A'\hat{C} = (A'A)^{-1}A'\tilde{C} - (A'A)^{-1}A'SA_2(A_2'SA_2)^{-1}A_2'\tilde{C},$$

which is identical to the expression previously obtained in (6.9), recalling that $A_1' = (A'A)^{-1}A'$.

Under normality of the ϵ_k, the ML estimator $\hat{B}$ given by (6.2) or (6.8) can be seen to be unbiased, i.e. $E(\hat{B}) = B$, using the property that $\tilde{C}$ and $\tilde{\Sigma}_{\epsilon\epsilon}$ are independently distributed and $E(\tilde{C}) = C \equiv AB$. The independence property can also be used to help establish that the covariance matrix of $\hat{B}$ is given by

$$\text{Cov}[\text{vec}(\hat{B})] = \frac{T-n-1}{T-n-1-(m-r)}(\boldsymbol{X}\boldsymbol{X}')^{-1} \otimes (A'\Sigma_{\epsilon\epsilon}^{-1}A)^{-1},$$

and an unbiased estimator of $(A'\Sigma_{\epsilon\epsilon}^{-1}A)^{-1}$ is $\frac{T}{T-n-(m-r)}(A'\tilde{\Sigma}_{\epsilon\epsilon}^{-1}A)^{-1}$. [For example, see Rao (1967) and Grizzle and Allen (1969) for details.] Although the first two moments of $\hat{B}$ are known, the complete distribution is not available in simple form. Gleser and Olkin (1966, 1970) and Kabe (1975) obtained expressions for the probability density of the ML estimator $\hat{B}$.

6.3 Likelihood Ratio Testing of Linear Hypotheses in Growth Curve Model

We now present and derive the LR test statistic for $H_0\colon F_1BG_2 = 0$, where F_1 and G_2 are known full-rank matrices of dimensions $r_1 \times r$ and $n \times n_2$, respectively, with $r_1 \leq r$ and $n_2 \leq n$. We also want to give the associated asymptotic distribution theory for the LR statistic. These results can be obtained from testing results for the classical multivariate linear regression model, as given in Section 1.3, applied to the conditional model (6.6). The hypothesis $H_0 : F_1BG_2 = 0$ for model (6.1) can be expressed in terms of the conditional model (6.6) as

$$H_0\colon F_1\,[B,\Theta]\begin{bmatrix} G_2 \\ 0 \end{bmatrix} \equiv F_1DG_2^* = 0$$

where $D = [B, \Theta]$ and $G_2^* = [G_2', 0']'$. Note that the conditional model (6.6) can be written as $Y_{1k} = [B, \Theta]X_k^* + \epsilon_{1k}^* = DX_k^* + \epsilon_{1k}^*$, where $X_k^* = [X_k', Y_{2k}']'$, which is in the framework of the classical multivariate regression model (1.1). So $H_0 : F_1 D G_2^* = 0$ can be tested using the results given in Section 1.3.

The unconstrained ML estimator of D is the LS estimator

$$\tilde{D} = \boldsymbol{Y}_1 \boldsymbol{X}^{*\prime}(\boldsymbol{X}^* \boldsymbol{X}^{*\prime})^{-1},$$

where $\boldsymbol{X}^{*\prime} = [\boldsymbol{X}', \boldsymbol{Y}_2']$, which yields the estimators $\hat{B}$ and $\hat{\Theta}$ discussed in Section 6.2. That is, $\tilde{D} = [\hat{B}, \hat{\Theta}]$ where $\hat{B}$ and $\hat{\Theta}$ are the (LS) estimators given in (6.7) and (6.8) of Section 6.2. The constrained ML estimator of D, and hence of B, under $H_0 : F_1 B G_2 \equiv F_1 D G_2^* = 0$ can be obtained from the expression in (1.18) of Section 1.3 with the appropriate changes in notation. The error sum of squares matrix from (unconstrained) LS estimation of the conditional model is $S_1^* = \boldsymbol{Y}_1[I - \boldsymbol{X}^{*\prime}(\boldsymbol{X}^* \boldsymbol{X}^{*\prime})^{-1}\boldsymbol{X}^*]\boldsymbol{Y}_1'$ where $\boldsymbol{Y}_1 = A_1'\boldsymbol{Y}$. However, notice that

$$\begin{aligned}
\boldsymbol{Y}_1[I - \boldsymbol{X}^{*\prime}(\boldsymbol{X}^* \boldsymbol{X}^{*\prime})^{-1}\boldsymbol{X}^*] &= \boldsymbol{Y}_1 - \tilde{D}\boldsymbol{X}^* = A_1'\boldsymbol{Y} - \hat{B}\boldsymbol{X} - \hat{\Theta}\boldsymbol{Y}_2 \\
&= [A_1' - (A_1'SA_2)(A_2'SA_2)^{-1}A_2']\boldsymbol{Y} - \hat{B}\boldsymbol{X} \\
&= (A'S^{-1}A)^{-1}A'S^{-1}\boldsymbol{Y} \\
&\qquad - (A'S^{-1}A)^{-1}A'S^{-1}\boldsymbol{Y}\boldsymbol{X}'(\boldsymbol{X}\boldsymbol{X}')^{-1}\boldsymbol{X} \\
&= (A'S^{-1}A)^{-1}A'S^{-1}\boldsymbol{Y}[I - \boldsymbol{X}'(\boldsymbol{X}\boldsymbol{X}')^{-1}\boldsymbol{X}],
\end{aligned}$$

again using the result from Lemma 2b of Rao (1967) in the same way as in (6.8). Hence the error sum of squares matrix can be expressed as

$$\begin{aligned}
S_1^* &= \boldsymbol{Y}_1[I - \boldsymbol{X}^{*\prime}(\boldsymbol{X}^* \boldsymbol{X}^{*\prime})^{-1}\boldsymbol{X}^*]\boldsymbol{Y}_1' \\
&= (A'S^{-1}A)^{-1}A'S^{-1}\boldsymbol{Y}[I - \boldsymbol{X}'(\boldsymbol{X}\boldsymbol{X}')^{-1}\boldsymbol{X}]\boldsymbol{Y}'A_1 \\
&= (A'S^{-1}A)^{-1}A'S^{-1}\,SA_1 = (A'S^{-1}A)^{-1}, \qquad (6.10)
\end{aligned}$$

recalling that $A'A_1 = I_r$. The hypothesis sum of squares matrix associated with $H_0 : F_1 B G_2 \equiv F_1 D G_2^* = 0$ is, from expression (1.19) of Section 1.3, given by

$$\begin{aligned}
H &= S_1^* F_1'(F_1 S_1^* F_1')^{-1}(F_1 \tilde{D} G_2^*)[G_2^{*\prime}(\boldsymbol{X}^* \boldsymbol{X}^{*\prime})^{-1}G_2^*]^{-1} \\
&\qquad \times (F_1 \tilde{D} G_2^*)'(F_1 S_1^* F_1')^{-1} F_1 S_1^* \\
&= S_1^* F_1'(F_1 S_1^* F_1')^{-1}(F_1 \hat{B} G_2)[G_2' \Omega G_2]^{-1} \\
&\qquad \times (F_1 \hat{B} G_2)'(F_1 S_1^* F_1')^{-1} F_1 S_1^* \qquad (6.11)
\end{aligned}$$

where $\Omega = [I_n, 0](\boldsymbol{X}^* \boldsymbol{X}^{*\prime})^{-1}[I_n, 0]'$ is the leading $n \times n$ submatrix of $(\boldsymbol{X}^* \boldsymbol{X}^{*\prime})^{-1}$, from which it is known that

$$\Omega = \{\boldsymbol{X}\boldsymbol{X}' - \boldsymbol{X}\boldsymbol{Y}_2'(\boldsymbol{Y}_2\boldsymbol{Y}_2')^{-1}\boldsymbol{Y}_2\boldsymbol{X}'\}^{-1} \equiv (1/T)\,\hat{\Sigma}_{xx.y_2}^{-1}.$$

In addition, from a standard result on the inverse of a partitioned matrix [e.g., see Rao (1973, p. 33)], with $E = \boldsymbol{Y}_2\boldsymbol{Y}_2' - \boldsymbol{Y}_2\boldsymbol{X}'(\boldsymbol{X}\boldsymbol{X}')^{-1}\boldsymbol{X}\boldsymbol{Y}_2' \equiv A_2'SA_2$, we also find that

$$\begin{aligned}
\Omega &= (\boldsymbol{X}\boldsymbol{X}')^{-1} + (\boldsymbol{X}\boldsymbol{X}')^{-1}\boldsymbol{X}\boldsymbol{Y}_2'E^{-1}\boldsymbol{Y}_2\boldsymbol{X}'(\boldsymbol{X}\boldsymbol{X}')^{-1} \\
&= (\boldsymbol{X}\boldsymbol{X}')^{-1} + (\boldsymbol{X}\boldsymbol{X}')^{-1}\boldsymbol{X}\boldsymbol{Y}'A_2(A_2'SA_2)^{-1}A_2'\boldsymbol{Y}\boldsymbol{X}'(\boldsymbol{X}\boldsymbol{X}')^{-1} \\
&= (\boldsymbol{X}\boldsymbol{X}')^{-1} + (\boldsymbol{X}\boldsymbol{X}')^{-1}\boldsymbol{X}\boldsymbol{Y}' \\
&\qquad \times [S^{-1} - S^{-1}A(A'S^{-1}A)^{-1}A'S^{-1}]\boldsymbol{Y}\boldsymbol{X}'(\boldsymbol{X}\boldsymbol{X}')^{-1} \\
&= (\boldsymbol{X}\boldsymbol{X}')^{-1} + (\boldsymbol{X}\boldsymbol{X}')^{-1}\boldsymbol{X}\boldsymbol{Y}'S^{-1}\boldsymbol{Y}\boldsymbol{X}'(\boldsymbol{X}\boldsymbol{X}')^{-1} \\
&\qquad - \hat{B}'(A'S^{-1}A)\hat{B}, \qquad (6.12)
\end{aligned}$$

using the matrix result from Lemma 1 of Khatri (1966) that

$$A_2(A_2'SA_2)^{-1}A_2' = S^{-1} - S^{-1}A(A'S^{-1}A)^{-1}A'S^{-1}$$

since $A_2'A = 0$.

As in Section 1.3, the LR test of H_0 is based on the ordered eigenvalues $\hat{\lambda}_1^2 > \cdots > \hat{\lambda}_{r_1}^2$ of the $r_1 \times r_1$ matrix $S_*^{-1}H_*$ where

$$H_* = F_1HF_1' = (F_1\hat{B}G_2)[G_2'\Omega G_2]^{-1}(F_1\hat{B}G_2)' \qquad (6.13)$$

and $S_* = F_1S_1^*F_1' = F_1(A'S^{-1}A)^{-1}F_1'$. The LR test statistic is

$$\begin{aligned}
\mathcal{M} &= -[T-n-(m-r) + \frac{1}{2}(n_2-r_1-1)] \log(U) \\
&\equiv [T-n-(m-r) + \frac{1}{2}(n_2-r_1-1)] \sum_{i=1}^{r_1} \log(1+\hat{\lambda}_i^2),
\end{aligned}$$

which is asymptotically distributed as $\chi^2_{r_1n_2}$, where $U^{-1} = |I + S_*^{-1}H_*| = \prod_{i=1}^{r_1}(1+\hat{\lambda}_i^2)$. The distribution of S_* is Wishart, $W(F_1\Sigma_1^*F_1', T-n-(m-r))$, and under H_0 the conditional distribution of H_* is also Wishart with matrix $F_1\Sigma_1^*F_1'$ and n_2 degrees of freedom, where $\Sigma_1^* = (A'\Sigma_{\epsilon\epsilon}^{-1}A)^{-1}$. Since this conditional null distribution for H_* does not depend on $\boldsymbol{X}^*$ (i.e., on $\boldsymbol{Y}_2$) it is also the unconditional null distribution of H_*. The above form of the LR test of $H_0: F_1BG_2 = 0$ was first derived by Khatri (1966). Some additional aspects of the testing procedure were further examined by Grizzle and Allen (1969).

Recall that in the growth curve model (6.1), A represents the design matrix of the within-individual model and $\boldsymbol{X}$ is the usual across-individual design matrix. Various choices of the matrices F_1 and G_2 in $H_0: F_1BG_2 = 0$ lead to special cases of tests of hypothesis of particular interest. These include testing for the order of polynomial in the within-individual model, and testing for homogeneity or similarity in regression coefficients across different treatment groups. For instance, the within-individual model is often represented in terms of a polynomial regression over time points, of degree $r-1$, and suppose we wish to test for lower degree $r-2$ versus $r-1$. That is, we test for the significance of the highest order coefficient of the within-individual model, over all individuals. This can be represented as $H_0 : \beta'_{(r)} = 0$, where $\beta'_{(r)}$ denotes the rth row of B, and this is equivalent to $H_0 : F_1 B G_2 = 0$ with $F_1 = (0, \ldots, 0, 1)$ and $G_2 = I_n$. For this we see from (6.13) that $H_* = (F_1\hat{B})\Omega^{-1}(F_1\hat{B})' = \hat{\beta}'_{(r)}\Omega^{-1}\hat{\beta}_{(r)}$, where $\hat{\beta}'_{(r)}$ is the rth row of $\hat{B}$, and $S_* = F_1(A'S^{-1}A)^{-1}F_1'$ is the rth diagonal element of $(A'S^{-1}A)^{-1}$. Hence, since $r_1 = 1$, the LR test is based on the scalar $H_*/S_* = \hat{\beta}'_{(r)}\Omega^{-1}\hat{\beta}_{(r)}/S_*$ which is distributed as $n/(T-n-(m-r))$ times the F-distribution with n and $T-n-(m-r)$ degrees of freedom under H_0.

For another example, suppose that the across-individual design matrix $\boldsymbol{X}$ corresponds to n distinct treatment groups as in a one-way multivariate ANOVA model, and we wish to test the null hypothesis of homogeneity of the growth curve parameters (or at least of a subset of these parameters) over the n groups. This is equivalent to the condition that the n columns of B be equal (or at least the lower $r_1 \times n$ submatrix of B have equal columns). This hypothesis can be expressed in the form $F_1 B G_2 = 0$ with $F_1 = [0, I_{r_1}]$ and

$$G_2 = \begin{bmatrix} 1 & 0 & . & . & 0 & 0 \\ 0 & 1 & . & . & 0 & 0 \\ . & . & . & . & . & . \\ . & . & . & . & . & . \\ 0 & 0 & . & . & 0 & 1 \\ -1 & -1 & . & . & -1 & -1 \end{bmatrix}$$

of dimension $n \times (n-1)$. For example, in the case with only $n = 2$ groups, we have $G_2 = (1, -1)'$ and $G_2'\Omega G_2 = (1/T_1) + (1/T_2) + (1/T)\{\bar{Y}_1 - \bar{Y}_2 - A(\hat{\beta}_1 - \hat{\beta}_2)\}'\tilde{\Sigma}_{\epsilon\epsilon}^{-1}\{\bar{Y}_1 - \bar{Y}_2 - A(\hat{\beta}_1 - \hat{\beta}_2)\}$, and with $r_1 = r$ the test statistic for $H_0 : BG_2 = 0$ (i.e., $\beta_1 - \beta_2 = 0$ where $B = [\beta_1, \beta_2]$) is based on the single nonzero eigenvalue of $S_*^{-1}H_*$, that is,

$$(\hat{B}G_2)'(A'S^{-1}A)(\hat{B}G_2)[G_2'\Omega G_2]^{-1} = \frac{1}{T}(\hat{\beta}_1 - \hat{\beta}_2)'A'\tilde{\Sigma}_{\epsilon\epsilon}^{-1}A(\hat{\beta}_1 - \hat{\beta}_2)/[G_2'\Omega G_2].$$

Also notice that a test of goodness-of-fit or adequacy of specification of the within-individual model can be tested in the framework of the general multivariate linear model $Y_k = CX_k + \epsilon_k$, with C specified to be of the form $C = AB$ under the growth curve model (6.1). The LR test can be carried out using the standard methods discussed in Section 1.3 as a test of $H_0 : A_2'C = 0$ in the model $Y_k = CX_k + \epsilon_k$, since this hypothesis is equivalent to the hypothesis that C has the form AB with A specified (and A_2 is the $m \times (m-r)$ matrix such that $A_2'A = 0$). Thus, from Section 1.3

we find that the hypothesis sum of squares matrix for H_0 is

$$H_* = A_2' \tilde{C}(\boldsymbol{X}\boldsymbol{X}')\tilde{C}' A_2 = A_2' \boldsymbol{Y}\boldsymbol{X}'(\boldsymbol{X}\boldsymbol{X}')^{-1}\boldsymbol{X}\boldsymbol{Y}' A_2,$$

while the error sum of squares matrix is $S_* = A_2' S A_2$, with $S_* + H_* = A_2' \boldsymbol{Y}\boldsymbol{Y}' A_2$. The LR test procedure uses the statistic $\mathcal{M} = -[T-n+(n-(m-r)-1)/2]\log(U)$, where $U = |S_*|/|S_* + H_*| = 1/|I_{m-r} + S_*^{-1} H_*|$, with $\mathcal{M}$ having an approximate $\chi^2_{n(m-r)}$ distribution under H_0. The LR can also be expressed in a form that avoids the explicit use of the matrix A_2. Specifically, from (6.3) we see that the $m \times m$ error sum of squares matrix under H_0 is

$$\begin{aligned} S_1 &\equiv S + H = (\boldsymbol{Y} - A\hat{B}\boldsymbol{X})(\boldsymbol{Y} - A\hat{B}\boldsymbol{X})' \\ &= S + (I_m - A(A'S^{-1}A)^{-1}A'S^{-1})\boldsymbol{Y}\boldsymbol{X}'(\boldsymbol{X}\boldsymbol{X}')^{-1}\boldsymbol{X}\boldsymbol{Y}' \\ &\qquad \times (I_m - A(A'S^{-1}A)^{-1}A'S^{-1})'. \end{aligned}$$

Therefore it follows that

$$\begin{aligned} U &= |S|/|S_1| = 1/|I_m + S^{-1}H| \\ &= 1/|I_m + S^{-1}(I_m - A(A'S^{-1}A)^{-1}A')S^{-1}\, \boldsymbol{Y}\boldsymbol{X}'(\boldsymbol{X}\boldsymbol{X}')^{-1}\boldsymbol{X}\boldsymbol{Y}'|, \end{aligned}$$

so that U can be expressed in terms of the $(m-r)$ nonzero eigenvalues of the matrix $S^{-1}(I_m - A(A'S^{-1}A)^{-1}A')S^{-1}\boldsymbol{Y}\boldsymbol{X}'(\boldsymbol{X}\boldsymbol{X}')^{-1}\boldsymbol{X}\boldsymbol{Y}'$. It is also shown by Kshirsagar and Smith (1995, pp. 194–195) that U can be expressed equivalently as $U = \{\,|S|\,|A'S^{-1}A|\,\}/\{\,|\boldsymbol{Y}\boldsymbol{Y}'|\,|A'(\boldsymbol{Y}\boldsymbol{Y}')^{-1}A|\,\}$.

6.4 An Extended Model for Growth Curve Data

An extension of the basic growth curve (or generalized MANOVA) model (6.1) to allow a mixture of growth curve and MANOVA components has been considered by Chinchilli and Elswick (1985). In this extended model the mean of the response vector Y_k is represented by the sum of two components, a growth curve portion and a standard MANOVA portion. The model could be viewed as an analysis of covariance model that adjusts the growth curve structure for the influence of additional covariates for the response vector. Specifically, the model considered is

$$Y_k = ABX_k + DZ_k + \epsilon_k, \qquad k = 1, \ldots, T, \tag{6.14}$$

where the $m \times 1$ response vector Y_k depends on the $n_1 \times 1$ vector X_k in the same way as in the basic growth curve model (6.1), and also on the $n_2 \times 1$ vector Z_k of additional covariates. However, the second set of covariates is not specified to enter the model with growth curve structure but simply with the standard multivariate linear model structure. This model may have appeal when it seems appropriate to express the effects of X_k in terms of growth curve structure over time, whereas it may not be desirable to express the effects of the additional covariates Z_k in this manner. Notice that the form of model (6.14) is similar to the reduced-rank model (3.3) of Section 3.1, but in the present growth curve model context the matrix A is known. As usual, the error terms ϵ_k are assumed to follow a multivariate normal distribution with zero mean vector and positive-definite covariance matrix $\Sigma_{\epsilon\epsilon}$.

The maximum likelihood estimators of the unknown parameters B, D, and $\Sigma_{\epsilon\epsilon}$ of model (6.14) are derived by Chinchilli and Elswick (1985). We can use some of the tools that have been developed in Sections 3.1 and 6.2 to readily present the results. Let $C = AB$ and denote the LS estimates of C and D in the model $Y_k = CX_k + DZ_k + \epsilon_k$ by $\tilde{C} = \hat{\Sigma}_{yx.z}\hat{\Sigma}_{xx.z}^{-1}$ and $\tilde{D} = \hat{\Sigma}_{yz.x}\hat{\Sigma}_{zz.x}^{-1}$, respectively. The ML estimates of model (6.14) are obtained by minimizing $|W|$ or, equivalently, $|\tilde{\Sigma}_{\epsilon\epsilon}^{-1}W|$ where

$$W = (1/T)(\boldsymbol{Y} - AB\boldsymbol{X} - D\boldsymbol{Z})(\boldsymbol{Y} - AB\boldsymbol{X} - D\boldsymbol{Z})'$$

and $\tilde{\Sigma}_{\epsilon\epsilon} = (1/T)(\boldsymbol{Y} - \tilde{C}\boldsymbol{X} - \tilde{D}\boldsymbol{Z})(\boldsymbol{Y} - \tilde{C}\boldsymbol{X} - \tilde{D}\boldsymbol{Z})' \equiv \hat{\Sigma}_{yy.(x,z)}$ as given in (3.10). As shown in Section 3.1, we have the decomposition

$$\begin{aligned} W = \tilde{\Sigma}_{\epsilon\epsilon} &+ (\tilde{C} - AB)\hat{\Sigma}_{xx.z}(\tilde{C} - AB)' \\ &+ (\tilde{D}^* - D - C\hat{\Sigma}_{xz}\hat{\Sigma}_{zz}^{-1})\hat{\Sigma}_{zz}(\tilde{D}^* - D - C\hat{\Sigma}_{xz}\hat{\Sigma}_{zz}^{-1})' \end{aligned}$$

where $\tilde{D}^* = \tilde{D} + \tilde{C}\hat{\Sigma}_{xz}\hat{\Sigma}_{zz}^{-1} \equiv \hat{\Sigma}_{yz}\hat{\Sigma}_{zz}^{-1}$. For any C, the last component above can be uniformly minimized, as the zero matrix, by the choice of $\hat{D} = \tilde{D}^* - C\hat{\Sigma}_{xz}\hat{\Sigma}_{zz}^{-1}$. Thus, the ML estimation of minimizing $|\tilde{\Sigma}_{\epsilon\epsilon}^{-1}\, W|$ is reduced to minimizing

$$|\, I_m + \tilde{\Sigma}_{\epsilon\epsilon}^{-1}(\tilde{C} - AB)\hat{\Sigma}_{xx.z}(\tilde{C} - AB)'\,|$$

with respect to B. This is the same form of minimization problem as considered at the beginning of Section 6.2 for the basic growth curve model. Hence, from the same arguments as given there, it follows that $\hat{B} = (A'\tilde{\Sigma}_{\epsilon\epsilon}^{-1}A)^{-1}A'\tilde{\Sigma}_{\epsilon\epsilon}^{-1}\tilde{C}$, with $\tilde{C} = \hat{\Sigma}_{yx.z}\hat{\Sigma}_{xx.z}^{-1}$, is the ML estimator of B in model (6.14). The corresponding ML estimator of D then is

$$\begin{aligned} \hat{D} = \tilde{D}^* - A\,\hat{B}\hat{\Sigma}_{xz}\hat{\Sigma}_{zz}^{-1} &= \hat{\Sigma}_{yz}\hat{\Sigma}_{zz}^{-1} - A\,\hat{B}\hat{\Sigma}_{xz}\hat{\Sigma}_{zz}^{-1} \\ &\equiv \tilde{D} + (\tilde{C} - A\hat{B})\hat{\Sigma}_{xz}\hat{\Sigma}_{zz}^{-1}. \qquad (6.15) \end{aligned}$$

It follows from the above decomposition that the ML estimate of the error covariance matrix is

$$\begin{aligned} \hat{\Sigma}_{\epsilon\epsilon} &= \tilde{\Sigma}_{\epsilon\epsilon} + (\tilde{C} - A\hat{B})\hat{\Sigma}_{xx.z}(\tilde{C} - A\hat{B})' \\ &= \tilde{\Sigma}_{\epsilon\epsilon} + (I_m - A(A'\tilde{\Sigma}_{\epsilon\epsilon}^{-1}A)^{-1}A'\tilde{\Sigma}_{\epsilon\epsilon}^{-1})\tilde{C}\hat{\Sigma}_{xx.z}\tilde{C}' \\ &\qquad \times\, (I_m - A(A'\tilde{\Sigma}_{\epsilon\epsilon}^{-1}A)^{-1}A'\tilde{\Sigma}_{\epsilon\epsilon}^{-1})'. \end{aligned}$$

As can be observed from the above results, the main differences between ML estimates of the parameters of the basic growth curve model (6.1) and ML estimates of the parameters of the extended model (6.14) are that the estimate $\tilde{\Sigma}_{\epsilon\epsilon}$ of the error covariance matrix in (6.14) is obtained from the LS regression of Y_k on X_k and Z_k and the estimate of B involves the LS regression coefficient matrix $\tilde{C} = \hat{\Sigma}_{yx.z}\hat{\Sigma}_{xx.z}^{-1}$ which is obtained after adjusting for the effect of Z_k on Y_k and X_k.

It is also possible to develop the conditional model approach for (6.14) in a similar way to model (6.6) for the basic growth curve model. Hypothesis testing under the extended model (6.14) can be developed along the same lines as in Section 6.3 for the model (6.1). In addition to the adjustments that were mentioned at the end of the last paragraph, the LR tests will involve an adjustment in the error degrees of freedom associated with the inclusion of the additional covariates Z_k in the model. We will not pursue the details for LR testing of linear hypotheses here. However, we do note that there are two separate tests concerning the issue of model adequacy that might be of interest for the extended model (6.14). The first type of test of hypothesis is similar to that mentioned at the end of Section 6.3, namely, a test of the adequacy of the growth curve feature (involving the term ABX_k) in the extended model (6.14) relative to a standard multivariate regression model $Y_k = CX_k + DZ_k + \epsilon_k$. The second type of test of interest would be the adequacy of a "combined" growth curve structure model of the form $Y_k = A(BX_k + B^* Z_k) + \epsilon_k \equiv A[B, B^*][X_k', Z_k']' + \epsilon_k$ relative to the extended model (6.14) which is a "mixture" of growth curve structure and standard multivariate regression.

6.5 Modification of Basic Growth Curve Model to Reduced-Rank Model

In order to relate the concepts in the growth curve model context to the reduced-rank modeling of Chapter 2, we need to recall the role of the matrix components A and B that appear in the basic growth curve model (6.1). As emphasized previously, the $m \times r$ matrix A is known or specified in the growth curve context and is taken to be the design matrix of the within-individual model. Because the elements of Y_k consist of repeated measurements of the same response variable over time, the columns of A are specified as known functions of time, typically polynomial terms. The $r \times n$ matrix B represents the regression coefficients associated with the known functions of time. The $n \times T$ matrix $\boldsymbol{X}$ is the usual across-individual design matrix. When the individuals are from n different groups and there are no additional explanatory variables, then $\boldsymbol{X}$ is simply a matrix of indicator variables as in the one-way ANOVA model.

To focus further on this specific situation, suppose there are n different experimental or treatment groups and T_i vector observations Y_{ij} are available for the ith group, for $j = 1, \ldots, T_i$. With observations in each group assumed to be measured at the same m time points, $t_1 < t_2 < \cdots < t_m$, the basic (polynomial) growth curve model can be written as $\boldsymbol{Y} = AB\boldsymbol{X} + \boldsymbol{\epsilon}$, where $\boldsymbol{Y} = [Y_{11}, \ldots, Y_{1T_1}, \ldots, Y_{n1}, \ldots, Y_{nT_n}]$,

$$A = \begin{bmatrix} 1 & t_1 & . & . & t_1^{r-1} \\ 1 & t_2 & . & . & t_2^{r-1} \\ . & . & . & . & . \\ . & . & . & . & . \\ 1 & t_m & . & . & t_m^{r-1} \end{bmatrix}, \qquad B = \begin{bmatrix} \beta_{11} & . & . & . & \beta_{1n} \\ \beta_{21} & . & . & . & \beta_{2n} \\ . & . & . & . & . \\ . & . & . & . & . \\ \beta_{r1} & . & . & . & \beta_{rn} \end{bmatrix},$$

$$\boldsymbol{X} = \begin{bmatrix} \mathbf{1}'_{T_1} & 0' & 0' & . & . & 0' \\ 0' & \mathbf{1}'_{T_2} & 0' & . & . & 0' \\ . & . & . & . & . & . \\ . & . & . & . & . & . \\ 0' & 0' & 0' & . & . & \mathbf{1}'_{T_n} \end{bmatrix},$$

and $\mathbf{1}_T$ denotes a $T \times 1$ vector of ones.

In Section 6.3, we indicated how various forms of hypotheses about the matrix B might be of interest for the analysis of the growth curve model. The degree $r-1$ of the polynomial and the number n of different treatment groups both play roles in determining the number of parameters $(= rn)$ to be estimated and compared in the growth curve analysis. If the growth curves could be adequately expressed by a lower-degree polynomial and if the number of distinct groupings could be reduced on the basis of similarity of growth curve features, a more parsimonious model would emerge. In practice, the choice of degree $r-1$ might be made initially on the basis of plots of mean growth curves of each of the n groups, and would subsequently be more formally chosen on the basis of goodness-of-fit LR tests as discussed in Section 6.3. However, in some situations a relatively low degree polynomial may not be adequate to represent the growth curve patterns over time. Equivalently, we may not want to specify the mathematical form of the growth curve function over time as a specific degree polynomial or any other specified function of time. For such situations, in this section we explore how the reduced-rank regression model and methods could be applied to build a parsimonious model.

In the reduced-rank regression model framework, $\boldsymbol{Y} = AB\boldsymbol{X} + \boldsymbol{\epsilon}$, of Chapter 2, both matrices A and B are unknown and the rank restrictions on $C = AB$ are meant to imply certain consequences. When considered for use in the growth curve data setting, the roles of the component matrices A and B can not be viewed in exactly the same way as the matrices A and B in the basic growth curve model (6.1). However, certain similarities and correspondences do exist. For the reduced-rank model, the $m \times r$ matrix A

is unknown but the r columns of A could be viewed as a (low-dimensional) set of basic curves over time (with general unspecified forms) such that each group's growth curve over time can be represented adequately by some linear combination of the basic set of curves. The coefficients in the ith column of the $r \times n$ matrix B then correspond to the coefficients of the linear combination of basic curves for the ith treatment group. The across-individual model could also be parameterized in terms of an overall mean growth curve and $n-1$ among-group contrasts, with a corresponding design matrix $\boldsymbol{X}$ whose first row has all elements equal to one. That is, for example, we can take

$$\boldsymbol{X} = \begin{bmatrix} \mathbf{1}'_{T_1} & \mathbf{1}'_{T_2} & \mathbf{1}'_{T_3} & . & . & \mathbf{1}'_{T_{n-1}} & \mathbf{1}'_{T_n} \\ \mathbf{1}'_{T_1} & 0' & 0' & . & . & 0' & -\mathbf{1}'_{T_n} \\ 0' & \mathbf{1}'_{T_2} & 0' & . & . & 0' & -\mathbf{1}'_{T_n} \\ . & . & . & . & . & . & . \\ . & . & . & . & . & . & . \\ 0' & 0' & 0' & . & . & \mathbf{1}'_{T_{n-1}} & -\mathbf{1}'_{T_n} \end{bmatrix}$$

with $B = [\,\bar{\boldsymbol{\beta}}, \boldsymbol{\beta}_1 - \bar{\boldsymbol{\beta}}, \ldots, \boldsymbol{\beta}_{n-1} - \bar{\boldsymbol{\beta}}]$, where $\boldsymbol{\beta}_i$ denotes the ith column of the original matrix B, that is, the $r \times 1$ vector of growth curve parameters for the ith group, and $\bar{\boldsymbol{\beta}} = n^{-1} \sum_{i=1}^{n} \boldsymbol{\beta}_i$. Then in this parameterization the first column of AB gives the general mean growth curve over all groups, and the remaining columns provide for various deviations or contrasts of the group means from the overall mean growth curve. Thus, the methodology of the reduced-rank regression modeling of Chapter 2 not only seeks for a basic set of curves over time, and the associated coefficients, it also searches for possible reductions in the number of basic curves required for adequate representation of all the groups growth curves. However, it does not seem that the unknown matrix A in the reduced-rank model should be viewed as a substitute for the known matrix A of polynomial terms in the basic growth curve model. Use of the reduced-rank modeling methodology might be considered when specification of the exact mathematical form of the growth curves, as polynomials of a certain degree, is not desirable or adequate, as mentioned.

We now recall the fundamental results from Chapter 2 concerning estimation of the reduced-rank regression model $Y_k = ABX_k + \epsilon_k$, $k = 1, \ldots, T$, for application to the growth curve data situation. The ML estimators of the component matrices A and B are as given in Section 2.3,

$$\hat{A} = \tilde{\Sigma}_{\epsilon\epsilon}^{1/2} \hat{V}, \qquad \hat{B} = \hat{V}' \tilde{\Sigma}_{\epsilon\epsilon}^{-1/2} \hat{\Sigma}_{yx} \hat{\Sigma}_{xx}^{-1}, \tag{6.16}$$

where $\hat{V} = [\hat{V}_1, \ldots, \hat{V}_r]$ and $\hat{V}_j$ is the normalized eigenvector corresponding to the jth largest eigenvalue $\hat{\lambda}_j^2$ of the matrix $\tilde{\Sigma}_{\epsilon\epsilon}^{-1/2} \hat{\Sigma}_{yx} \hat{\Sigma}_{xx}^{-1} \hat{\Sigma}_{xy} \tilde{\Sigma}_{\epsilon\epsilon}^{-1/2}$. The estimates satisfy the normalizations $\hat{A}' \tilde{\Sigma}_{\epsilon\epsilon}^{-1} \hat{A} = I_r$ and $\hat{B} \hat{\Sigma}_{xx} \hat{B}' = \hat{\Lambda}^2$. The ML estimate of the error covariance matrix $\Sigma_{\epsilon\epsilon}$ can be expressed as

$$\hat{\Sigma}_{\epsilon\epsilon} = \tilde{\Sigma}_{\epsilon\epsilon} + (\tilde{C} - \hat{C}) \hat{\Sigma}_{xx} (\tilde{C} - \hat{C})'$$

where $\hat{C} = \hat{A}\hat{B}$ and $\tilde{C} = \hat{\Sigma}_{yx}\hat{\Sigma}_{xx}^{-1}$. A "goodness-of-fit" test for the adequacy of the model with reduced rank r, that is, of H_0: rank$(C) = r$ against the general alternative of a full-rank regression matrix C is provided by the LR test statistic given in Section 2.6,

$$\mathcal{M} = -[T - n + (n - m - 1)/2] \sum_{j=r+1}^{m} \log(1 - \hat{\rho}_j^2),$$

where $\hat{\rho}_1^2 > \cdots \geq \hat{\rho}_m^2 \geq 0$ are the ordered eigenvalues of the matrix

$$\hat{\Sigma}_{yy}^{-1/2}\hat{\Sigma}_{yx}\hat{\Sigma}_{xx}^{-1}\hat{\Sigma}_{xy}\hat{\Sigma}_{yy}^{-1/2}.$$

The asymptotic distribution of the ML estimators $\hat{A}$ and $\hat{B}$ of the reduced-rank component matrices is given in Theorem 2.4 of Section 2.5. For the growth curve setting, one may also want to consider the extended version of the reduced-rank model as discussed in Section 3.1, $Y_k = ABX_k + DZ_k + \epsilon_k$, where Z_k represents a portion or subset of the total set of explanatory variables that are not postulated to enter the model with a growth curve or reduced-rank structure. As noted earlier, this model is in the same spirit as the extended growth curve model discussed in Section 6.4. For example, in the ANOVA situation with n treatment groups, it may be desired to separate the overall mean response curve from the growth curve structure and incorporate growth curve structure only for the remaining $n-1$ among-group contrasts. This corresponds to a case with $Z_k \equiv 1$ for all k, and X_k are $(n-1)$-dimensional vectors of group indicators.

6.6 Reduced-Rank Growth Curve Models

For longitudinal or "growth curve" data Y_k, as we have seen in earlier parts of this chapter it is often appropriate to specify a growth curve model structure,

$$Y_k = ABX_k + \epsilon_k, \qquad k = 1, \ldots, T,$$

where A is a known or specified $m \times r$ matrix and the $r \times n$ matrix B is unknown. The matrix B plays the role of summarizing the mean structure of the response variables over time (i.e., the growth curve structure), typically in the form of regression coefficients of polynomial functions over time. For example, BX_k can be viewed as the mean vector of the r-dimensional summary of the response vector, $A_1'Y_k = (A'A)^{-1}A'Y_k$, and the "complete" response mean vector can be expressed in terms of this through $E(Y_k) = A(BX_k)$. In this way, the mean vectors of the Y_k are specified to have a reduced-rank structure of a specified form. However, the $r \times n$ regression coefficient matrix B, which summarizes the mean vector structure, could itself also have reduced rank. Thus, in this section we consider the possibility of reduced-rank structure for B in the basic growth curve model, and consider the assumption that $\text{rank}(B) = r_1 < \min(r, n)$. As in previous chapters, this assumption of reduced rank yields the matrix decomposition as

$$B = B_1 B_2$$

where B_1 and B_2 are $r \times r_1$ and $r_1 \times n$ matrices, respectively, of unknown parameters. Therefore, the reduced-rank growth curve model can be written as

$$Y_k = A\, B_1 B_2\, X_k + \epsilon_k, \qquad k = 1, \ldots, T. \tag{6.17}$$

The reduced rank of B implies that the vectors of growth curve parameters exist in r_1-dimensional space. The r_1 columns of B_1 provide a basis for this space; the elements in the ith column of B_2 provide the coefficients in the linear combination of basis growth curve parameters for the ith treatment group, in the n-group ANOVA situation, for example. The model also retains the usual reduced-dimension interpretation that the $r_1 \times 1$ vector of linear combinations, $X_k^* = B_2 X_k$, is sufficient to represent the linear relationships between Y_k and X_k. The analysis of model (6.17), including estimation of the unknown parameters B_1, B_2, and $\Sigma_{\epsilon\epsilon}$, and extensions and applications of this model, will be the focus of this section.

To derive the ML estimates of B_1 and B_2, we again will utilize the estimation results of Sections 6.2 and 2.3. The ML estimates are obtained by minimizing $|(\boldsymbol{Y} - AB_1B_2\boldsymbol{X})(\boldsymbol{Y} - AB_1B_2\boldsymbol{X})'|$ with respect to B_1 and B_2, where A is a known matrix. Following the derivation given in Section 6.2 for the ML estimate of B in the model (6.1) where B is assumed to be of full rank, we obtain the relation

$$\begin{aligned}
&|(\boldsymbol{Y} - AB_1B_2\boldsymbol{X})(\boldsymbol{Y} - AB_1B_2\boldsymbol{X})'| \\
&\quad = |S|\; |I_n + \hat{\Sigma}_{xx}\Psi|\; |I_n + (\hat{\Sigma}_{xx}^{-1} + \Psi)^{-1}(\tilde{B} - B)'(A'\tilde{\Sigma}_{\epsilon\epsilon}^{-1}A)(\tilde{B} - B)| \\
&\quad = |S|\; |I_n + \hat{\Sigma}_{xx}\Psi| \\
&\qquad \times |I_r + (A'\tilde{\Sigma}_{\epsilon\epsilon}^{-1}A)(\tilde{B} - B_1B_2)(\hat{\Sigma}_{xx}^{-1} + \Psi)^{-1}(\tilde{B} - B_1B_2)'|,
\end{aligned} \tag{6.18}$$

where the "full-rank" ML estimate of B is now denoted as

$$\tilde{B} = (A'\tilde{\Sigma}_{\epsilon\epsilon}^{-1}A)^{-1}A'\tilde{\Sigma}_{\epsilon\epsilon}^{-1}\tilde{C}$$

with $\tilde{C} = \boldsymbol{Y}\boldsymbol{X}'(\boldsymbol{X}\boldsymbol{X}')^{-1}$, and $\Psi = (\tilde{C} - A\tilde{B})'\tilde{\Sigma}_{\epsilon\epsilon}^{-1}(\tilde{C} - A\tilde{B})$. Thus, ML estimation for B_1 and B_2 is equivalent to minimizing the last term on the right side of (6.18). Note that this term has the same form as the quantity involved in the expressions following (2.17) in Section 2.3, and that the minimization is achieved by simultaneous minimization, with respect to B_1 and B_2, of all the eigenvalues of the matrix

$$(A'\tilde{\Sigma}_{\epsilon\epsilon}^{-1}A)^{1/2}(\tilde{B} - B_1B_2)(\hat{\Sigma}_{xx}^{-1} + \Psi)^{-1}(\tilde{B} - B_1B_2)'(A'\tilde{\Sigma}_{\epsilon\epsilon}^{-1}A)^{1/2}, \tag{6.19}$$

where $\tilde{\Sigma}_{\epsilon\epsilon} = (1/T)S$. The matrix in (6.19) can be expressed as $(S_{(r)} - P_{(r)})(S_{(r)} - P_{(r)})'$, where

$$\begin{aligned} S_{(r)} &= (A'\tilde{\Sigma}_{\epsilon\epsilon}^{-1}A)^{1/2}\tilde{B}(\hat{\Sigma}_{xx}^{-1} + \Psi)^{-1/2} \\ &= (A'\tilde{\Sigma}_{\epsilon\epsilon}^{-1}A)^{-1/2}A'\tilde{\Sigma}_{\epsilon\epsilon}^{-1}\tilde{C}(\hat{\Sigma}_{xx}^{-1} + \Psi)^{-1/2} \end{aligned}$$

and $P_{(r)} = (A'\tilde{\Sigma}_{\epsilon\epsilon}^{-1}A)^{1/2}B_1B_2(\hat{\Sigma}_{xx}^{-1} + \Psi)^{-1/2}$. From the same arguments as in Section 2.3, using the result of Lemma 2.2 in particular, the minimizing matrix $P_{(r)}$ is chosen as the rank-r_1 approximation of $S_{(r)}$ obtained through the singular value decomposition of $S_{(r)}$ as $P_{(r)} = \hat{V}_{(r_1)}\hat{V}_{(r_1)}'S_{(r)}$, where $\hat{V}_{(r_1)} = [\hat{V}_1, \ldots, \hat{V}_{r_1}]$ and $\hat{V}_j$ is the normalized eigenvector of the matrix

$$S_{(r)}S_{(r)}' = (A'\tilde{\Sigma}_{\epsilon\epsilon}^{-1}A)^{-1/2}A'\tilde{\Sigma}_{\epsilon\epsilon}^{-1}\tilde{C}(\hat{\Sigma}_{xx}^{-1} + \Psi)^{-1}\tilde{C}'\tilde{\Sigma}_{\epsilon\epsilon}^{-1}A(A'\tilde{\Sigma}_{\epsilon\epsilon}^{-1}A)^{-1/2} \quad (6.20)$$

corresponding to the jth largest eigenvalue $\hat{\lambda}_j^2$. The ML estimates of B_1 and B_2 can then be written explicitly, in a similar form as (2.17) of Section 2.3, as

$$\hat{B}_1 = (A'\tilde{\Sigma}_{\epsilon\epsilon}^{-1}A)^{-1/2}\hat{V}_{(r_1)}, \quad (6.21a)$$

$$\hat{B}_2 = \hat{V}_{(r_1)}'(A'\tilde{\Sigma}_{\epsilon\epsilon}^{-1}A)^{1/2}\tilde{B} = \hat{V}_{(r_1)}'(A'\tilde{\Sigma}_{\epsilon\epsilon}^{-1}A)^{-1/2}A'\tilde{\Sigma}_{\epsilon\epsilon}^{-1}\tilde{C}. \quad (6.21b)$$

The ML estimate of $B = B_1B_2$ under the reduced-rank assumption is therefore given by

$$\begin{aligned} \hat{B} = \hat{B}_1\hat{B}_2 &= (A'\tilde{\Sigma}_{\epsilon\epsilon}^{-1}A)^{-1/2}\hat{V}_{(r_1)}\hat{V}_{(r_1)}'(A'\tilde{\Sigma}_{\epsilon\epsilon}^{-1}A)^{1/2}\tilde{B} \\ &= (A'\tilde{\Sigma}_{\epsilon\epsilon}^{-1}A)^{-1/2}\hat{V}_{(r_1)}\hat{V}_{(r_1)}'(A'\tilde{\Sigma}_{\epsilon\epsilon}^{-1}A)^{-1/2}A'\tilde{\Sigma}_{\epsilon\epsilon}^{-1}\tilde{C}. \end{aligned}$$

Since the $\hat{V}_j$ are normalized eigenvectors, implicit in the construction of the estimators $\hat{B}_1$ and $\hat{B}_2$ are the usual normalizations

$$\hat{B}_2(\hat{\Sigma}_{xx}^{-1} + \Psi)^{-1}\hat{B}_2' = \hat{\Lambda}^2 \equiv \text{diag}\{\hat{\lambda}_1^2, \ldots, \hat{\lambda}_{r_1}^2\}, \qquad \hat{B}_1'(A'\tilde{\Sigma}_{\epsilon\epsilon}^{-1}A)\hat{B}_1 = I_{r_1}.$$

The ML estimator of $\Sigma_{\epsilon\epsilon}$ for model (6.17) can be written as

$$\begin{aligned} \hat{\Sigma}_{\epsilon\epsilon} &= \frac{1}{T}(\boldsymbol{Y} - A\hat{B}_1\hat{B}_2\boldsymbol{X})(\boldsymbol{Y} - A\hat{B}_1\hat{B}_2\boldsymbol{X})' \\ &= \tilde{\Sigma}_{\epsilon\epsilon} + (\tilde{C} - A\hat{B}_1\hat{B}_2)\hat{\Sigma}_{xx}(\tilde{C} - A\hat{B}_1\hat{B}_2)' \\ &= \tilde{\Sigma}_{\epsilon\epsilon} + (I_m - P)(\tilde{C}\hat{\Sigma}_{xx}\tilde{C}')(I_m - P)' \end{aligned}$$

where $P = A(A'\tilde{\Sigma}_{\epsilon\epsilon}^{-1}A)^{-1/2}\hat{V}_{(r_1)}\hat{V}_{(r_1)}'(A'\tilde{\Sigma}_{\epsilon\epsilon}^{-1}A)^{-1/2}A'\tilde{\Sigma}_{\epsilon\epsilon}^{-1}$.

It is easy to see that the results derived for the reduced-rank growth curve model (6.17) simplify to the results for the standard reduced-rank regression model discussed in Chapter 2. These are obtained by setting $A = I_m$ with $r = m$, so that $\tilde{B} \equiv \tilde{C}$ and $\Psi = 0$. The matrix $S_{(r)}S'_{(r)}$ which is involved in the reduced-rank growth curve model simplifies to $\hat{R} = \tilde{\Sigma}_{\epsilon\epsilon}^{-1/2}\hat{\Sigma}_{yx}\hat{\Sigma}_{xx}^{-1}\hat{\Sigma}_{xy}\tilde{\Sigma}_{\epsilon\epsilon}^{-1/2}$, the matrix through which the standard reduced-rank model calculations were made in Chapter 2. We also readily see that the above expression for $\hat{\Sigma}_{\epsilon\epsilon}$ simplifies to (2.18) when $A = I_m$ for the standard reduced-rank model.

The same estimation results can be derived using the conditional model approach as presented in Section 6.2. The conditional model (6.6) when B is assumed to be of reduced rank can be written as

$$Y_{1k} = B_1 B_2 X_k + \Theta Y_{2k} + \epsilon^*_{1k}, \tag{6.22}$$

which is in the form of the reduced-rank model (3.3) of Section 3.1. From (3.11) it follows that the ML estimates of B_1 and B_2 from this model are given by

$$\hat{B}_1 = (A'\tilde{\Sigma}_{\epsilon\epsilon}^{-1}A)^{-1/2}\hat{V}, \qquad \hat{B}_2 = \hat{V}'(A'\tilde{\Sigma}_{\epsilon\epsilon}^{-1}A)^{1/2}\hat{\Sigma}_{y_1x.y_2}\hat{\Sigma}_{xx.y_2}^{-1},$$

recalling from (6.10) that $\tilde{\Sigma}_1^* = (A'\tilde{\Sigma}_{\epsilon\epsilon}^{-1}A)^{-1}$ is the "unrestricted" ML estimate of the error covariance matrix $\Sigma_1^* = \text{Cov}(\epsilon^*_{1k})$ in model (6.6), and $\hat{V} = [\hat{V}_1, \ldots, \hat{V}_{r_1}]$ where $\hat{V}_j$ is the normalized eigenvector of the matrix

$$(A'\tilde{\Sigma}_{\epsilon\epsilon}^{-1}A)^{1/2}\hat{\Sigma}_{y_1x.y_2}\hat{\Sigma}_{xx.y_2}^{-1}\hat{\Sigma}_{xy_1.y_2}(A'\tilde{\Sigma}_{\epsilon\epsilon}^{-1}A)^{1/2} \tag{6.23}$$

corresponding to the jth largest eigenvalue. However, from (6.8) we know that $\hat{\Sigma}_{y_1x.y_2}\hat{\Sigma}_{xx.y_2}^{-1} \equiv \tilde{B} \equiv (A'\tilde{\Sigma}_{\epsilon\epsilon}^{-1}A)^{-1}A'\tilde{\Sigma}_{\epsilon\epsilon}^{-1}\tilde{C}$, and it has also been established in Section 6.3 that $\hat{\Sigma}_{xx.y_2} = (\hat{\Sigma}_{xx}^{-1} + \Psi)^{-1}$ [noting the correspondence $\hat{\Sigma}_{xx}^{-1} + \Psi \equiv T\,\Omega = \hat{\Sigma}_{xx.y_2}^{-1}$ from (6.12)]. So we see that the matrix in (6.23) is identical to the matrix $S_{(r)}S'_{(r)}$ in (6.20) that is involved in ML estimation through the previous direct approach. Hence it immediately follows that the above expressions for $\hat{B}_1$ and $\hat{B}_2$, obtained from the conditional model approach, can be expressed in an identical form as in (6.21). In addition, note that the "unrestricted" ML estimate of the error covariance matrix Σ_1^* in model (6.6) can be expressed in the form $\tilde{\Sigma}_1^* = (A'\tilde{\Sigma}_{\epsilon\epsilon}^{-1}A)^{-1} \equiv \hat{\Sigma}_{y_1y_1.y_2} - \hat{\Sigma}_{y_1x.y_2}\hat{\Sigma}_{xx.y_2}^{-1}\hat{\Sigma}_{xy_1.y_2}$, and that

$$\begin{aligned}\hat{\Sigma}_{y_1y_1.y_2} &= \frac{1}{T}[A_1'\boldsymbol{Y}\boldsymbol{Y}'A_1 - A_1'\boldsymbol{Y}\boldsymbol{Y}'A_2(A_2'\boldsymbol{Y}\boldsymbol{Y}'A_2)^{-1}A_2'\boldsymbol{Y}\boldsymbol{Y}'A_1] \\ &= \frac{1}{T}(A'(\boldsymbol{Y}\boldsymbol{Y}')^{-1}A)^{-1} \equiv (A'\hat{\Sigma}_{yy}^{-1}A)^{-1},\end{aligned}$$

with $\hat{\Sigma}_{yy} = \frac{1}{T}\boldsymbol{Y}\boldsymbol{Y}'$, using Lemma 1 of Khatri (1966). Hence, we see that

$$\begin{aligned}\hat{\Sigma}_{y_1x.y_2}\hat{\Sigma}_{xx.y_2}^{-1}\hat{\Sigma}_{xy_1.y_2} &= \hat{\Sigma}_{y_1y_1.y_2} - (A'\tilde{\Sigma}_{\epsilon\epsilon}^{-1}A)^{-1} \\ &= (A'\hat{\Sigma}_{yy}^{-1}A)^{-1} - (A'\tilde{\Sigma}_{\epsilon\epsilon}^{-1}A)^{-1},\end{aligned}$$

so it immediately follows that another equivalent form to represent the matrix in (6.23) is as $(A'\tilde{\Sigma}_{\epsilon\epsilon}^{-1}A)^{1/2}[(A'\hat{\Sigma}_{yy}^{-1}A)^{-1} - (A'\tilde{\Sigma}_{\epsilon\epsilon}^{-1}A)^{-1}](A'\tilde{\Sigma}_{\epsilon\epsilon}^{-1}A)^{1/2}$, which is similar to the form given by Albert and Kshirsagar (1993) in related work.

The formulation of the reduced-rank growth curve model (6.17) in the conditional model form (6.22) leads more directly to LR testing for the rank of the matrix B. Because (6.22) is in the framework of the reduced-rank model (3.3) of Section 3.1, a LR test of the rank of B can be obtained through the eigenvalues $\hat{\lambda}_j^2$ of the matrix in (6.23), which is equivalent to the matrix $S_{(r)}S_{(r)}'$ in (6.20) as noted. These eigenvalues are also directly related, in the usual way, to the sample partial canonical correlations between Y_{1k} and X_k after adjusting for the effects of Y_{2k}. Following the developments in Section 3.1, the LR test for H_0: rank$(B) = r_1$ is based on the test statistic

$$\mathcal{M} = [T - n - (m-r) + (n-r-1)/2] \sum_{j=r_1+1}^{r} \log(1 + \hat{\lambda}_j^2), \qquad (6.24)$$

whose null distribution is approximated by the χ^2-distribution with $(r-r_1)(n-r_1)$ degrees of freedom.

Note that LR testing for the rank of the matrix C $(= AB)$ in the standard reduced-rank regression model, $Y_k = CX_k + \epsilon_k = ABX_k + \epsilon_k$, ignoring the knowledge that the matrix A is known in the growth curve model, still could serve as a valid initial testing procedure, for convenience. Under the assumed reduced-rank growth curve model (6.17), the additional rank deficiency of $C = AB = AB_1B_2$ is due to the component $B = B_1B_2$, since A is specified to be of full rank r. Hence, LR testing in the standard regression model (as summarized in Section 6.5) would be valid to suggest a possible rank for the matrix B, initially. However, this procedure would not be as efficient as the LR testing procedure discussed in the preceding paragraph, since it does not take into account the known specified value of the component matrix A in model (6.17).

We comment briefly on the relationship between the LR test for specified linear constraints on the coefficient matrix B and the LR test results above for the rank of B. The LR test for $H_0 : FB = 0$, where F is a specified $(r-r_1) \times r$ full-rank matrix, is derived in Section 6.3 to have the same form as the statistic $\mathcal{M}$ in (6.24), with r replaced by $(r-r_1)$, and the statistic is asymptotically distributed as $\chi^2_{(r-r_1)n}$. In this context, the $\hat{\lambda}_j^2$ are the eigenvalues of the matrix

$$S_*^{-1}H_* \equiv \{F(A'\tilde{\Sigma}_{\epsilon\epsilon}^{-1}A)^{-1}F'\}^{-1}F(A'\tilde{\Sigma}_{\epsilon\epsilon}^{-1}A)^{-1/2}S_{(r)}S_{(r)}'(A'\tilde{\Sigma}_{\epsilon\epsilon}^{-1}A)^{-1/2}F'.$$

When B is of reduced rank $r_1 < r$, this is equivalent to the condition related to H_0 above that B satisfies $FB = 0$ for an unknown $(r-r_1) \times r$ full-rank matrix F. The LR test statistic for the rank of B is given by (6.24) and has an asymptotic $\chi^2_{(r-r_1)(n-r_1)}$ distribution. Thus we see the similarity between the two testing situations, with the matrix $S_{(r)}S'_{(r)}$ playing a main role in both cases. Because the constraints are unknown in the reduced-rank case and need to be estimated, the degrees of freedom $(r-r_1)n$ in the known constraints case is decreased to $(r-r_1)(n-r_1)$ for the reduced-rank case.

6.6.1 Extensions of the Reduced-Rank Growth Curve Model

In the same spirit as the extended model (6.14) considered in Section 6.4, the reduced-rank growth curve model (6.17) can be extended to allow for the effect of some predictor variables on the response without growth curve structure. That is, we consider the model

$$Y_k = A\,B_1 B_2\,X_k + DZ_k + \epsilon_k, \qquad k = 1, \ldots, T, \tag{6.25}$$

where all terms are similar to those in (6.14) and (6.17). Maximum likelihood estimation of the component matrices B_1 and B_2, as well as D, can be easily carried out along the same lines as the relevant calculations for model (6.17). The main difference is that the basic calculations now will be performed after adjustment for the additional predictor variables Z_k, as discussed in Section 6.4. Hence the procedures are similar to the preceding ones except that various sample covariance matrices are replaced by sample partial covariance matrices, adjusting for Z_k. In particular, ML estimation of B_1 and B_2, similar to the expressions in (6.21), is now based on $\tilde{B} = (A'\tilde{\Sigma}_{\epsilon\epsilon}^{-1}A)^{-1}A'\tilde{\Sigma}_{\epsilon\epsilon}^{-1}\tilde{C}$ with $\tilde{C} = \hat{\Sigma}_{yx.z}\hat{\Sigma}_{xx.z}^{-1}$, and the matrix similar to that in (6.19) now has $\hat{\Sigma}_{xx.z}^{-1}$ in place of $\hat{\Sigma}_{xx}^{-1}$.

Another extension of the model (6.17) concerns the inclusion of the additional predictor variables Z_k in the growth curve structure, but without any reduced-rank feature. This is equivalent to representing D in model (6.25) in the form $D = AB^*$, so that the resulting model can be represented as

$$\begin{aligned} Y_k &= A\,B_1 B_2\,X_k + AB^* Z_k + \epsilon_k \\ &\equiv A[B_1 B_2 X_k + B^* Z_k] + \epsilon_k, \qquad k = 1, \ldots, T. \end{aligned} \tag{6.26}$$

This can be viewed as analogous to the reduced-rank model (3.3) of Section 3.1, but where the reduced-rank and "standard" regression structures are in terms of the growth curve parameters of the responses Y_k rather than in terms of the response vectors Y_k themselves. A special case of model (6.26) was the focus of work by Albert and Kshirsagar (1993). Their study was in the context of a multivariate one-way ANOVA situation involving n treatment groups, with $Z_k \equiv 1$ and X_k an $(n-1)$-dimensional vector of indicators. They considered the model of the form of (6.26) for the purpose of combining the features of growth curve with the application of reduced-rank regression to linear discriminant analysis. In Section 3.2 we discussed the application of the reduced-rank regression model for data from a multivariate one-way ANOVA situation, and indicated its relation to linear discriminant analysis in the case of m-dimensional unstructured mean vectors. The above model (6.26) can be applied to growth curve structures in a similar way, with emphasis on discriminant analysis through use of the r-dimensional growth curve parameter vectors of each group to summarize their mean structure.

We will first discuss the ML estimation of parameters in model (6.26), and then examine in more detail the special case of the multivariate one-way ANOVA situation. For model (6.26), let $D^* = D + C\hat{\Sigma}_{xz}\hat{\Sigma}_{zz}^{-1}$ with $C = AB = AB_1B_2$ and $D = AB^*$, so that $D^* = AB^* + AB_1B_2\hat{\Sigma}_{xz}\hat{\Sigma}_{zz}^{-1} = A(B^* + B_1B_2\hat{\Sigma}_{xz}\hat{\Sigma}_{zz}^{-1}) \equiv AB^{**}$. Also let $\tilde{D}^* = \hat{\Sigma}_{yz}\hat{\Sigma}_{zz}^{-1}$ and $\tilde{C} = \hat{\Sigma}_{yx.z}\hat{\Sigma}_{xx.z}^{-1}$ denote the unrestricted LS estimators of D^* and C. For ML estimation of (6.26), we need to minimize $|W|$, where $W = (1/T)(\boldsymbol{Y} - AB_1B_2\boldsymbol{X} - AB^*\boldsymbol{Z})(\boldsymbol{Y} - AB_1B_2\boldsymbol{X} - AB^*\boldsymbol{Z})'$. As in previous cases, ML estimation and LR testing results for the extended reduced-rank growth curve model (6.26) can be derived using the conditional model approach. Analogous to (6.22), the conditional model corresponding to (6.26) can be written as

$$\begin{aligned} Y_{1k} &= B_1B_2X_k + B^*Z_k + \Theta Y_{2k} + \epsilon_{1k}^* \\ &= B_1B_2X_k + \Theta^*Y_{2k}^* + \epsilon_{1k}^*, \end{aligned} \tag{6.27}$$

where $\Theta^* = [\,B^*, \Theta]$ and $Y_{2k}^* = [\,Z_k', Y_{2k}']'$. This model is in the form of the reduced-rank model (3.3) of Section 3.1. From (3.11) and recalling from (6.10) that $\tilde{\Sigma}_1^* = (A'\tilde{\Sigma}_{\epsilon\epsilon}^{-1}A)^{-1}$ is the "unrestricted" ML estimate of the error covariance matrix $\Sigma_1^* = \text{Cov}(\epsilon_{1k}^*)$ in the conditional models (6.6) and (6.27), it follows that the ML estimates of B_1 and B_2 from this model are given by

$$\hat{B}_1 = (A'\tilde{\Sigma}_{\epsilon\epsilon}^{-1}A)^{-1/2}\hat{V}_{(r_1)}^*, \qquad \hat{B}_2 = \hat{V}_{(r_1)}^{*\prime}(A'\tilde{\Sigma}_{\epsilon\epsilon}^{-1}A)^{1/2}\hat{\Sigma}_{y_1x.y_2^*}\hat{\Sigma}_{xx.y_2^*}^{-1}, \tag{6.28}$$

where $\hat{V}_{(r_1)}^* = [\hat{V}_1^*, \ldots, \hat{V}_{r_1}^*]$ and $\hat{V}_j^*$ is the normalized eigenvector of the matrix

$$(A'\tilde{\Sigma}_{\epsilon\epsilon}^{-1}A)^{1/2}\hat{\Sigma}_{y_1x.y_2^*}\hat{\Sigma}_{xx.y_2^*}^{-1}\hat{\Sigma}_{xy_1.y_2^*}(A'\tilde{\Sigma}_{\epsilon\epsilon}^{-1}A)^{1/2}$$

corresponding to the jth largest eigenvalue $\hat{\lambda}_j^{*2}$. However, from previous results associated with the reduced-rank growth curve model (6.17), it follows that $\hat{\Sigma}_{y_1x^*.y_2}\hat{\Sigma}_{x^*x^*.y_2}^{-1} = (A'\tilde{\Sigma}_{\epsilon\epsilon}^{-1}A)^{-1}A'\tilde{\Sigma}_{\epsilon\epsilon}^{-1}\tilde{G} \equiv \tilde{B}_*$, with $\tilde{G} = [\,\tilde{C}, \tilde{D}]$, and that $\hat{\Sigma}_{x^*x^*.y_2} = (\hat{\Sigma}_{x^*x^*}^{-1} + \Psi_*)^{-1}$, where $\boldsymbol{X}^* = [\boldsymbol{X}', \boldsymbol{Z}']'$ and $\Psi_* = (\tilde{G} - A\tilde{B}_*)'\tilde{\Sigma}_{\epsilon\epsilon}^{-1}(\tilde{G} - A\tilde{B}_*)$. From these results it can readily be established that $\hat{\Sigma}_{y_1x.y_2^*}\hat{\Sigma}_{xx.y_2^*}^{-1} = \tilde{B} \equiv (A'\tilde{\Sigma}_{\epsilon\epsilon}^{-1}A)^{-1}A'\tilde{\Sigma}_{\epsilon\epsilon}^{-1}\tilde{C}$, and that $\hat{\Sigma}_{xx.y_2^*} = (\hat{\Sigma}_{xx.z}^{-1} + \Psi)^{-1}$, where $\Psi = (\tilde{C} - A\tilde{B})'\tilde{\Sigma}_{\epsilon\epsilon}^{-1}(\tilde{C} - A\tilde{B})$. So we find that, equivalently, the eigenvectors $\hat{V}_j^*$ involved in the ML estimates $\hat{B}_1$ and $\hat{B}_2$ in

(6.28) are those obtained from the matrix

$$S^*_{(r)}S^{*\prime}_{(r)} = (A'\tilde{\Sigma}_{\epsilon\epsilon}^{-1}A)^{1/2}\tilde{B}\,(\hat{\Sigma}_{xx.z}^{-1} + \Psi)^{-1}\,\tilde{B}'(A'\tilde{\Sigma}_{\epsilon\epsilon}^{-1}A)^{1/2} \tag{6.29}$$

with $\tilde{B} \equiv (A'\tilde{\Sigma}_{\epsilon\epsilon}^{-1}A)^{-1}A'\tilde{\Sigma}_{\epsilon\epsilon}^{-1}\tilde{C}$. Similar to the arguments following equation (6.23), the matrix in (6.29) can be expressed equivalently in the form $(A'\tilde{\Sigma}_{\epsilon\epsilon}^{-1}A)^{1/2}[(A'\hat{\Sigma}_{yy.z}^{-1}A)^{-1} - (A'\tilde{\Sigma}_{\epsilon\epsilon}^{-1}A)^{-1}](A'\tilde{\Sigma}_{\epsilon\epsilon}^{-1}A)^{1/2}$.

The ML estimate of $B = B_1B_2$ under the reduced-rank assumption is thus given by

$$\begin{aligned}\hat{B} = \hat{B}_1\hat{B}_2 &= (A'\tilde{\Sigma}_{\epsilon\epsilon}^{-1}A)^{-1/2}\hat{V}^*_{(r_1)}\hat{V}^{*\prime}_{(r_1)}(A'\tilde{\Sigma}_{\epsilon\epsilon}^{-1}A)^{1/2}\tilde{B}\\ &= (A'\tilde{\Sigma}_{\epsilon\epsilon}^{-1}A)^{-1/2}\hat{V}^*_{(r_1)}\hat{V}^{*\prime}_{(r_1)}(A'\tilde{\Sigma}_{\epsilon\epsilon}^{-1}A)^{-1/2}A'\tilde{\Sigma}_{\epsilon\epsilon}^{-1}\tilde{C}.\end{aligned}$$

In addition, from results in Section 3.1 for the model (3.3), we obtain the ML estimate of $\Theta^* = [\,B^*, \Theta]$ from the conditional model (6.27) as $\hat{\Theta}^* = \hat{\Sigma}_{y_1y_2^*}\hat{\Sigma}_{y_2^*y_2^*}^{-1} - \hat{B}\hat{\Sigma}_{xy_2^*}\hat{\Sigma}_{y_2^*y_2^*}^{-1}$. From this, we can show that the ML estimate of the component B^* is obtained as

$$\hat{B}^* = \tilde{B}^{**} - \hat{B}_1\hat{B}_2\{\,\hat{\Sigma}_{xz}\hat{\Sigma}_{zz}^{-1} - \hat{\Sigma}_{xy.z}\hat{\Sigma}_{yy.z}^{-1}(\tilde{D}^* - A\tilde{B}^{**})\,\} \tag{6.30}$$

where $\tilde{D}^* = \hat{\Sigma}_{yz}\hat{\Sigma}_{zz}^{-1}$ and $\tilde{B}^{**} = (A'\hat{\Sigma}_{yy.z}^{-1}A)^{-1}A'\hat{\Sigma}_{yy.z}^{-1}\tilde{D}^*$. Since $\tilde{C} = \hat{\Sigma}_{yx.z}\hat{\Sigma}_{xx.z}^{-1}$, $\tilde{\Sigma}_{\epsilon\epsilon} = \hat{\Sigma}_{yy.z} - \hat{\Sigma}_{yx.z}\hat{\Sigma}_{xx.z}^{-1}\hat{\Sigma}_{xy.z}$, and $\tilde{C}\hat{\Sigma}_{xz}\hat{\Sigma}_{zz}^{-1} = \tilde{D}^* - \tilde{D}$, where $\tilde{D} = \hat{\Sigma}_{yz.x}\hat{\Sigma}_{zz.x}^{-1}$, it follows that

$$\begin{aligned}&(A'\tilde{\Sigma}_{\epsilon\epsilon}^{-1}A)^{-1}A'\tilde{\Sigma}_{\epsilon\epsilon}^{-1}\tilde{C}\{\,\hat{\Sigma}_{xz}\hat{\Sigma}_{zz}^{-1} - \hat{\Sigma}_{xy.z}\hat{\Sigma}_{yy.z}^{-1}(\tilde{D}^* - A\tilde{B}^{**})\,\}\\ &= (A'\tilde{\Sigma}_{\epsilon\epsilon}^{-1}A)^{-1}A'\tilde{\Sigma}_{\epsilon\epsilon}^{-1}\{\,\tilde{D}^* - \tilde{D} + (\tilde{\Sigma}_{\epsilon\epsilon} - \hat{\Sigma}_{yy.z})\hat{\Sigma}_{yy.z}^{-1}(\tilde{D}^* - A\tilde{B}^{**})\,\}\\ &= \tilde{B}^{**} - (A'\tilde{\Sigma}_{\epsilon\epsilon}^{-1}A)^{-1}A'\tilde{\Sigma}_{\epsilon\epsilon}^{-1}\tilde{D}.\end{aligned}$$

Therefore, the ML estimator $\hat{B}^*$ in (6.30) can also be expressed in the equivalent form

$$\hat{B}^* = \tilde{B}^{**} + (A'\tilde{\Sigma}_{\epsilon\epsilon}^{-1}A)^{-1/2}\hat{V}^*_{(r_1)}\hat{V}^{*\prime}_{(r_1)}(A'\tilde{\Sigma}_{\epsilon\epsilon}^{-1}A)^{1/2}(\tilde{B}^* - \tilde{B}^{**}), \tag{6.31}$$

where $\tilde{B}^* = (A'\tilde{\Sigma}_{\epsilon\epsilon}^{-1}A)^{-1}A'\tilde{\Sigma}_{\epsilon\epsilon}^{-1}\tilde{D}$. Finally, the ML estimate of the error covariance matrix $\Sigma_{\epsilon\epsilon}$ is

$$\begin{aligned}\hat{\Sigma}_{\epsilon\epsilon} &= \frac{1}{T}(\boldsymbol{Y} - A\hat{B}_1\hat{B}_2\boldsymbol{X} - A\hat{B}^*\boldsymbol{Z})(\boldsymbol{Y} - A\hat{B}_1\hat{B}_2\boldsymbol{X} - A\hat{B}^*\boldsymbol{Z})'\\ &= \tilde{\Sigma}_{\epsilon\epsilon} + \frac{1}{T}(\tilde{C}\boldsymbol{X} + \tilde{D}\boldsymbol{Z} - A\hat{B}_1\hat{B}_2\boldsymbol{X} - A\hat{B}^*\boldsymbol{Z})\\ &\qquad\qquad \times\,(\tilde{C}\boldsymbol{X} + \tilde{D}\boldsymbol{Z} - A\hat{B}_1\hat{B}_2\boldsymbol{X} - A\hat{B}^*\boldsymbol{Z})'\\ &= \tilde{\Sigma}_{\epsilon\epsilon} + (\tilde{C} - A\hat{B}_1\hat{B}_2)\hat{\Sigma}_{xx.z}(\tilde{C} - A\hat{B}_1\hat{B}_2)'\\ &\qquad + (I_m - A\hat{B}_1\hat{B}_2\hat{\Sigma}_{xy.z}\hat{\Sigma}_{yy.z}^{-1})(\tilde{D}^* - A\tilde{B}^{**})\hat{\Sigma}_{zz}\\ &\qquad\qquad \times\,(\tilde{D}^* - A\tilde{B}^{**})'(I_m - A\hat{B}_1\hat{B}_2\hat{\Sigma}_{xy.z}\hat{\Sigma}_{yy.z}^{-1})'.\end{aligned}$$

The formulation of the extended reduced-rank growth curve model (6.26) in the conditional model form (6.27) also leads directly to LR testing for the rank of the matrix B. Because (6.27) is in the framework of the reduced-rank model (3.3) of Section 3.1, a LR test of the rank of B can be obtained through the eigenvalues $\hat{\lambda}_j^{*2}$ of the matrix $S_{(r)}^* S_{(r)}^{*\prime}$ in (6.29). These eigenvalues are also directly related, in the usual way, to the sample partial canonical correlations between Y_{1k} and X_k after adjusting for the effects of Y_{2k}^*. Following the developments in Section 3.1, the LR test for $H_0: \text{rank}(B) = r_1$ is based on the test statistic

$$\mathcal{M} = [T - n - (m-r) + (n_1 - r - 1)/2] \sum_{j=r_1+1}^{r} \log(1 + \hat{\lambda}_j^{*2}), \qquad (6.32)$$

whose null distribution is asymptotically the χ^2-distribution with $(r - r_1)(n_1 - r_1)$ degrees of freedom.

We now elaborate on the multivariate one-way ANOVA case, and consider the situation and details similar to those from Section 3.2. Suppose we have independent random samples $Y_{i1}, \dots, Y_{iT_i}$, $i = 1, \dots, n$, from n multivariate normal distributions $N(\mu_i, \Sigma)$, and set $T = T_1 + \cdots + T_n$. We suppose the mean vectors have the growth curve structure $\mu_i = A\beta_i$, where A is the known $m \times r$ within-subject design matrix and β_i is the $r \times 1$ vector of unknown growth curve parameters for the ith group. It will be assumed that the β_i lie in some r_1-dimensional linear subspace with $r_1 \leq \min(n-1, r)$. This is equivalent to the condition that the matrix of contrasts $[\beta_1 - \beta_n, \dots, \beta_{n-1} - \beta_n]$ is of reduced rank r_1. As mentioned above, the model (6.26) accommodates this special situation where the model is expressed as $Y_{ij} = A[BX_i + B^* Z_i] + \epsilon_{ij}$, with $B = B_1 B_2$ and the choices of B and B^* as $B = [\beta_1 - \beta_n, \dots, \beta_{n-1} - \beta_n]$ and $B^* = \beta_n$, and $Z_i \equiv 1$. The LS estimates of $C = AB$ and $D = AB^*$ are as in (3.14), $\tilde{C} = [\bar{Y}_1 - \bar{Y}_n, \dots, \bar{Y}_{n-1} - \bar{Y}_n]$ and $\tilde{D} = \bar{Y}_n$, and the corresponding estimate of $\Sigma_{\epsilon\epsilon}$ is $\tilde{\Sigma}_{\epsilon\epsilon} = \frac{1}{T} \sum_{i=1}^{n} \sum_{j=1}^{T_i} (Y_{ij} - \bar{Y}_i)(Y_{ij} - \bar{Y}_i)'$. The (full-rank) estimates of B and B^* are obtained directly from the results of Section 6.2 as

$$\tilde{B} = (A'\tilde{\Sigma}_{\epsilon\epsilon}^{-1} A)^{-1} A'\tilde{\Sigma}_{\epsilon\epsilon}^{-1} \tilde{C}, \qquad \tilde{B}^* = (A'\tilde{\Sigma}_{\epsilon\epsilon}^{-1} A)^{-1} A'\tilde{\Sigma}_{\epsilon\epsilon}^{-1} \tilde{D}. \qquad (6.33)$$

The calculations related to the reduced-rank growth curve aspect, that is, ML estimation of the component matrices B_1 and B_2, follow from the preceding general results. From (6.28) and (6.29), the ML estimates of the component matrices are based on the normalized eigenvectors of

$$(A'\tilde{\Sigma}_{\epsilon\epsilon}^{-1} A)^{-1/2} A'\tilde{\Sigma}_{\epsilon\epsilon}^{-1} \tilde{C} \, (\hat{\Sigma}_{xx.z}^{-1} + \Psi)^{-1} \, \tilde{C}'\tilde{\Sigma}_{\epsilon\epsilon}^{-1} A (A'\tilde{\Sigma}_{\epsilon\epsilon}^{-1} A)^{-1/2}. \qquad (6.34)$$

Let $\hat{V}_{(r_1)}^*$ denote the $r \times r_1$ matrix whose r_1 columns are the normalized eigenvectors of the matrix above corresponding to its r_1 largest eigenvalues. Then it follows from (6.28) that the ML estimates of the component matrices are

$$\hat{B}_1 = (A'\tilde{\Sigma}_{\epsilon\epsilon}^{-1}A)^{-1/2}\hat{V}^*_{(r_1)}, \qquad \hat{B}_2 = \hat{V}^{*\prime}_{(r_1)}(A'\tilde{\Sigma}_{\epsilon\epsilon}^{-1}A)^{-1/2}A'\tilde{\Sigma}_{\epsilon\epsilon}^{-1}\tilde{C}. \quad (6.35)$$

We set $\hat{V}^{**}_{(r_1)} = (A'\tilde{\Sigma}_{\epsilon\epsilon}^{-1}A)^{1/2}\hat{V}^*_{(r_1)}$ so that it satisfies the normalization $\hat{V}^{**\prime}_{(r_1)}(A'\tilde{\Sigma}_{\epsilon\epsilon}^{-1}A)^{-1}\hat{V}^{**}_{(r_1)} = I_{r_1}$. Then we can express the ML estimates in (6.35) as $\hat{B}_1 = (A'\tilde{\Sigma}_{\epsilon\epsilon}^{-1}A)^{-1}\hat{V}^{**}_{(r_1)}$ and $\hat{B}_2 = \hat{V}^{**\prime}_{(r_1)}(A'\tilde{\Sigma}_{\epsilon\epsilon}^{-1}A)^{-1}A'\tilde{\Sigma}_{\epsilon\epsilon}^{-1}\tilde{C}$. The corresponding ML reduced-rank estimate of B is

$$\hat{B} = \hat{B}_1\,\hat{B}_2 = (A'\tilde{\Sigma}_{\epsilon\epsilon}^{-1}A)^{-1}\hat{V}^{**}_{(r_1)}\hat{V}^{**\prime}_{(r_1)}(A'\tilde{\Sigma}_{\epsilon\epsilon}^{-1}A)^{-1}A'\tilde{\Sigma}_{\epsilon\epsilon}^{-1}\tilde{C}. \quad (6.36)$$

The ML estimate of $B^* = \beta_n$ is, from (6.31), given by

$$\begin{aligned}\hat{\beta}_n &= \tilde{B}^{**} + (A'\tilde{\Sigma}_{\epsilon\epsilon}^{-1}A)^{-1/2}\hat{V}^*_{(r_1)}\hat{V}^{*\prime}_{(r_1)}(A'\tilde{\Sigma}_{\epsilon\epsilon}^{-1}A)^{1/2}(\tilde{B}^* - \tilde{B}^{**})\\ &= \tilde{B}^{**} + (A'\tilde{\Sigma}_{\epsilon\epsilon}^{-1}A)^{-1}\hat{V}^{**}_{(r_1)}\hat{V}^{**\prime}_{(r_1)}(\tilde{B}^* - \tilde{B}^{**})\end{aligned}$$

where $\tilde{B}^* = (A'\tilde{\Sigma}_{\epsilon\epsilon}^{-1}A)^{-1}A'\tilde{\Sigma}_{\epsilon\epsilon}^{-1}\bar{Y}_n$ and $\tilde{B}^{**} = (A'\hat{\Sigma}_{yy.z}^{-1}A)^{-1}A'\hat{\Sigma}_{yy.z}^{-1}\bar{Y}$, since $\tilde{D}^* = \hat{\Sigma}_{yz}\hat{\Sigma}_{zz}^{-1} = \bar{Y}$, with $\hat{\Sigma}_{yy.z} = (1/T)\sum_{i=1}^n\sum_{j=1}^{T_i}(Y_{ij}-\bar{Y})(Y_{ij}-\bar{Y})'$. Because the ith column of $\tilde{C}$ is $\bar{Y}_i - \bar{Y}_n$ and the ML reduced-rank growth curve estimate is $\hat{B} \equiv \hat{B}_1\hat{B}_2 = [\hat{\beta}_1 - \hat{\beta}_n, \ldots, \hat{\beta}_{n-1} - \hat{\beta}_n]$, we also have

$$(\hat{\beta}_i - \hat{\beta}_n) = (A'\tilde{\Sigma}_{\epsilon\epsilon}^{-1}A)^{-1}\hat{V}^{**}_{(r_1)}\hat{V}^{**\prime}_{(r_1)}(A'\tilde{\Sigma}_{\epsilon\epsilon}^{-1}A)^{-1}A'\tilde{\Sigma}_{\epsilon\epsilon}^{-1}(\bar{Y}_i - \bar{Y}_n). \quad (6.37)$$

Combining the last two expressions we have the ML reduced-rank estimate of β_i as

$$\begin{aligned}\hat{\beta}_i &= (A'\hat{\Sigma}_{yy.z}^{-1}A)^{-1}A'\hat{\Sigma}_{yy.z}^{-1}\bar{Y} + (A'\tilde{\Sigma}_{\epsilon\epsilon}^{-1}A)^{-1}\hat{V}^{**}_{(r_1)}\hat{V}^{**\prime}_{(r_1)}\\ &\quad\times\{\,(A'\tilde{\Sigma}_{\epsilon\epsilon}^{-1}A)^{-1}A'\tilde{\Sigma}_{\epsilon\epsilon}^{-1}\bar{Y}_i - (A'\hat{\Sigma}_{yy.z}^{-1}A)^{-1}A'\hat{\Sigma}_{yy.z}^{-1}\bar{Y}\,\}. \quad (6.38)\end{aligned}$$

Thus, the $\beta_i - \beta_n$ are estimated using only the r_1 linear combinations $\hat{V}^{**\prime}_{(r_1)}(A'\tilde{\Sigma}_{\epsilon\epsilon}^{-1}A)^{-1}A'\tilde{\Sigma}_{\epsilon\epsilon}^{-1}(\bar{Y}_i - \bar{Y}_n)$ of the differences in the sample mean vectors. Observe that when there is no growth curve structure, hence $r = m$ and $A = I_m$, the result in (6.38) reduces to the result (3.18) in Section 3.2.

Note that the matrix in (6.34), whose eigenvectors and eigenvalues are needed, involves the matrix $\tilde{C}(\hat{\Sigma}_{xx.z}^{-1} + \Psi)^{-1}\tilde{C}'$, where

$$\Psi = (\tilde{C} - A\tilde{B})'\tilde{\Sigma}_{\epsilon\epsilon}^{-1}(\tilde{C} - A\tilde{B})$$

and $(\tilde{C} - A\tilde{B}) = (I_m - P)\tilde{C}$ with $P = A(A'\tilde{\Sigma}_{\epsilon\epsilon}^{-1}A)^{-1}A'\tilde{\Sigma}_{\epsilon\epsilon}^{-1}$. Now the matrix $\tilde{C}\hat{\Sigma}_{xx.z}\tilde{C}' = \frac{1}{T}S_B$ from (3.15), where $S_B = \sum_{i=1}^n T_i(\bar{Y}_i - \bar{Y})(\bar{Y}_i - \bar{Y})'$ is the "between-group" sum of squares matrix. Therefore, using the following standard matrix inversion result (Rao, 1973, p. 33),

$$(\hat{\Sigma}_{xx.z}^{-1}+\Psi)^{-1}=\hat{\Sigma}_{xx.z}-\hat{\Sigma}_{xx.z}(\tilde{C}-A\tilde{B})' \\ \times\{(\tilde{C}-A\tilde{B})\hat{\Sigma}_{xx.z}(\tilde{C}-A\tilde{B})'+\tilde{\Sigma}_{\epsilon\epsilon}\}^{-1}(\tilde{C}-A\tilde{B})\hat{\Sigma}_{xx.z},$$

it follows that the matrix involved in (6.34) can be expressed in a more explicit form as

$$\begin{aligned}&\tilde{C}\,(\hat{\Sigma}_{xx.z}^{-1}+\Psi)^{-1}\,\tilde{C}' \\ &=\frac{1}{T}\,S_B-\frac{1}{T}S_B(I_m-P)' \\ &\qquad\times\left\{(I_m-P)\left[\frac{1}{T}S_B\right](I_m-P)'+\tilde{\Sigma}_{\epsilon\epsilon}\right\}^{-1}(I_m-P)\frac{1}{T}S_B.\end{aligned}$$

Finally, we briefly discuss aspects of linear discriminant analysis within the reduced-rank growth curve context for the multivariate one-way ANOVA growth curve model, $Y_{ij}=\mu_i+\epsilon_{ij}\equiv A\beta_i+\epsilon_{ij}$, $j=1,\ldots,T_i$, $i=1,\ldots,n$. Under the reduced-rank assumptions above for the growth curve parameters β_i, for $i=1,\ldots,n-1$ we have

$$\beta_i=\beta_n+(\beta_i-\beta_n)=\beta_n+B_1B_2^{(i)} \qquad \text{and} \qquad \mu_i=A\beta_i=A\beta_n+AB_1B_2^{(i)},$$

where $B_2^{(i)}$ is the ith column of the matrix B_2. Let $B_{1*}=[B_1,B_*]$ be an $r\times r$ full-rank augmentation of the matrix B_1, normalized by $B_{1*}'(A'\tilde{\Sigma}_{\epsilon\epsilon}^{-1}A)B_{1*}=I_r$, and set $L_*'=[\ell_1,\ldots,\ell_r]'=B_{1*}'A'\tilde{\Sigma}_{\epsilon\epsilon}^{-1}$. Then considering the r linear combinations

$$L_*'\mu_i=B_{1*}'A'\tilde{\Sigma}_{\epsilon\epsilon}^{-1}(A\beta_n+AB_1B_2^{(i)})=L_*'A\beta_n+\begin{bmatrix}I_{r_1}\\0\end{bmatrix}B_2^{(i)}, \quad i=1,\ldots,n\text{–}1,$$

we see that only the first r_1 vectors $\ell_1,\ldots,\ell_{r_1}$ yield different mean vectors among the n groups. So these first r_1 vectors represent the set of potentially useful information for discrimination among the n groups and thus may be regarded as an adequate set of linear discriminant functions. The ML estimate of these r_1 discriminant vectors $L'=[\ell_1,\ldots,\ell_{r_1}]'=B_1'A'\tilde{\Sigma}_{\epsilon\epsilon}^{-1}$ is

$$\hat{L}'=\hat{B}_1'A'\tilde{\Sigma}_{\epsilon\epsilon}^{-1}=\hat{V}_{(r_1)}^{**\prime}(A'\tilde{\Sigma}_{\epsilon\epsilon}^{-1}A)^{-1}A'\tilde{\Sigma}_{\epsilon\epsilon}^{-1},$$

where $\hat{B}_1=(A'\tilde{\Sigma}_{\epsilon\epsilon}^{-1}A)^{-1}\hat{V}_{(r_1)}^{**}$ is the ML estimate of B_1. Also notice that the ML estimates of the corresponding linear combinations of mean vectors, $L'\mu_i$, are obtained using (6.38) as

$$\begin{aligned}\hat{L}'\hat{\mu}_i &= \hat{L}'A\hat{\beta}_i=\hat{V}_{(r_1)}^{**\prime}\hat{\beta}_i \\ &= \hat{V}_{(r_1)}^{**\prime}(A'\tilde{\Sigma}_{\epsilon\epsilon}^{-1}A)^{-1}A'\tilde{\Sigma}_{\epsilon\epsilon}^{-1}\bar{Y}_i\equiv\hat{V}_{(r_1)}^{**\prime}\tilde{\beta}_i, \qquad i=1,\ldots,n, \qquad (6.39)\end{aligned}$$

where $\tilde{\beta}_i = (A'\tilde{\Sigma}_{\epsilon\epsilon}^{-1}A)^{-1}A'\tilde{\Sigma}_{\epsilon\epsilon}^{-1}\bar{Y}_i$ is the "full-rank" ML estimate of the growth curve parameters for the ith group. So the r_1 column vectors of $\hat{V}_{(r_1)}^{**}$ may be referred to as the discriminant vectors for the space of growth curve parameters (e.g., Albert and Kshirsagar, 1993, p. 351). Thus for an individual with response vector Y_{ij}, $\tilde{\beta}_{ij} = (A'\tilde{\Sigma}_{\epsilon\epsilon}^{-1}A)^{-1}A'\tilde{\Sigma}_{\epsilon\epsilon}^{-1}Y_{ij}$ can be regarded as a weighted LS estimate of its individual growth curve parameters, and the ML estimates of its linear discriminants are $\hat{L}'Y_{ij} = \hat{V}_{(r_1)}^{**\prime}(A'\tilde{\Sigma}_{\epsilon\epsilon}^{-1}A)^{-1}A'\tilde{\Sigma}_{\epsilon\epsilon}^{-1}Y_{ij} \equiv \hat{V}_{(r_1)}^{**\prime}\tilde{\beta}_{ij}$, r_1 linear combinations of its estimated growth curve parameters.

6.7 A Numerical Example

In this section we present details for a numerical example on multivariate growth curve and reduced-rank growth curve analysis for some bioassay data, to illustrate some of the methods and results described in the previous sections of this chapter.

Example 6.1. We consider the basic multivariate growth curve model analysis methods of the previous sections applied to some bioassay data taken from a study by Volund (1980). The data consist of measurements on blood sugar concentration of 36 rabbits at 0, 1, 2, 3, 4, and 5 hours after administration of an insulin dose. The 36 rabbits comprise four groups of nine animals according to a 2×2 factorial with two types of insulin preparation, a 'standard' and a 'test' preparation, crossed with two dose levels, 0.75 and 1.50 units per rabbit, based on an assumed potency of 26 units per mg insulin. In fact, the 'standard' preparation was a preparation of MC porcine insulin and the 'test' was a preparation from another batch of MC porcine insulin, so we will not expect any substantial "type of treatment" effect. Figure 6.1 shows the average blood sugar concentrations for each of the four groups at each time point.

We consider the vector $Y = (y_1, y_2, \ldots, y_5)'$ of measurements at the times 1, 2, 3, 4, and 5 hours as the $m = 5$ dimension response vector, and use as predictor variables the indicator variable for insulin type ($x_1 = \pm 1$), the indicator variable for dose level ($x_2 = \pm 1$), the initial blood sugar concentration at time 0 minus 100 ($x_3 \equiv y_0 - 100$), and a term for "interaction" between dose level and initial concentration ($x_4 = x_2 x_3$). Thus, the initial multivariate linear regression model that we consider for the kth vector of responses is of the form

$$Y_k = c_0 + c_1 x_{1k} + c_2 x_{2k} + c_3 x_{3k} + c_4 x_{4k} + \epsilon_k \equiv C X_k + \epsilon_k, \qquad k = 1, \ldots, T,$$

where $C = [c_0, c_1, \ldots, c_4]$ and $X_k = (1, x_{1k}, \ldots, x_{4k})'$, and $T = 36$. With $\boldsymbol{Y}$ and $\boldsymbol{X}$ denoting 5×36 data matrices, respectively, the least squares estimate $\tilde{C} = \boldsymbol{Y}\boldsymbol{X}'(\boldsymbol{X}\boldsymbol{X}')^{-1}$ of the regression coefficient matrix C is obtained

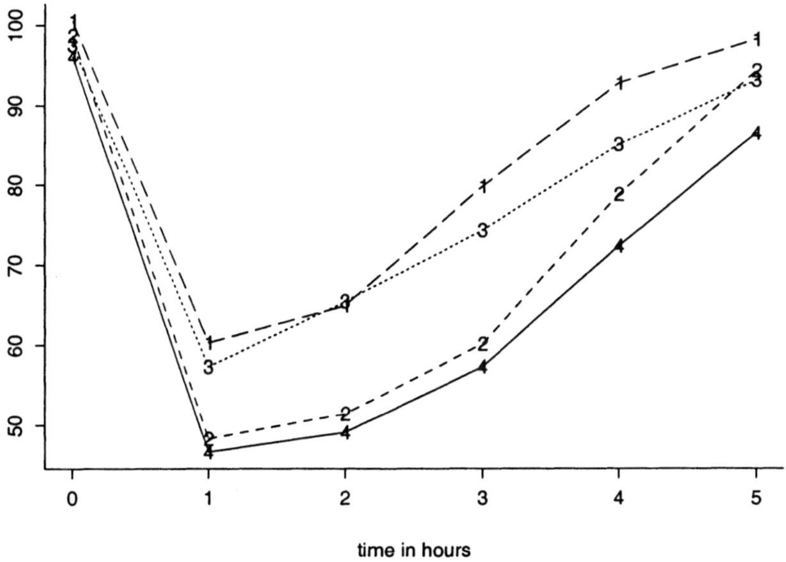

Figure 6.1. Average blood sugar concentrations over time for each of four treatment groups. Different groups indicated by symbols: 1, standard treatment at low dose; 2, standard treatment at high dose; 3, test treatment at low dose; 4, test treatment at high dose

as

$$\tilde{C} = \begin{bmatrix} 54.5405 & -0.0500 & -5.0428 & 0.7879 & -0.0062 \\ (1.4840) & (1.4530) & (1.4820) & (0.2300) & (0.2242) \\ 59.0460 & 0.6836 & -7.0757 & 0.8073 & -0.1451 \\ (1.5700) & (1.5380) & (1.5680) & (0.2434) & (0.2372) \\ 69.8572 & -0.6324 & -8.1554 & 1.0494 & 0.0948 \\ (1.9110) & (1.8710) & (1.9080) & (0.2962) & (0.2887) \\ 85.1540 & -1.7673 & -4.4398 & 1.3019 & 0.6732 \\ (2.1070) & (2.0630) & (2.1040) & (0.3266) & (0.3183) \\ 95.9062 & -1.6904 & -0.5283 & 1.1409 & 0.6809 \\ (1.8180) & (1.7810) & (1.8160) & (0.2818) & (0.2747) \end{bmatrix},$$

with estimated standard deviations given in parentheses below the corresponding estimate. As was anticipated, it is seen that the insulin type "treatment" effect variable x_1 is not a significant factor for these data, whereas the dose variable (x_2) and initial concentration (x_3) are quite important as expected, and the interaction (x_4) of the dose variable with initial concentration is a relevant factor mainly for the later responses at times of 4 and 5 hours. The ML estimate of the 5×5 covariance matrix of the errors ϵ_k (with correlations shown above the diagonal) is given by

$$\tilde{\Sigma}_{\epsilon\epsilon} = \frac{1}{36}\hat{\epsilon}\hat{\epsilon}' = \begin{bmatrix} 62.1769 & 0.8552 & 0.5617 & 0.3149 & 0.1569 \\ 56.2715 & 69.6356 & 0.7907 & 0.5422 & 0.3865 \\ 44.9707 & 66.9940 & 103.0968 & 0.8106 & 0.5045 \\ 27.8012 & 50.6523 & 92.1435 & 125.3396 & 0.7572 \\ 11.9579 & 31.1633 & 49.5009 & 81.9090 & 93.3698 \end{bmatrix}$$

where $\hat{\epsilon} = \boldsymbol{Y} - \tilde{C}\boldsymbol{X}$, with $\log(|\tilde{\Sigma}_{\epsilon\epsilon}|) = 17.7096$.

We next allow for a growth curve model structure of the form of (6.1), $Y_k = ABX_k + \epsilon_k$, to represent the pattern of mean blood sugar concentrations over the times 1 through 5 hours. With the columns of the within-subject design matrix A specified as orthogonal polynomial terms, it is found that a polynomial model of degree 3 ($r = 4$) is required for an adequate fit. The goodness-of-fit chi-squared test statistic, discussed in Section 6.3, for adequacy of this model gives the value of $\mathcal{M} = 6.22$ with 5 degrees of freedom. For the matrix A as shown, we obtain the ML estimate for the growth curve coefficient matrix B as follows:

$$A' = \begin{bmatrix} 1 & 1 & 1 & 1 & 1 \\ -2 & -1 & 0 & 1 & 2 \\ 2 & -1 & -2 & -1 & 2 \\ -1 & 2 & 0 & -2 & 1 \end{bmatrix},$$

$$\hat{B} = \begin{bmatrix} 72.9036 & -0.6908 & -5.0451 & 1.0176 & 0.2599 \\ 10.6675 & -0.6092 & 0.9131 & 0.1137 & 0.1932 \\ 1.0211 & -0.1128 & 0.9673 & -0.0307 & 0.0220 \\ -1.0303 & 0.3352 & -0.0117 & -0.0620 & -0.0884 \end{bmatrix},$$

where $\hat{B} = (A'\tilde{\Sigma}_{\epsilon\epsilon}^{-1}A)^{-1}A'\tilde{\Sigma}_{\epsilon\epsilon}^{-1}\tilde{C}$, with ML error covariance matrix estimate $\hat{\Sigma}_{\epsilon\epsilon}$ such that $\log(|\hat{\Sigma}_{\epsilon\epsilon}|) = 17.9010$. It is found that the growth curve model reproduces the mean blood sugar concentration curves over time, in the sense that the estimates of mean concentration obtained from the growth curve model, for each group, are quite close to the mean concentrations estimated from the LS regression results.

Next we explore the possibility of a reduced-rank regression model form, without specification of any specific polynomial growth curve model structure. That is, as suggested in Section 6.5, we consider models of the form of Chapter 2, $Y_k = ABX_k + \epsilon_k$, where the multivariate regression coefficient matrix $C = AB$ is assumed to have reduced rank r and A and B are both unknown (unspecified) matrices of parameters to be estimated (subject to normalization constraints, of course). LR testing procedures for the rank of C, as described in Section 2.6, indicate that a coefficient matrix of rank $r = 2$ will be adequate, with the LR statistic for testing $H_0 : \text{rank}(C) \leq 2$ having the value $\mathcal{M} = 8.62$ with $(5-2)(5-2) = 9$ degrees of freedom. The

ML estimates of the component matrices A and B are obtained as in (2.17) of Section 2.3 [also (6.16) in Section 6.5], and are given by

$$\hat{A}' = \begin{bmatrix} 4.6510 & 5.0493 & 5.9560 & 7.2172 & 8.1490 \\ -4.2176 & -6.3067 & -6.5796 & -2.1946 & 1.3839 \end{bmatrix},$$

$$\hat{B} = \begin{bmatrix} 11.7598 & -0.2109 & -0.2252 & 0.1454 & 0.0656 \\ 0.0601 & -0.2346 & 0.9531 & -0.0078 & 0.0770 \end{bmatrix},$$

with ML error covariance matrix estimate $\hat{\Sigma}_{\epsilon\epsilon}$ having $\log(|\hat{\Sigma}_{\epsilon\epsilon}|) = 17.9922$. Notice, first, that the two column vectors of the estimated matrix $\hat{A}$ can be viewed as the two "basis" functions over time needed to represent the mean blood sugar concentration curves over time for each of the groups, and these provide essentially as good a representation as the third degree polynomial growth curve model. Also, taking into account the scales of the variables x_{ik}, the first linear combination $x^*_{1k} = \hat{\beta}'_1 X_k$ of predictor variables in $X^*_k = \hat{B} X_k$ consists mainly of a constant term and contributions from the initial concentration (x_3) and interaction (x_4) variables, whereas the second linear combination $x^*_{2k} = \hat{\beta}'_2 X_k$ is dominantly the dose level variable (x_2). We also mention that although a rank 2 model provides an acceptable fit, a rank 3 model produces an excellent fit with $\log(|\hat{\Sigma}_{\epsilon\epsilon}|) = 17.7404$.

We can also consider a reduced-rank structure within the growth curve model framework, as detailed in Section 6.6. Thus, we consider a model in the form of (6.17), $Y_k = A\, B_1 B_2\, X_k + \epsilon_k$, with the within-subject growth curve design matrix A specified to have the third degree orthogonal polynomial terms as in the previous full-rank growth curve model. The LR testing procedure for the rank associated with this model again indicates that a rank 2 model would be adequate, with the LR test statistic from Section 6.6 for a test of $H_0 : \text{rank}(B) \leq 2$ giving the value $\mathcal{M} = 7.74$ with $(4-2)(5-2) = 6$ degrees of freedom. The ML estimates of the component growth curve coefficient matrices B_1 and B_2 are obtained from (6.21) and are given as

$$\hat{B}'_1 = \begin{bmatrix} 6.2299 & 0.8919 & 0.0813 & -0.0785 \\ -3.7281 & 1.3797 & 0.9835 & -0.2211 \end{bmatrix},$$

$$\hat{B}_2 = \begin{bmatrix} 11.7553 & -0.2105 & -0.2389 & 0.1453 & 0.0644 \\ 0.1224 & -0.2746 & 0.8787 & -0.0094 & 0.0696 \end{bmatrix},$$

with ML error covariance matrix estimate $\hat{\Sigma}_{\epsilon\epsilon}$ having $\log(|\hat{\Sigma}_{\epsilon\epsilon}|) = 18.1590$. Notice that the two estimated linear combinations of predictor variables which occur in this model, $x^*_{ik} = \hat{\beta}'_i X_k$, $i = 1, 2$, are very similar to those estimated for the previous reduced-rank regression model with rank of 2 (especially the first linear combination x^*_{1k}). In addition, it is found that the factor $A\hat{B}_1$ from the reduced-rank growth curve model above is fairly similar to the estimated left coefficient matrix $\hat{A}$ in the previous reduced-rank regression model (especially the first column), with

$$(A\hat{B}_1)' = \begin{bmatrix} 4.6871 & 5.0997 & 6.0674 & 7.1976 & 8.0977 \\ -4.2995 & -6.5335 & -5.6950 & -2.8896 & 0.7772 \end{bmatrix}.$$

Finally, we can also consider the "extended" versions of the reduced-rank regression and reduced-rank growth curve models, as discussed in Sections 3.1 and 6.6, respectively. The extended reduced-rank regression model we considered is $Y_k = C_1 X_{1,k} + DZ_k + \epsilon_k$, with $C_1 = AB$, where $Z_k \equiv 1$ corresponds to the constant terms in the model and $X_{1,k} = (x_{1k}, x_{2k}, x_{3k}, x_{4k})'$ contains the remaining predictor variables. LR testing procedures for the appropriate rank of C_1, through canonical correlation analysis calculations as described in Section 3.1, indicate that a rank of $r_1 = 2$ would be appropriate for C_1, with a LR test statistic for H_0 : rank$(C_1) \leq 2$ having the value $\mathcal{M} = 1.93$ on $(5-2)(4-2) = 6$ degrees of freedom. The ML estimates of the component matrices A and B, obtained from equation (3.11) as described in Section 3.1, are given by

$$\hat{A}' = \begin{bmatrix} 5.3545 & 5.7591 & 8.0997 & 9.2386 & 6.9661 \\ -3.3313 & -5.4329 & -5.0752 & -0.3182 & 2.5203 \end{bmatrix},$$

$$\hat{B} = \begin{bmatrix} -0.2229 & -0.4256 & 0.1489 & 0.0643 \\ -0.2865 & 0.8867 & 0.0192 & 0.0917 \end{bmatrix},$$

with $\hat{D} = (54.4812, 58.9653, 69.9291, 85.1844, 95.8072)'$, and ML error covariance matrix estimate $\hat{\Sigma}_{\epsilon\epsilon}$ having $\log(|\hat{\Sigma}_{\epsilon\epsilon}|) = 17.7738$. Note that the ML estimate $\hat{D}$ in this model is nearly identical to the first column (corresponding to the constant terms) of the LS estimate $\tilde{C}$ in the multivariate regression model. In addition, notice that the estimated coefficients in $\hat{B}$, especially those in the first row, that determine the two linear combinations of the predictor variables $X_{1,k}$ are fairly similar to the corresponding values in the previous reduced-rank model, which did not separate out the constant term from the remaining predictor variables in formulating the reduced-rank structure. The current extended model does provide for a slightly better fitting model, however.

The extended version of the reduced-rank growth curve model, $Y_k = ABX_{1,k} + AB^* Z_k + \epsilon_k$, with $B = B_1 B_2$, was also considered. A rank of 2 was selected for the growth curve coefficient matrix B (LR statistic for testing H_0 : rank$(B) \leq 2$ of $\mathcal{M} = 1.15$ with 4 degrees of freedom). The ML estimates were obtained from (6.28) as

$$\hat{B}_1' = \begin{bmatrix} 7.0023 & 0.7475 & -0.4158 & -0.5541 \\ -2.8220 & 1.4915 & 0.8697 & -0.3841 \end{bmatrix},$$

$$\hat{B}_2 = \begin{bmatrix} -0.2312 & -0.3852 & 0.1498 & 0.0684 \\ -0.3156 & 0.8303 & 0.0099 & 0.0810 \end{bmatrix},$$

with $\hat{B}^* = (72.8995, 10.6707, 0.9950, -1.0610)'$, and ML error covariance matrix estimate $\hat{\Sigma}_{\epsilon\epsilon}$ having $\log(|\hat{\Sigma}_{\epsilon\epsilon}|) = 17.9400$. Once again, the implied two linear combinations of predictor variables determined under this model are similar to those estimated in the previous reduced-rank models. In addition, the ML estimate $\hat{B}^*$ in this model is very similar to the first column of the estimate $\hat{B}$ in the previous "full-rank" growth curve model.

A summary of the fits of several multivariate regression and growth curve models considered for the blood sugar concentration data is presented in Table 6.1. Given there are values of $\log|\hat{\Sigma}_{\epsilon\epsilon}|$ and of the AIC and BIC information criteria for each model, where AIC $= \log|\hat{\Sigma}_{\epsilon\epsilon}| + 2n^*/T$ and BIC $= \log|\hat{\Sigma}_{\epsilon\epsilon}| + n^* \log(T)/T$, and n^* denotes the number of estimated regression parameters in the model. From this summary of results, it can be seen that the models with reduced ranks offer desirable improvements (and simplifications) over the full-rank models. Thus, we have illustrated that more parsimonious models can be developed in the growth curve context through the use of reduced-rank regression methods.

Table 6.1 Summary of fitting results of several multivariate regression and growth curve models estimated for the blood sugar concentrations data.

Type of model	Rank of C or B	Number of parameters, n^*	$\log\|\hat{\Sigma}_{\epsilon\epsilon}\|$	AIC $= \log\|\hat{\Sigma}_{\epsilon\epsilon}\| + 2n^*/T$	BIC $= \log\|\hat{\Sigma}_{\epsilon\epsilon}\| + n^* \log T/T$
Regression, full-rank	5	25	17.7096	19.0985	20.1981
Regression, reduced-rank	2	16	17.9922	18.8811	19.5849
Regression, extended reduced-rank	2	19	17.7738	18.8293	19.6651
Growth curve, full-rank	4	20	17.9010	19.0121	19.8918
Growth curve, reduced-rank	2	14	18.1590	18.9368	19.5526
Growth curve, extended reduced-rank	2	16	17.9400	18.8289	19.5327

7

Seemingly Unrelated Regressions Models With Reduced Ranks

7.1 Introduction and the Seemingly Unrelated Regressions Model

The classical multivariate linear regression model discussed in Chapter 1 can be generalized by allowing the different response variables y_{ik} to have different input or predictor variables $X_{ik} = (x_{i1k}, \ldots, x_{ink})'$ for different i, so that $y_{ik} = X_{ik}' C_{(i)} + \epsilon_{ik}$, $i = 1, \ldots, m$, and the errors ϵ_{ik} are contemporaneously correlated across the different response variables. Multivariate linear regression models of this form were considered by Zellner (1962), who referred to them as *seemingly unrelated regression* (SUR) equations models. In experimental design situations, the model is also referred to as the multiple design multivariate linear model (e.g., Srivastava, 1967; Roy, Gnanadesikan, and Srivastava, 1971). In the notation of Section 1.2, the linear regression model for the $T \times 1$ vector of values for the ith response variable is

$$\boldsymbol{Y}_{(i)} = \boldsymbol{X}_i' C_{(i)} + \epsilon_{(i)}, \qquad \text{for} \quad i = 1, \ldots, m, \tag{7.1}$$

where $\boldsymbol{X}_i = [X_{i1}, \ldots, X_{iT}]$ is $n \times T$, while $\boldsymbol{Y}_{(i)} = (y_{i1}, \ldots, y_{iT})'$, $\epsilon_{(i)} = (\epsilon_{i1}, \ldots, \epsilon_{iT})'$, and $C_{(i)}$ are the same as in the previous model of Chapter 1. Stacking the m vector equations of (7.1) together, the SUR model can be expressed in vector form as

$$y = \overline{\boldsymbol{X}} c + e, \tag{7.2}$$

where $\overline{\boldsymbol{X}} = \text{Diag}(\boldsymbol{X}_1', \boldsymbol{X}_2', \ldots, \boldsymbol{X}_m')$, while the vectors $c = (C_{(1)}', \ldots, C_{(m)}')'$, $y = (\boldsymbol{Y}_{(1)}', \ldots, \boldsymbol{Y}_{(m)}')'$, and $e = (\epsilon_{(1)}', \ldots, \epsilon_{(m)}')'$ are the same as in Section 1.3, and hence $\text{Cov}(e) = \Omega = \Sigma_{\epsilon\epsilon} \otimes I_T = [(\sigma_{ij}\, I_T)]$.

Before we proceed with discussion of statistical properties associated with analysis of the SUR model, we want to motivate the usefulness of such models. The dependence of the m equations in (7.1) through the contemporaneous correlations of the error terms $\epsilon_{(i)}$ is a special feature of the SUR model. As will be demonstrated below, use of this additional infoı-mation across the equations can improve the statistical properties of the estimators of the regression coefficients. The econometrics literature contains many applications of the SUR model [e.g., see Srivastava and Giles (1987, pp. 2-3)]. The first example used by Zellner (1962) is based on data that appear in Boot and de Wit (1960). For two large corporations, General Electric and Westinghouse, both from the same type of industry, data were taken on the investment functions of the firms annually from 1935 through 1954. It seems reasonable to assume that the error terms associated with the investment functions of the two companies may be contemporaneously correlated because of the influence of common market factors. Another example where the SUR specification is used is in the context of explaining certain economic activities in different geographic regions. Giles and Hampton (1984) estimated production functions for several regions in New Zealand assuming the possibility of inter-regional dependency through the correlations in the error terms. These examples suggest that the SUR model is appropriate and useful for a wide range of applications.

The least squares (LS) estimators of the $C_{(i)}$ in model (7.1) are similar to those in (1.8),

$$\tilde{C}_{(i)} = (\boldsymbol{X}_i \boldsymbol{X}_i')^{-1} \boldsymbol{X}_i \boldsymbol{Y}_{(i)}, \qquad i = 1, \ldots, m.$$

The LS estimators have the properties that $E(\tilde{C}_{(i)}) = C_{(i)}$ and

$$\text{Cov}(\tilde{C}_{(i)}, \tilde{C}_{(j)}) = \sigma_{ij} (\boldsymbol{X}_i \boldsymbol{X}_i')^{-1} \boldsymbol{X}_i \boldsymbol{X}_j' (\boldsymbol{X}_j \boldsymbol{X}_j')^{-1}, \qquad i, j = 1, \ldots, m.$$

In particular, we have the usual result that $\text{Cov}(\tilde{C}_{(i)}) = \sigma_{ii} (\boldsymbol{X}_i \boldsymbol{X}_i')^{-1}$. However, unlike the previous classical multivariate linear model, in the SUR model the LS estimators are not the same as the generalized least squares (GLS) estimator, which is given by

$$\hat{c} = \{\overline{\boldsymbol{X}}' (\Sigma_{\epsilon\epsilon}^{-1} \otimes I_T)\, \overline{\boldsymbol{X}}\}^{-1} \overline{\boldsymbol{X}}' (\Sigma_{\epsilon\epsilon}^{-1} \otimes I_T) y. \tag{7.3}$$

Hence, the GLS estimator for this model does depend on the covariance matrix $\Sigma_{\epsilon\epsilon}$ and does not coincide with the individual LS estimators of the m regression models in (7.1), unless $\overline{\boldsymbol{X}} = I_m \otimes \boldsymbol{X}'$, that is, $\boldsymbol{X}_1 = \boldsymbol{X}_2 = \cdots = \boldsymbol{X}_m$, so that the m linear regression models each involve the same design matrix $\boldsymbol{X}$ of input variables. The GLS estimator in (7.3) has mean vector $E(\hat{c}) = c$ and covariance matrix given by

$$\mathrm{Cov}(\hat{c}) = \{\overline{X}'(\Sigma_{\epsilon\epsilon}^{-1} \otimes I_T)\,\overline{X}\}^{-1}. \tag{7.4}$$

The difficulty with implementing the GLS estimator in (7.3) is that the covariance matrix $\Sigma_{\epsilon\epsilon}$ will typically be unknown. A two-step estimation procedure was originally proposed (e.g., Zellner, 1962) in which an estimate of $\Sigma_{\epsilon\epsilon}$ is obtained from the residuals from the LS estimators $\tilde{C}_{(i)}$ as $\tilde{\Sigma}_{\epsilon\epsilon} = T^{-1}\hat{\boldsymbol{\epsilon}}\hat{\boldsymbol{\epsilon}}'$, where $\hat{\boldsymbol{\epsilon}} = [\hat{\boldsymbol{\epsilon}}_{(1)}, \dots, \hat{\boldsymbol{\epsilon}}_{(m)}]'$ with $\hat{\boldsymbol{\epsilon}}_{(i)} = \boldsymbol{Y}_{(i)} - \boldsymbol{X}_i'\tilde{C}_{(i)}$, $i = 1, \dots, m$, being the vectors of residuals from the LS estimation of the individual linear regression models in (7.1). Then, the GLS estimator of the form of (7.3) is obtained, but with $\Sigma_{\epsilon\epsilon}$ replaced by the estimate $\tilde{\Sigma}_{\epsilon\epsilon}$, that is,

$$\tilde{c} = \{\overline{X}'(\tilde{\Sigma}_{\epsilon\epsilon}^{-1} \otimes I_T)\,\overline{X}\}^{-1}\overline{X}'(\tilde{\Sigma}_{\epsilon\epsilon}^{-1} \otimes I_T)y. \tag{7.5}$$

The covariance matrix of the estimator $\tilde{c}$ is then approximated by $\hat{\mathrm{Cov}}(\tilde{c}) = \{\overline{X}'(\tilde{\Sigma}_{\epsilon\epsilon}^{-1} \otimes I_T)\,\overline{X}\}^{-1}$. The two-step procedure can also be iterated, with the current GLS estimator $\tilde{c}$ in (7.5) at any given stage used to form the estimate $\tilde{\Sigma}_{\epsilon\epsilon}$, and this will lead to the maximum likelihood estimators of c and $\Sigma_{\epsilon\epsilon}$ under the normality assumption of the errors $\boldsymbol{\epsilon}_{(i)}$ in (7.1).

In principle, the GLS estimator $\hat{c} = (\hat{C}_{(1)}', \dots, \hat{C}_{(m)}')'$ in (7.3) offers gains in efficiency, in terms of reduced variance, over the LS estimators. The largest gains in efficiency can be expected for situations when the design matrices satisfy $\boldsymbol{X}_i\boldsymbol{X}_j' \approx 0$ and the squared correlations $\rho_{ij}^2 = \sigma_{ij}^2/(\sigma_{ii}\sigma_{jj})$ of the errors are large [e.g., see Zellner (1962, 1963); Revankar (1974)]. For illustration, consider the case with $m = 2$ equations. Then, from (7.4), the covariance matrix of the GLS estimator can be written as

$$\mathrm{Cov}(\hat{c}) = \mathrm{Cov}\begin{bmatrix} \hat{C}_{(1)} \\ \hat{C}_{(2)} \end{bmatrix} = \begin{bmatrix} \sigma^{11}\boldsymbol{X}_1\boldsymbol{X}_1' & \sigma^{12}\boldsymbol{X}_1\boldsymbol{X}_2' \\ \sigma^{12}\boldsymbol{X}_2\boldsymbol{X}_1' & \sigma^{22}\boldsymbol{X}_2\boldsymbol{X}_2' \end{bmatrix}^{-1},$$

where we denote $\Sigma_{\epsilon\epsilon}^{-1} = [(\sigma^{ij})]$, with $\sigma^{11} = \sigma_{22}/(\sigma_{11}\sigma_{22} - \sigma_{12}^2)$, $\sigma^{22} = \sigma_{11}/(\sigma_{11}\sigma_{22} - \sigma_{12}^2)$, and $\sigma^{12} = -\sigma_{12}/(\sigma_{11}\sigma_{22} - \sigma_{12}^2)$. In particular, using a standard matrix inversion result, we have that

$$\begin{aligned} \mathrm{Cov}(\hat{C}_{(1)}) &= \{\sigma^{11}\boldsymbol{X}_1\boldsymbol{X}_1' - [(\sigma^{12})^2/\sigma^{22}]\boldsymbol{X}_1\boldsymbol{X}_2'(\boldsymbol{X}_2\boldsymbol{X}_2')^{-1}\boldsymbol{X}_2\boldsymbol{X}_1'\}^{-1} \\ &= \sigma_{11}(1 - \rho_{12}^2)\{\boldsymbol{X}_1\boldsymbol{X}_1' - \rho_{12}^2\boldsymbol{X}_1\boldsymbol{X}_2'(\boldsymbol{X}_2\boldsymbol{X}_2')^{-1}\boldsymbol{X}_2\boldsymbol{X}_1'\}^{-1}, \end{aligned}$$

noting that $1/\sigma^{11} = (\sigma_{11}\sigma_{22} - \sigma_{12}^2)/\sigma_{22} = \sigma_{11}(1 - \rho_{12}^2)$ and $(\sigma^{12})^2/(\sigma^{11}\sigma^{22}) = (\sigma_{12})^2/(\sigma_{11}\sigma_{22}) = \rho_{12}^2$. Compared to the covariance matrix of the LS estimator, $\mathrm{Cov}(\tilde{C}_{(1)}) = \sigma_{11}(\boldsymbol{X}_1\boldsymbol{X}_1')^{-1}$, we readily see that the largest reduction in variance for the GLS estimator will occur when $\boldsymbol{X}_1\boldsymbol{X}_2' = 0$ and ρ_{12}^2 is large. In practice, for the GLS estimator in (7.5), the use of the estimated error covariance matrix $\tilde{\Sigma}_{\epsilon\epsilon}$ will cause some increase in the covariance matrix of the resulting GLS estimator above the ideal result in (7.4).

7.2 Relation of Growth Curve Model to the Seemingly Unrelated Regressions Model

It has been noted by Stanek and Koch (1985) that the basic growth curve (or GMANOVA) model considered in Chapter 6 can be interpreted as a special case of a SUR model. Recognition of this relationship between the growth curve and SUR models can be useful for better appreciation of estimation procedures under the two models, and for methods of extension to more flexible growth curve models. Thus, we briefly discuss the relationship in this section.

Consider from Section 6.2 the basic growth curve model in matrix form, $\boldsymbol{Y} = AB\boldsymbol{X} + \boldsymbol{\epsilon}$, and let $H = [A_1, A_2]$ where $A_1 = A(A'A)^{-1}$ and A_2 is $m \times (m-r)$ such that $A_2'A = 0$. Then under the transformation of the model as in (6.5), with $\boldsymbol{Y}^* = H'\boldsymbol{Y} = [\boldsymbol{Y}^*_{(1)}, \ldots, \boldsymbol{Y}^*_{(m)}]'$, we have

$$\boldsymbol{Y}^* = H'AB\boldsymbol{X} + H'\boldsymbol{\epsilon} = P'B\boldsymbol{X} + \boldsymbol{\epsilon}^* = \begin{bmatrix} B\boldsymbol{X} \\ 0 \end{bmatrix} + \boldsymbol{\epsilon}^*, \tag{7.6}$$

where $P' = H'A = [I_r, 0]'$ and $\boldsymbol{\epsilon}^* = H'\boldsymbol{\epsilon} = [\boldsymbol{\epsilon}^*_{(1)}, \ldots, \boldsymbol{\epsilon}^*_{(m)}]'$. Set $\Sigma^* = \mathrm{Cov}(\epsilon^*_k) = H'\Sigma_{\epsilon\epsilon}H$, and write the $r \times n$ matrix B of unknown growth curve coefficients as $B = [B_{(1)}, \ldots, B_{(r)}]'$. The transformed growth curve model (7.6) can then be expressed as

$$\boldsymbol{Y}^*_{(i)} = \boldsymbol{X}'B_{(i)} + \boldsymbol{\epsilon}^*_{(i)}, \qquad i = 1, \ldots, r, \tag{7.7a}$$

and

$$\boldsymbol{Y}^*_{(i)} = \boldsymbol{\epsilon}^*_{(i)}, \qquad i = r+1, \ldots, m. \tag{7.7b}$$

The (transformed) growth curve model, expressed in the form (7.7), can be seen to be a particular version of a SUR model (7.1). The model can be written in vector form similar to (7.2) as

$$y^* = \overline{\boldsymbol{X}}\, b + e^* = (P' \otimes \boldsymbol{X}')b + e^*, \tag{7.8}$$

with $\overline{\boldsymbol{X}} = (P' \otimes \boldsymbol{X}') = [I_r \otimes \boldsymbol{X}, 0]'$, $b = \mathrm{vec}(B') = (B'_{(1)}, \ldots, B'_{(r)})'$, and $\mathrm{Cov}(e^*) = \Sigma^* \otimes I_T$. For this model, a two-stage GLS estimator can be constructed by using an estimate of Σ^* based on LS estimation of each of the m regression equations separately, as discussed at the end of Section 7.1, with the common design matrix $\boldsymbol{X}$. The resulting estimator of Σ^* is $\tilde{\Sigma}^* = (1/T)\boldsymbol{Y}^*[I_T - \boldsymbol{X}'(\boldsymbol{X}\boldsymbol{X}')^{-1}\boldsymbol{X}]\boldsymbol{Y}^* = H'\tilde{\Sigma}_{\epsilon\epsilon}H$. From the last relation and the nonsingularity of H, it follows that $H\tilde{\Sigma}^{*-1}H' = \tilde{\Sigma}_{\epsilon\epsilon}^{-1}$. Thus, when GLS estimation is applied to model (7.8), with the estimate $\tilde{\Sigma}^*$, we obtain

$$\begin{aligned} \tilde{b} &= [(P \otimes \boldsymbol{X})(\tilde{\Sigma}^{*-1} \otimes I_T)(P' \otimes \boldsymbol{X}')]^{-1}(P \otimes \boldsymbol{X})(\tilde{\Sigma}^{*-1} \otimes I_T)y^* \\ &= [(P\tilde{\Sigma}^{*-1}P')^{-1}P\tilde{\Sigma}^{*-1} \otimes (\boldsymbol{X}\boldsymbol{X}')^{-1}\boldsymbol{X}]y^*. \end{aligned}$$

In matrix form, using the relations $P = A'H$ and $H\tilde{\Sigma}^{*-1}H' = \tilde{\Sigma}_{\epsilon\epsilon}^{-1}$, we find that this GLS estimator can be written equivalently as

$$\begin{aligned} \tilde{B} &= (P\tilde{\Sigma}^{*-1}P')^{-1}P\tilde{\Sigma}^{*-1}\boldsymbol{Y}^*\boldsymbol{X}'(\boldsymbol{X}\boldsymbol{X}')^{-1} \\ &= (A'\tilde{\Sigma}_{\epsilon\epsilon}^{-1}A)^{-1}A'\tilde{\Sigma}_{\epsilon\epsilon}^{-1}\boldsymbol{Y}\boldsymbol{X}'(\boldsymbol{X}\boldsymbol{X}')^{-1}. \end{aligned} \tag{7.9}$$

Notice that this last expression in (7.9) is identical to the ML estimator of B for the growth curve model (6.1), as given in (6.2) of Section 6.2. Hence, we see that the basic growth curve model can be transformed and reexpressed as a particular SUR model (7.7), and that a corresponding two-step GLS estimation for the SUR model yields the ML estimate for the growth curve model. Since SUR models can be of much more general form than in (7.7), the connection displayed above between growth curve models and SUR models suggests that, perhaps, more general and flexible growth curve models could be formulated and analyzed through the methods of SUR models.

7.3 Reduced-Rank Coefficient in Seemingly Unrelated Regressions Model

Before we formally introduce the reduced-rank aspects for the SUR model (7.1), we want to discuss the need and usefulness of such a feature particularly in modeling large scale data sets. From the previous discussion thus far, it can be seen that the main idea is that the m regression equations in (7.1) are indeed related to each other and this relationship is captured through the correlations between the error terms. The aspect of reduced rank that we will propose here captures additional relationship in a more 'structured' way. To expand on this, consider the investment function example used by Zellner (1962). Denote I_t as the current gross investment of the firm, C_{t-1} as the firm's beginning of the year capital stock, and F_{t-1} as the value of the firm's outstanding shares at the beginning of the year. These variables are measured for the two electric companies annually from 1935 through 1954. Observe that each company has its own values of the predictor variables C_{t-1} and F_{t-1}, but they are conceptually the same type of variables. General Electric's gross investment, on the average, is about twice as large as that of Westinghouse. For these data, the GLS estimates obtained from (7.5) are given as

$$\text{General Electric}: \; I_{1t} = 0.139C_{1,t-1} + 0.038F_{1,t-1} - 27.7$$

$$\text{Westinghouse}: \; I_{2t} = 0.064C_{2,t-1} + 0.058F_{2,t-1} - 1.3\,.$$

The estimated correlation between the errors in the two equations is approximately 0.73. The standard errors of the regression coefficients from the SUR model approach are generally smaller than the standard errors of the ordinary LS coefficient estimates. Thus the SUR model approach provides more efficient estimates.

From the above estimated equations, we observe that although the response variable I_t is of different magnitude for the two companies, the regression coefficients of the predictor variables are roughly of the same magnitude. This raises the possibility that apart from the constant terms, the coefficients of certain of the predictor variables, for example F_{t-1}, could be constrained to be the same or more generally, the matrix formed from the coefficients of each SUR equation could be of reduced rank. The possible relations or constraints in regression coefficients across equations is somewhat suggested even in the two equation example, but we might expect such features to be more prominent when the number of SUR equations is larger. Thus the dependence among the SUR equations can be more directly modeled through dependency of the regression coefficient vectors from the different equations because the predictor variables are conceptually similar. While this dependency can be estimated mainly through the available data on the observed response and predictor variables, the dependency through correlations among the error terms as formulated by Zellner (1962) might be thought of as representing the unknown or unobservable common factors, possibly not accounted for by the available data or the regression component of the model (7.1). The approach we will take here has the features of the reduced-rank regression model first discussed in Chapter 2. It provides an opportunity for dimension reduction as well as a way to make interpretations of linear combinations of predictor variables as canonical variables in a certain sense. These features will be emphasized as we develop the model.

As a starting point for the development of the reduced-rank model, we restate the SUR model (7.1) as

$$\boldsymbol{Y}_{(i)} = \boldsymbol{X}_i' C_{(i)} + \epsilon_{(i)}, \qquad \text{for} \quad i = 1, \ldots, m.$$

Arranging the regression coefficient vectors $C_{(i)}$ in matrix form, we have $C = [C_{(1)}, C_{(2)}, \ldots, C_{(m)}]'$ as an $m \times n$ matrix. As noted previously, the model (7.1) becomes equivalent to the standard multivariate regression model when the predictor variables are the same for all m equations, that is, $\boldsymbol{X}_1 = \boldsymbol{X}_2 = \cdots = \boldsymbol{X}_m = \boldsymbol{X}$, with the equations $\boldsymbol{Y}'_{(i)} = C'_{(i)}\boldsymbol{X} + \epsilon'_{(i)}$ then being stacked to yield the matrix equation $\boldsymbol{Y} = C\boldsymbol{X} + \epsilon$ as in (1.4), where $\boldsymbol{Y} = [\boldsymbol{Y}_{(1)}, \ldots, \boldsymbol{Y}_{(m)}]'$ and $\epsilon = [\epsilon_{(1)}, \ldots, \epsilon_{(m)}]'$.

For the SUR equations model we shall assume that the $m \times n$ matrix C is of reduced rank. More formally, we assume that

$$\text{rank}(C) = r \leq \min(m, n). \tag{7.10}$$

Recall that an implication of this assumption is that C can be written as $C = AB$, where A is a full-rank matrix of dimension $m \times r$ and B is full rank of dimension $r \times n$. Then, because $C_{(i)} = B'a_{(i)}$ where $A' = [a_{(1)}, \ldots, a_{(m)}]$ and each $a_{(i)}$ is of dimension $r \times 1$, we can write the reduced-rank version of the SUR model (7.1) as

$$\boldsymbol{Y}_{(i)} = \boldsymbol{X}_i' B' a_{(i)} + \epsilon_{(i)}, \qquad i = 1, \ldots, m. \tag{7.11}$$

From the discussion in Chapter 2 it may be recalled that when the matrices $\boldsymbol{X}_i$ are the same, the linear combinations of $\boldsymbol{X}$ that are most useful for modeling the relationships with the response variables $\boldsymbol{Y}$ are given by $B\boldsymbol{X}$. Depending on the dimensions involved, substantial reduction in parameterization is possible, and $B\boldsymbol{X}$ also can be related to canonical variables. We see from (7.11) that such an idea can be extended to the more general multivariate regression system (7.1) and the coefficients (the matrix B) of a type of canonical variables are fixed to be the same for all m seemingly unrelated data sets. The distinctions among these regression equations are reflected in the coefficients $a_{(i)}$ which are allowed to differ among the regression equations in (7.1). Thus the reduced-rank regression approach to analyze the regression models in (7.1) provides a canonical method to link these models through the structure of the coefficients in addition to taking into account the correlations among the error terms.

We will focus on the estimation of the component matrices A and B and also on the subsequent estimation of the matrix C $(= AB)$. Similar to the previous reduced-rank model situations, we need to impose normalization conditions as in (2.14) to uniquely identify the parameters A and B. Because of the different matrices $\boldsymbol{X}_i$ that appear in model (7.1), natural normalization conditions for B are not obvious. Recall that we earlier argued that dimension reduction through reduced rank seemed sensible because of the 'similarity' of the predictor variable sets among the m different regression equations. Hence, for the normalization condition it appears reasonable to use the average of the second moment matrices of the predictor variables as follows:

$$A'\Gamma A = I_r \qquad \text{and} \qquad B\Sigma_{xx}B' = \Lambda_r^2, \tag{7.12}$$

where Γ is a positive-definite matrix which will typically be specified as $\Gamma = \Sigma_{\epsilon\epsilon}^{-1}$ and

$$\Sigma_{xx} = \lim_{T\to\infty} \frac{1}{mT} \sum_{i=1}^{m} \boldsymbol{X}_i \boldsymbol{X}_i'.$$

Note that these conditions correspond to (2.14) when $\boldsymbol{X}_i = \boldsymbol{X}$ for all i.

7.4 Maximum Likelihood Estimators for Reduced-Rank Model

We now consider derivation of efficient estimators of the coefficient matrices A and B in the reduced-rank SUR model under the normality assumption for the error terms in (7.1). Let $\theta = (\alpha', \beta')'$, where $\alpha = \text{vec}(A')$ and $\beta = \text{vec}(B')$. The criterion to be minimized can be stated as

$$S_T(\theta) = \frac{1}{2T} e'(\Gamma \otimes I_T) e \tag{7.13}$$

where $e = y - \overline{\boldsymbol{X}}\, c$, subject to the normalization conditions as in (7.12) that are based on sample versions of the normalizing matrices. We take $\Gamma = \tilde{\Sigma}_{\epsilon\epsilon}^{-1}$, the inverse of the error covariance matrix estimate based on LS residuals, and Σ_{xx} is replaced by $\hat{\Sigma}_{xx} = \frac{1}{mT} \sum_{i=1}^{m} \boldsymbol{X}_i \boldsymbol{X}_i'$. It appears that it is not possible to obtain an explicit solution for A and B simultaneously. Therefore, we consider the first order conditions for minimization of (7.13) from which one can solve for either one of the component matrices when the other is known.

Since $C = AB$, we have $c = \text{vec}(C') = (A \otimes I_n)\text{vec}(B') = (I_m \otimes B')\text{vec}(A')$, and so $e = y - \overline{\boldsymbol{X}}(A \otimes I_n)\beta = y - \overline{\boldsymbol{X}}(I_m \otimes B')\alpha$. Therefore, the first order conditions from (7.13) are given as

$$\frac{\partial S_T(\theta)}{\partial \alpha} = -\frac{1}{T}(I_m \otimes B)\overline{\boldsymbol{X}}'(\Gamma \otimes I_T)[\, y - \overline{\boldsymbol{X}}\,(I_m \otimes B')\alpha\,], \tag{7.14a}$$

$$\frac{\partial S_T(\theta)}{\partial \beta} = -\frac{1}{T}(A' \otimes I_n)\overline{\boldsymbol{X}}'(\Gamma \otimes I_T)[\, y - \overline{\boldsymbol{X}}\,(A \otimes I_n)\beta\,]. \tag{7.14b}$$

Introduce the notations $\overline{\boldsymbol{X}}(B) = \overline{\boldsymbol{X}}(I_m \otimes B') = \text{Diag}\{\boldsymbol{X}_1' B', \ldots, \boldsymbol{X}_m' B'\}$ and $\overline{\boldsymbol{X}}(A) = \overline{\boldsymbol{X}}(A \otimes I_n)$. Then the solutions to (7.14) for α and β, each in terms of the other parameter, can be expressed as

$$\hat{\alpha} = [\overline{\boldsymbol{X}}(B)'(\Gamma \otimes I_T)\overline{\boldsymbol{X}}(B)]^{-1}\overline{\boldsymbol{X}}(B)'(\Gamma \otimes I_T)y\,, \tag{7.15a}$$

$$\hat{\beta} = [\overline{\boldsymbol{X}}(A)'(\Gamma \otimes I_T)\overline{\boldsymbol{X}}(A)]^{-1}\overline{\boldsymbol{X}}(A)'(\Gamma \otimes I_T)y\,. \tag{7.15b}$$

It is easy to see that when $\boldsymbol{X}_1 = \cdots = \boldsymbol{X}_m = \boldsymbol{X}$, $\overline{\boldsymbol{X}}' = (I_m \otimes \boldsymbol{X})$ and the above equations reduce to the partial least squares equation results for the standard multivariate regression model case as given in Section 2.3. That is, we then have $\overline{\boldsymbol{X}}(B) = (I_m \otimes \boldsymbol{X}'B')$ and $\overline{\boldsymbol{X}}(A) = (A \otimes \boldsymbol{X}')$, so that the solutions in (7.15) reduce to $\hat{\alpha} = [I_m \otimes (B\boldsymbol{X}\boldsymbol{X}'B')^{-1}B\boldsymbol{X}]y$ and $\hat{\beta} = [(A'\Gamma A)^{-1}A'\Gamma \otimes (\boldsymbol{X}\boldsymbol{X}')^{-1}\boldsymbol{X}]y$, equivalently, $\hat{A} = \boldsymbol{Y}\boldsymbol{X}'B'(B\boldsymbol{X}\boldsymbol{X}'B')^{-1}$ and $\hat{B} = (A'\Gamma A)^{-1}A'\Gamma\boldsymbol{Y}\boldsymbol{X}'(\boldsymbol{X}\boldsymbol{X}')^{-1}$ with $y = \text{vec}(\boldsymbol{Y}')$.

In the above expressions we will take $\Gamma = \tilde{\Sigma}_{\epsilon\epsilon}^{-1}$, where $\tilde{\Sigma}_{\epsilon\epsilon}$ is obtained from the LS residuals as defined in Section 7.1. As noted in previous chapters, the first order equations in (7.14) refer only to necessary conditions for a minimum to occur, and so do not guarantee the global minimum. We suggest two computational schemes for solving the first order equations to obtain the efficient estimator.

The partial least squares procedure is perhaps computationally the easiest to implement. For a given estimate $\hat{\beta}$ or $\hat{B}$, the equation in (7.15a) is used to compute $\hat{\alpha}$ or $\hat{A}$ and similarly, given the estimate $\hat{\alpha}$, the equation (7.15b) is used to arrive at an estimate $\hat{\beta}$. Once these are computed, we use the equations

$$\hat{A}'\tilde{\Sigma}_{\epsilon\epsilon}^{-1}\hat{A} = I_r \qquad \text{and} \qquad \hat{B}\hat{\Sigma}_{xx}\hat{B}' = \Lambda_r^2 \tag{7.16}$$

to normalize the estimates, with $\hat{\Sigma}_{xx} = \frac{1}{mT}\sum_{i=1}^{m} \boldsymbol{X}_i\boldsymbol{X}_i'$. To perform the normalizations based on given unnormalized estimates $\hat{A}$ and $\hat{B}$, we first obtain the normalized eigenvectors $V_1, \ldots, V_r$ of $\hat{A}'\tilde{\Sigma}_{\epsilon\epsilon}^{-1}\hat{A}$ with corresponding eigenvalues $d_1, \ldots, d_r$, and define the $r \times r$ matrices $V = [V_1, \ldots, V_r]$ and $D = \mathrm{diag}(d_1, \ldots, d_r)$. Next we obtain the normalized eigenvectors $U_1, \ldots, U_r$ of the matrix $D^{1/2}V'\hat{B}\hat{\Sigma}_{xx}\hat{B}'VD^{1/2}$, with corresponding eigenvalues $\lambda_1^2, \ldots, \lambda_r^2$, and set $U = [U_1, \ldots, U_r]$. Then the normalized versions of the estimates of A and B are given by $\hat{A}_* = \hat{A}VD^{-1/2}U$ and $\hat{B}_* = U'D^{1/2}V'\hat{B}$, such that $\hat{C} \equiv \hat{A}\hat{B} = \hat{A}_*\hat{B}_*$, and $\hat{A}_*$ and $\hat{B}_*$ satisfy the normalization conditions (7.16). The partial least squares procedure then continues to iterate between the two solutions $\hat{\alpha}$ and $\hat{\beta}$, and at each step the above normalizations are imposed.

The second computational procedure is essentially a Newton–Raphson method based on the Taylor expansion of the first order equations. Thus, it is based on the same relations as given in (3.29) of Section 3.5.1 and in (4.11) of Section 4.3. Similar to the procedures given in Sections 3.5.1 and 4.4, the ith step of the iterative scheme is given by

$$\begin{bmatrix} \hat{\theta}_i \\ \hat{\lambda}_i \end{bmatrix} = \begin{bmatrix} \hat{\theta}_{i-1} \\ 0 \end{bmatrix} - \begin{bmatrix} W_{i-1} & Q_{i-1} \\ Q_{i-1}' & -I_{r^2} \end{bmatrix} \begin{bmatrix} \partial S_T(\theta)/\partial\theta|_{\hat{\theta}_{i-1}} \\ h(\theta)|_{\hat{\theta}_{i-1}} \end{bmatrix}, \tag{7.17}$$

where

$$W = (B_\theta + H_\theta H_\theta')^{-1}B_\theta(B_\theta + H_\theta H_\theta')^{-1} \tag{7.18}$$

and $Q = -(B_\theta + H_\theta H_\theta')^{-1}H_\theta$, with $H_\theta = \{\partial h_j(\theta)/\partial\theta_i\}$ where $h(\theta) = 0$ denotes the $r^2 \times 1$ vector of normalization conditions (7.12). In (7.18), the matrix B_θ is $B_\theta = \lim_{T\to\infty} \partial^2 S_T(\theta)/\partial\theta\partial\theta'$ as defined below in equation (7.20) in Theorem 7.1. Notice also that the vector of first partial derivatives in (7.17) can be expressed as

$$\frac{\partial S_T(\theta)}{\partial\theta} = -\frac{1}{T}M'\overline{\boldsymbol{X}}'(\Sigma_{\epsilon\epsilon}^{-1} \otimes I_T)e, \tag{7.19}$$

where $M = [(I_m \otimes B'), (A \otimes I_n)]$. The iterations are carried out until successive changes in $\hat{\theta}_i$ or $\partial S_T(\hat{\theta}_i)/\partial\theta$ are reasonably small.

We now state a result on the asymptotic properties of the efficient estimator. For this we take $\Gamma = \Sigma_{\epsilon\epsilon}^{-1}$ as known in (7.13) and (7.12), for convenience of presentation. Again, the theoretical results follow from the work of Robinson (1973) and Silvey (1959), and the proof is essentially of the same lines as those of Theorems 3.2 and 4.1. Therefore, details of the proof are omitted; we note only that the main asymptotic distributional result needed in the proof is that $\frac{1}{\sqrt{T}}\overline{\boldsymbol{X}}'(\Sigma_{\epsilon\epsilon}^{-1} \otimes I_T)e \xrightarrow{d} N(0, V)$ as $T \to \infty$ where V is defined below in Theorem 7.1 [e.g., see Srivastava and Giles (1987, p. 28)].

Theorem 7.1 Consider the model (7.1) under the stated assumptions, with $C = AB$, where A and B satisfy the normalization conditions (7.12), and let $h(\theta) = 0$ denote the $r^2 \times 1$ vector of normalization conditions obtaining by stacking the conditions (7.12) without duplication. Assume the limits $\lim_{T\to\infty} \frac{1}{T}\boldsymbol{X}_i\boldsymbol{X}_j' = \lim_{T\to\infty} \frac{1}{T}\sum_{k=1}^{T} X_{ik}X_{jk}' = \Omega_{ij}$, $i, j = 1, \dots, m$, exist almost surely and $\Sigma_{\epsilon\epsilon}$ is a positive-definite matrix. Let $\hat{\theta}$ be the estimator that minimizes the criterion (7.13) subject to conditions (7.12) for the choice $\Gamma = \Sigma_{\epsilon\epsilon}^{-1}$. Then as $T \to \infty$, $\hat{\theta} \to \theta$ almost surely and $\sqrt{T}(\hat{\theta} - \theta)$ has a limiting multivariate normal distribution with zero mean vector and (singular) covariance matrix given by the matrix $W = (B_\theta + H_\theta H_\theta')^{-1}B_\theta(B_\theta + H_\theta H_\theta')^{-1}$, as in (7.18), where

$$B_\theta = \lim_{T\to\infty} \frac{\partial^2 S_T(\theta)}{\partial\theta\partial\theta'} = M'VM, \qquad M = [(I_m \otimes B'), (A \otimes I_n)], \tag{7.20}$$

and $V = \{(V_{ij})\}$ is $mn \times mn$, where $V_{ij} = \sigma^{ij}\Omega_{ij}$ is an $n \times n$ matrix, σ^{ij} denotes the (i, j)th element of $\Sigma_{\epsilon\epsilon}^{-1}$, and $H_\theta = \{\partial h_j(\theta)/\partial\theta_i\}$ is of dimension $r(m + n) \times r^2$.

The asymptotic covariance matrix W can be used to make inferences on the parameters of the component matrices A and B based on the efficient estimators $\hat{A}$ and $\hat{B}$. The inferential aspects associated with the overall reduced-rank regression coefficient matrix estimator $\hat{C} = \hat{A}\hat{B}$ are also of interest. Because $\hat{C} - C = (\hat{A} - A)B + A(\hat{B} - B) + o_p(T^{-1/2})$, it follows as in previous cases that $\sqrt{T}\text{vec}(\hat{C}' - C') \xrightarrow{d} N(0, MWM')$, and $MWM' = M(M'VM)^{-}M'$. Inferences related to $\hat{C}$ can be based on this asymptotic distribution result.

7.5 An Alternate Estimator and Its Properties

The computation of the ML estimator in the reduced-rank SUR model requires iterative procedures, as noted in the previous section, and for good convergence of the procedures a reasonably accurate starting value is needed. Therefore, initial estimates of the component matrices A and B should be chosen with some care. In this section, we seek an alternate to the ML estimator that is computationally simpler and that can be used to specify initial values for the component matrices in the ML procedures. In addition, such an alternate estimator can be used in a more computationally convenient way than the ML estimator to initially specify or identify the appropriate rank of the coefficient matrix C through hypothesis testing procedures. As in the approach discussed for the model considered in Chapter 4, a full-rank estimator can be first computed and it may be used to help reveal any possible reduced-rank structure in the true coefficient matrix. Hence, we will construct an alternate estimator for the component matrices A and B based on the full-rank GLS estimator of the regression coefficient matrix C for model (7.1).

The motivation for the alternate estimator in the reduced-rank SUR model (7.1) is somewhat similar to the reasoning offered for the alternate estimator considered in Section 4.5 for the model of Chapter 4. Observe that the criterion (7.13), with the choice $\Gamma = \tilde{\Sigma}_{\epsilon\epsilon}^{-1}$, can be written as

$$S_T(\theta) = \frac{1}{2T}(y - \overline{\boldsymbol{X}}\,\tilde{c})'(\tilde{\Sigma}_{\epsilon\epsilon}^{-1} \otimes I_T)(y - \overline{\boldsymbol{X}}\,\tilde{c}) + \frac{1}{2T}(\tilde{c} - c)'\overline{\boldsymbol{X}}'(\tilde{\Sigma}_{\epsilon\epsilon}^{-1} \otimes I_T)\,\overline{\boldsymbol{X}}\,(\tilde{c} - c), \tag{7.21}$$

where $\tilde{c} = \{\overline{\boldsymbol{X}}'(\tilde{\Sigma}_{\epsilon\epsilon}^{-1} \otimes I_T)\,\overline{\boldsymbol{X}}\}^{-1}\overline{\boldsymbol{X}}'(\tilde{\Sigma}_{\epsilon\epsilon}^{-1} \otimes I_T)y$ is the full-rank GLS estimator in (7.5). Minimizing $S_T(\theta)$ with respect to A and B (with $C = AB$) is equivalent to minimizing the second term above subject to the normalization conditions in (7.16). As discussed in the previous section, the computational procedures to solve for the efficient estimates of the component matrices are iterative. However, in the special case of the classical multivariate regression setup, we have seen in Chapter 2 that the ML estimates of the component matrices can be calculated simultaneously and explicitly through computation of certain eigenvectors and eigenvalues. In this special case, this is possible because we then have $\overline{\boldsymbol{X}}' = (I_m \otimes \boldsymbol{X})$ and so the second term in (7.21) reduces to

$$\frac{1}{2T}(\tilde{c} - c)'(\tilde{\Sigma}_{\epsilon\epsilon}^{-1} \otimes \boldsymbol{X}\boldsymbol{X}')(\tilde{c} - c) = \frac{1}{2T}\text{tr}[\,\tilde{\Sigma}_{\epsilon\epsilon}^{-1}(\tilde{C} - AB)\boldsymbol{X}\boldsymbol{X}'(\tilde{C} - AB)'\,],$$

where $\tilde{c} = \text{vec}(\tilde{C}')$ and $\tilde{C} = \boldsymbol{Y}\boldsymbol{X}'(\boldsymbol{X}\boldsymbol{X}')^{-1}$. The Householder–Young Theorem can be used to arrive at explicit solutions for A and B, for example, as given in (2.17) of Section 2.3. These computational simplifications are possible mainly because in the criterion the matrix $\tilde{\Sigma}_{\epsilon\epsilon}^{-1} \otimes \boldsymbol{X}\boldsymbol{X}'$, which acts as a "weights matrix" for the vector $\tilde{c} - c$ that indicates the distance between the (full-rank) GLS estimator and the true parameters, is of Kronecker product form, whereas such simplifications do not occur in the SUR model since the matrix $\overline{\boldsymbol{X}}'(\tilde{\Sigma}_{\epsilon\epsilon}^{-1} \otimes I_T)\,\overline{\boldsymbol{X}}$ is of more complicated form.

For an alternate estimation procedure, we suggest that this latter matrix be replaced in the criterion by a weights matrix of similar form to that which appears in the classical multivariate regression setup. Specifically, in place of $\frac{1}{T}\boldsymbol{X}\boldsymbol{X}'$ in the classical case, we choose the moment matrix of the averages of the predictor variable sets,

$$\hat{\Sigma}_{\bar{x}\bar{x}} = \frac{1}{T}\bar{\boldsymbol{X}}\bar{\boldsymbol{X}}' = \frac{1}{m^2T}\sum_{i=1}^{m}\sum_{j=1}^{m}\boldsymbol{X}_i\boldsymbol{X}_j',$$

where $\bar{\boldsymbol{X}} = \frac{1}{m}\sum_{i=1}^{m}\boldsymbol{X}_i$, and consider use of the following criterion to be minimized:

$$\frac{1}{2}(\tilde{c}-c)'(\tilde{\Sigma}_{\epsilon\epsilon}^{-1}\otimes\hat{\Sigma}_{\bar{x}\bar{x}})(\tilde{c}-c) = \frac{1}{2}\text{tr}[\,\tilde{\Sigma}_{\epsilon\epsilon}^{-1}(\tilde{C}-AB)\,\hat{\Sigma}_{\bar{x}\bar{x}}\,(\tilde{C}-AB)'\,]. \quad (7.22)$$

It follows from the Householder–Young Theorem and related results in Chapter 2 that the alternate estimates $\tilde{A}$ and $\tilde{B}$, which are obtained by minimizing (7.22) and which satisfy similar normalization conditions as in (7.16) for the ML estimates, can be computed in the following way. The columns of $\hat{\Sigma}_{\bar{x}\bar{x}}^{1/2}\tilde{B}'$ are chosen to be the eigenvectors corresponding to the r largest eigenvalues of the matrix $\hat{\Sigma}_{\bar{x}\bar{x}}^{1/2}\tilde{C}'\tilde{\Sigma}_{\epsilon\epsilon}^{-1}\tilde{C}\hat{\Sigma}_{\bar{x}\bar{x}}^{1/2}$, and $\tilde{A} = \tilde{C}\hat{\Sigma}_{\bar{x}\bar{x}}\tilde{B}'\tilde{\Lambda}^{-2}$, where $\tilde{\Lambda}$ is a diagonal matrix with positive diagonal elements equal to the square roots of the eigenvalues of the matrix. Equivalently,

$$\tilde{A} = \tilde{\Sigma}_{\epsilon\epsilon}^{1/2}[\tilde{V}_1,\ldots,\tilde{V}_r] \quad \text{and} \quad \tilde{B} = [\tilde{V}_1,\ldots,\tilde{V}_r]'\tilde{\Sigma}_{\epsilon\epsilon}^{-1/2}\tilde{C},$$

where the $\tilde{V}_j$ are normalized eigenvectors of the matrix $\tilde{\Sigma}_{\epsilon\epsilon}^{-1/2}\tilde{C}\hat{\Sigma}_{\bar{x}\bar{x}}\tilde{C}'\tilde{\Sigma}_{\epsilon\epsilon}^{-1/2}$ corresponding to its r largest eigenvalues.

It is expected that this alternate procedure might result in estimators of A and B that have fairly high efficiency in cases where the matrices of regressors $\boldsymbol{X}_i$ have somewhat similar characteristics, in the sense that the terms $\frac{1}{T}\boldsymbol{X}_i\boldsymbol{X}_j'$ are moderately similar for all i and j, since then the "weights matrix" $\frac{1}{T}\overline{\boldsymbol{X}}'(\tilde{\Sigma}_{\epsilon\epsilon}^{-1}\otimes I_T)\,\overline{\boldsymbol{X}}$ might be well approximated by $\tilde{\Sigma}_{\epsilon\epsilon}^{-1}\otimes\hat{\Sigma}_{\bar{x}\bar{x}}$. Otherwise, the alternate estimators $\tilde{A}$ and $\tilde{B}$ might tend to not have good efficiency, but they still might have reasonable use as initial estimators. We also suggest the following improvement to this initial estimation procedure. Specifically, once the initial estimate $\tilde{B}$ has been obtained as indicated, the estimate of A is obtained by a (one-step) GLS estimation of the SUR model (7.11) for the $\boldsymbol{Y}_{(i)}$ on the constructed regressors $(\tilde{B}\boldsymbol{X}_i)'$, $i = 1,\ldots,m$. That is, we obtain the initial estimate of A using (7.15a), based on the initial estimate $\tilde{B}$, as

$$\tilde{\alpha} = [\overline{\boldsymbol{X}}(\tilde{B})'(\tilde{\Sigma}_{\epsilon\epsilon}^{-1}\otimes I_T)\overline{\boldsymbol{X}}(\tilde{B})]^{-1}\overline{\boldsymbol{X}}(\tilde{B})'(\tilde{\Sigma}_{\epsilon\epsilon}^{-1}\otimes I_T)y\,,$$

with $\overline{\boldsymbol{X}}(\tilde{B}) = \text{Diag}\{\boldsymbol{X}_1'\tilde{B}',\ldots,\boldsymbol{X}_m'\tilde{B}'\}$ and $\tilde{\alpha} = \text{vec}(\tilde{A}') = (\tilde{a}_{(1)}',\ldots,\tilde{a}_{(m)}')'$.

Our main interest in developing the alternate estimator is to provide initial values for the component matrices in computing the ML estimates, and also for use in preliminary specification of the rank, as will be discussed in Section 7.6. However, the asymptotic distribution of the alternate estimator might be of some interest for comparison with the efficient estimator. Because the details are similar to Theorem 7.1 and to results from Chapter 4, we only briefly mention the result. We let $\tilde{\theta} = (\tilde{\alpha}', \tilde{\beta}')'$, where $\tilde{\alpha} = \text{vec}(\tilde{A}')$ and $\tilde{\beta} = \text{vec}(\tilde{B}')$ with the alternate estimators $\tilde{A}$ and $\tilde{B}$ as given above. It can then be established that $\sqrt{T}(\tilde{\theta} - \theta)$ has a limiting multivariate normal distribution with zero mean vector and covariance matrix given by the leading $r(m+n)$-order square submatrix of the $r(m+n+r)$-order matrix

$$\begin{bmatrix} M'(\Sigma_{\epsilon\epsilon}^{-1} \otimes \Sigma_{\bar{x}\bar{x}})M & -H_\theta \\ -H_\theta' & 0 \end{bmatrix}^{-1} \times \begin{bmatrix} M'(\Sigma_{\epsilon\epsilon}^{-1} \otimes \Sigma_{\bar{x}\bar{x}})V^{-1}(\Sigma_{\epsilon\epsilon}^{-1} \otimes \Sigma_{\bar{x}\bar{x}})M & 0 \\ 0 & 0 \end{bmatrix} \times \begin{bmatrix} M'(\Sigma_{\epsilon\epsilon}^{-1} \otimes \Sigma_{\bar{x}\bar{x}})M & -H_\theta \\ -H_\theta' & 0 \end{bmatrix}^{-1},$$

where M, V, and H_θ are as defined before in Theorem 7.1, and

$$\Sigma_{\bar{x}\bar{x}} = \lim_{T\to\infty} \frac{1}{m^2 T} \sum_{i=1}^{m}\sum_{j=1}^{m} \boldsymbol{X}_i \boldsymbol{X}_j' = \frac{1}{m^2} \sum_{i=1}^{m}\sum_{j=1}^{m} \Omega_{ij}.$$

This covariance matrix is expressible more explicitly as

$$W^* = (B_\theta^* + H_\theta H_\theta')^{-1} M'(\Sigma_{\epsilon\epsilon}^{-1} \otimes \Sigma_{\bar{x}\bar{x}})V^{-1}(\Sigma_{\epsilon\epsilon}^{-1} \otimes \Sigma_{\bar{x}\bar{x}})M(B_\theta^* + H_\theta H_\theta')^{-1},$$

where $B_\theta^* = M'(\Sigma_{\epsilon\epsilon}^{-1} \otimes \Sigma_{\bar{x}\bar{x}})M$. Note that the matrix V is the limiting form of T times the covariance matrix in (7.4). Although the distinction among the different matrices of predictor variables $\boldsymbol{X}_i$ is not fully accounted for in construction of the component matrix estimates under the alternate procedure, the asymptotic covariance matrix of the alternate estimator $\tilde{\theta}$ does reflect the distinctness of the $\boldsymbol{X}_i$ through the matrix V.

7.6 Identification of Rank of the Regression Coefficient Matrix

To carry out the computations related to estimating the component matrices A and B through the maximum likelihood method as outlined in Section 7.4, the rank of the regression coefficient matrix C needs to be determined. In the classical multivariate linear regression model situation, where all $\boldsymbol{X}_i$ in model (7.1) are equal, the specification of rank could be conveniently carried out because of the correspondence between the number of nonzero canonical correlations and the rank of the regression matrix. For this case, we have used the LR statistics

$$-[(T-n)+\frac{1}{2}(n-m-1)]\sum_{j=r+1}^{m}\log(1-\hat{\rho}_j^2),$$

where the $\hat{\rho}_j^2$ are the squared sample canonical correlations. This form of test does not apply in the SUR model, since each response variable has its own set of predictor variables. We therefore need to develop some large sample procedures to test the hypothesis of the rank of C.

Before we further consider the problem of testing for the rank of the matrix C, we want to briefly comment on procedures for testing usual linear hypotheses for coefficients in the SUR model (7.1). Typically, the model is considered in the form given by (7.2) and the linear hypothesis constraints are stated as $Rc = 0$, where R is a known matrix. A particular application of interest is testing for aggregation bias as discussed by Zellner (1962). Recall that the model (7.1) may often refer to data relating to several different micro-units. Simple aggregation of data from these units into macro-unit data, and subsequent estimation of the aggregate model, will not lead to any aggregation bias if the hypothesis $H_0 : C_{(1)} = C_{(2)} = \cdots = C_{(m)}$ holds, or equivalently, $Rc = 0$ where R is an appropriately defined known $(m-1)n \times mn$ matrix. The usual LR procedures can be applied to test such hypotheses. The linear constraints imposed by a reduced-rank condition are more general than the above equality of coefficients hypothesis, however. If the above hypothesis holds, it implies that the rank of the matrix C is one; the reverse implications are somewhat more general than the coefficient vectors being identical (i.e., they are proportional). Examining the investment function example described in Section 7.3 for aggregation bias, based on the LR test Zellner (1962) concluded that the hypothesis of coefficient vector equality is rejected. Note in our earlier discussion of this example that the coefficients of some predictor variables, not necessarily all, might be constrained to be the same implying that the regression coefficient matrix could be of lower rank. Thus, interest in testing for the rank of the regression coefficient matrix may stem from other than purely statistical modeling considerations. We now discuss procedures for testing the rank of the matrix C of regression coefficients.

We can still consider the LR statistic for testing $H_0 : \text{rank}(C) \leq r$. In general terms, this is given as

$$M = [(T-n)+\frac{1}{2}(n-m-1)]\log(|\hat{\Sigma}_{\epsilon\epsilon}^{(r)}|/|\hat{\Sigma}_{\epsilon\epsilon}|), \tag{7.23}$$

where $\hat{\Sigma}_{\epsilon\epsilon}$ denotes the unrestricted ML estimate of $\Sigma_{\epsilon\epsilon}$ and $\hat{\Sigma}_{\epsilon\epsilon}^{(r)}$ denotes the restricted estimate based on the reduced-rank ML estimate $\hat{C} = \hat{A}\hat{B}$ of C, subject to the condition that $\text{rank}(C) \leq r$. The statistic M asymptotically will follow a $\chi^2_{(m-r)(n-r)}$ distribution under the null hypothesis. The test statistic could be computed for various values of r and used to test the validity of each rank assumption. Because this procedure would require the somewhat involved calculations for the ML estimates $\hat{A}$ and $\hat{B}$ to obtain $\hat{\Sigma}_{\epsilon\epsilon}^{(r)}$, for a range of possible ranks r, at the initial specification stage it would not be a convenient procedure for selection of an appropriate rank for C, prior to computing the ML estimate of C. Therefore, for initial specification of possible rank, we also suggest using an alternate testing procedure based on the alternate estimator of C described in Section 7.5. The numerical example presented in the next section indicates that the alternate estimation procedure is somewhat useful for initial testing of rank, as well as for obtaining initial values in iterative computation of the ML estimates of A and B.

The alternate test statistic that we consider is thus $M^* = [(T-n) + \frac{1}{2}(n-m-1)] \log(|\tilde{\Sigma}_{\epsilon\epsilon}^{(r)}|/|\tilde{\Sigma}_{\epsilon\epsilon}|)$, where $\tilde{\Sigma}_{\epsilon\epsilon}^{(r)}$ is the error covariance matrix estimate constructed, in the usual way, from the residual vectors $\hat{\epsilon}_{(i)} = \boldsymbol{Y}_{(i)} - \boldsymbol{X}_i'\tilde{C}_{(i)}^{(r)}$, $i = 1, \ldots, m$, where $\tilde{C}^{(r)} = [\tilde{C}_{(1)}^{(r)}, \ldots, \tilde{C}_{(m)}^{(r)}]' = \tilde{A}^{(r)}\tilde{B}^{(r)}$ and $\tilde{A}^{(r)}$ and $\tilde{B}^{(r)}$ are the alternate estimates of A and B described in Section 7.5. Another possible and convenient alternate test statistic that is motivated by the alternate estimation procedure in Section 7.5 is of the form $M^* = [(T-n) + \frac{1}{2}(n-m-1)] \sum_{j=r+1}^{m} \log(1+\tilde{\lambda}_j^2)$, where the $\tilde{\lambda}_j^2$ are the eigenvalues of the matrix $\tilde{\Sigma}_{\epsilon\epsilon}^{-1/2}\tilde{C}\hat{\Sigma}_{\bar{x}\bar{x}}\tilde{C}'\tilde{\Sigma}_{\epsilon\epsilon}^{-1/2}$, $\hat{\Sigma}_{\bar{x}\bar{x}} = \frac{1}{m^2T}\sum_{i=1}^{m}\sum_{j=1}^{m} \boldsymbol{X}_i\boldsymbol{X}_j'$, and $\tilde{C}$ is the full-rank GLS estimator obtained from (7.5). Although this test statistic does not involve any quantities that are canonical correlations, because of the analogue to the classical multivariate linear model case we may interpret $\tilde{\rho}_j^2 = \tilde{\lambda}_j^2/(1+\tilde{\lambda}_j^2)$ as 'canonical correlation' quantities. The values $\tilde{\rho}_j^2$ might be easier to interpret as descriptive tools than the values $\tilde{\lambda}_j^2$. We thus suggest using the above statistic M^* and the reference $\chi^2_{(m-r)(n-r)}$ distribution to initially test the hypothesis of the rank of C, for various values of r. This alternate 'approximate' test procedure will tend to be conservative, because the determinant of the estimated error covariance matrix obtained using the alternate estimator will be larger than that obtained using the ML estimator of A and B.

7.7 A Numerical Illustration With Scanner Data

We illustrate the procedures for the seemingly unrelated regressions model developed in this chapter with an application in the area of marketing. Typically, manufacturers and retailers have to make, on a continual basis, important marketing decisions regarding the promotion and pricing of products. They make these decisions based on elasticity estimates related to various promotional activities and the price discounts. Now that scanner data of weekly sales for retail chains are readily available these elasticities can be easily computed. But these extensive data do not guarantee reliable estimates for individual product brands and generally yield a large set of estimates that may be difficult to fully comprehend. We attempt to address these problems through the reduced-rank SUR models developed in this chapter.

The point-of-sale data are collected at the store level for all the brands in a product category for a weekly time interval. In large cities, a retail chain typically owns several stores. For our illustrative example, we aggregate the store level data into data for the chain and compute the elasticity estimates, because important marketing negotiations related to trade discounts and promotions are usually made between the manufacturers and the retail chains; the retail chains then decide on marketing plans at the store level.

The product whose data we consider in this analysis is a refrigerated product with several brands; the data are for a large chain located in a highly populated city in northeastern United States. The weekly data span the period from mid-September 1988 to mid-June 1990, covering 93 weeks. In addition to the number of units sold, for each brand we consider data on regular price, actual price, whether promotion was present and whether the brand was on display. With the information available, we consider the following model, similar to the model advocated by Blattberg and George (1991) and Blattberg, Kim, and Ye (1994), as appropriate for the store level data analysis:

$$S_t = \beta_0 + \beta_1 S_{t-1} + \beta_2 P_t + \beta_3 PD_t + \beta_4 PR_t + \beta_5 D_t + \beta_6 DD_t + \beta_7 CPD_t + \epsilon_t \tag{7.24}$$

where

S_t = logarithm of number of units sold in tth week
P_t = regular price
PD_t = price discount = (regular price − actual price)/regular price
PR_t = indicator for promotions
D_t = indicator for display
DD_t = deal decay code = k if deal is present in $(k+1)$st week
CPD_t = maximum price discount for competing brands in a chain

The 'semi-log' model (7.24) has also been used by Blattberg and Wisniewski (1989) and is known to fit better than log-log or linear models. While the regression coefficients of a log-log model can be directly interpreted as elasticities, the coefficients of a semi-log model can be converted into point elasticities.

There are some broad assertions from the marketing literature that can guide us to interpret the coefficients of the above model. Three promotional variables, PR_t, D_t, and DD_t, are included in the model. Although the promotion and display variables had several categories, we use them as binary indicator variables because of lack of variation in the actual occurrence of these categories. The deal decay index is supposed to reflect the waning effectiveness of promotions. It is postulated that the first weeks of promotions are far more effective than later weeks. Displays are expected to increase sales because the retail space available is limited and expensive; they are usually reserved for high volume items. Two variables are included to measure the price effect, the regular price and price discount. It is believed that deal elasticities are generally higher than price elasticities. The effect of competition can be measured through their promotional and price cross-elasticities but here we chose only the competitors' price discount as a variable that may play a significant role. There is some discussion on deciding on which brands compete with each other but it generally is agreed that brands with similar prices tend to compete with each other. Brand switching based on price discounts occurs mostly toward higher priced brands; if the lower-tier brand promotes, it does not usually attract customers from higher-tier brands. The lagged value of S_t is included in (7.24) to account for serial correlation.

Because we consider the chain level data, the variables in (7.24) are all aggregated over the stores within that chain. The response variable S_t at the chain level is the logarithm of total sales aggregated over the stores, P_t is the average price, PD_t is the average price discount, PR_t is the proportion of stores promoting, D_t is the proportion of stores displaying, DD_t is the deal decay code averaged over stores, and CPD_t is the average price discount offered by the competitors. To get some understanding of the nature of the data, Figure 7.1 displays time series plots of S_t, log sales for the $m = 8$ brands. For all but brand 6, we notice that the sales generally peak during the summer weeks, especially around July 4th. In Figure 7.2 we present plots of the predictor variables for brand 1 only, to avoid repetition. The regular price is a nondecreasing step function with relatively few jumps over the duration. The plots show that price discounts and promotions are fairly frequent, and their peaks tend to match peaks in log sales, whereas there was no display activity for brand 1 over this time period. Also, for most weeks, one or more competitors' brands offer price discounts.

All variables are adjusted by subtraction of sample means before we proceed with the analysis. The ordinary LS estimates of the regression model (7.24) were first computed for each of the $m = 8$ brands with $n = 7$ predictor variables, based on $T = 92$ observations, because of the lag term included in the model. The correlations of the residuals across equations are all positive and range from 0.1 to 0.6, indicating similarity of the sales of different brands over time in the product market. This suggests the possibility that SUR equations estimates might be desirable. Both the two-step and the ML estimates of the SUR model were computed. The determinants of the residual covariance matrices are 0.02101, 0.01244, and 0.01141 ($\times 10^{-8}$) for the LS, two-step SUR, and ML SUR estimation, respectively. The (full-rank) ML estimate of the coefficient matrix C is presented in Table 7.1, with expected signs of the coefficients for individual predictor variables indicated in parentheses in the column headings; an asterisk indicates that the coefficient estimate is significant at the 5% level.

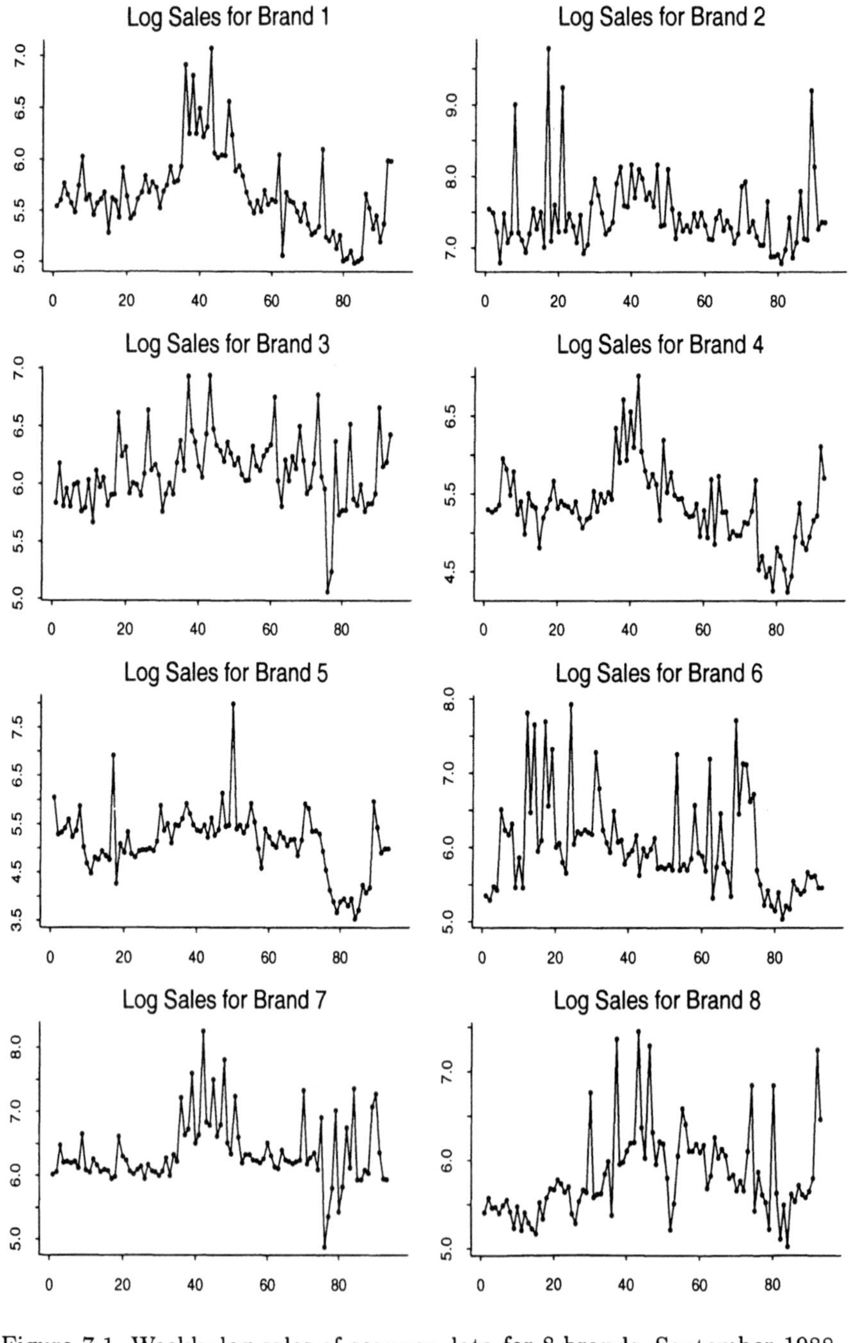

Figure 7.1. Weekly log sales of scanner data for 8 brands, September 1988 – June 1990

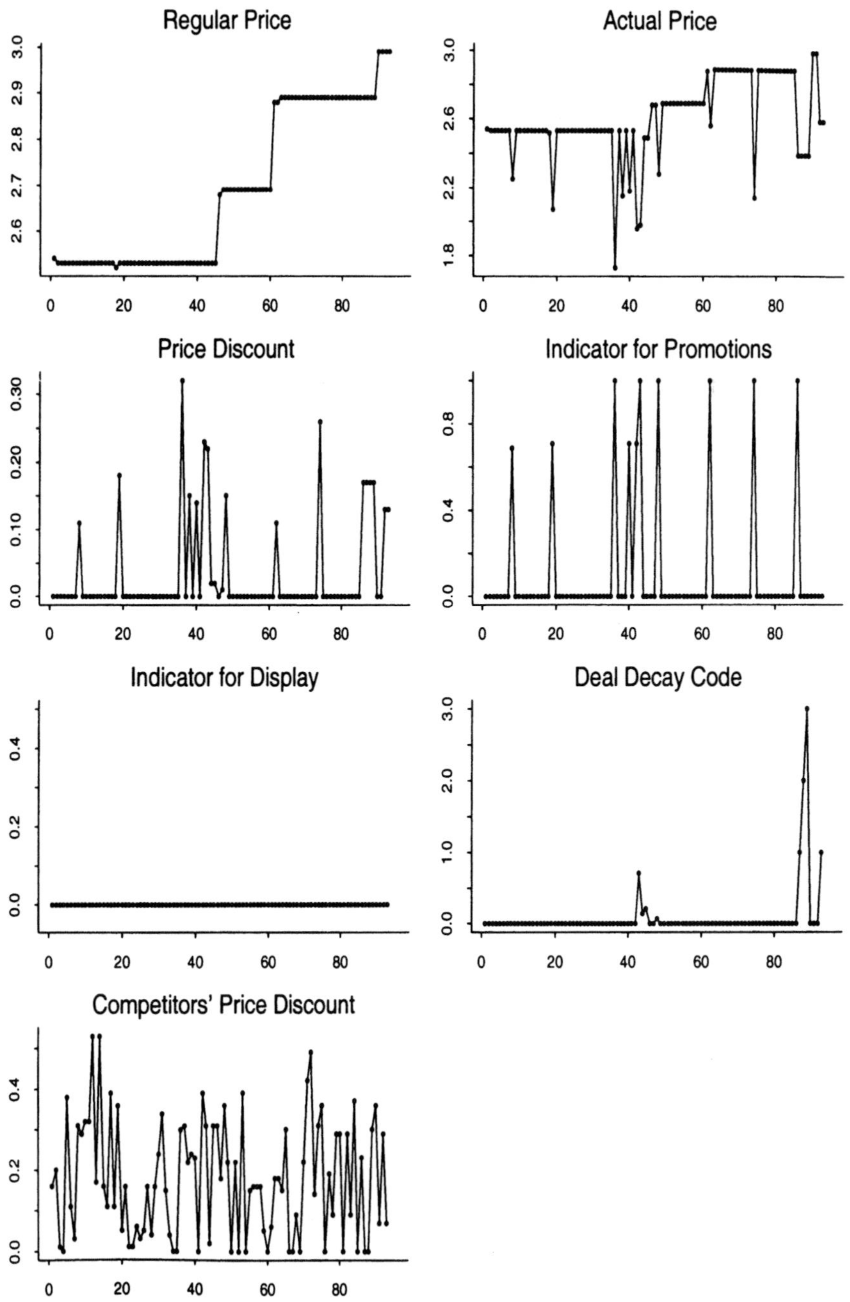

Figure 7.2. Weekly predictor variable values for log sales of scanner data for brand 1, September 1988 – June 1990

Table 7.1 Full-rank ML estimates of coefficient matrix C in SUR model for the scanner data.

Brand	Lag Sales (+)	Price (−)	Price Disc (+)	Prom (+)	Disp (+)	DD (−)	CPD (−)
1	0.358*	−0.795*	2.597*	0.213	—	−0.161*	−0.048
2	0.127*	−0.816	0.617	0.587*	2.814*	−0.104	0.304
3	0.041	−0.584*	−1.786*	0.699*	—	0.014	0.166
4	0.311*	−2.736*	2.753*	0.173	—	−0.924	0.150
5	0.244*	−4.696*	1.817*	0.372*	5.646*	−0.047	−0.058
6	0.040	−0.865*	2.367*	0.495*	7.403*	−0.263*	−0.642*
7	0.054	−1.973*	3.136*	0.211	3.111*	−0.116*	−0.164
8	0.292*	−0.621	2.262	0.709	4.633	−2.140	−0.537*

For brands 1, 3, and 4, because of no activity in display area, the coefficients for the display indicator D_t cannot be estimated and they are taken to be 'missing'. Each type of predictor variable exhibits significance for one or more brands, confirming the usefulness of including them in the model. The brands that have nonsignificant coefficients for lagged sales tend to be influenced by promotions. All significant coefficients have the correct expected signs except for the price discount for brand 3. We are not able to explain this anomaly, but observe that having only one 'unreliable' estimate among such a large number of estimates might be taken as satisfactory (see Blattberg and George, 1991, for related discussion).

To proceed with reduced-rank analysis under the SUR model framework, using the initial alternate estimation procedure, we need to have the full-rank estimate of C with no 'missing' elements. Because the display variable is significant for most brands and is potentially important from the marketing area point of view, we want to entertain the possibility of accommodating this predictor variable in the model for all brands. While there are various options to fill in the missing estimates, we need to be sure that the procedure does not unduly affect the estimation of component matrices in the reduced-rank model. We have considered two options: (1) substituting the average value of the available display coefficient estimates for the missing values; (2) because brands tend to compete within a price tier, substituting the available competitor's display variable coefficient estimate for the missing value. Specifically, brands 1 and 2 are in the top price tier, 3, 4, and 5 are in the mid price tier, and brands 6, 7, and 8 are in the low price tier. Since the estimation results were very similar using both options, we will follow option (2) and present the corresponding analysis and results.

As suggested in Section 7.6, we use the results from Section 7.5 that are related to the alternate estimator to obtain information for an initial specification of the rank of C and initial estimates of the coefficient matrices A and B in the reduced-rank SUR model. The nonzero eigenvalues $\tilde{\lambda}_j^2$ of $\tilde{\Sigma}_{\epsilon\epsilon}^{-1/2}\tilde{C}\hat{\Sigma}_{\bar{x}\bar{x}}\tilde{C}'\tilde{\Sigma}_{\epsilon\epsilon}^{-1/2}$ are found to be 2.740, 1.784, 0.691, 0.183, 0.118, 0.067, and 0.005. The corresponding squared 'canonical correlation' quantities $\tilde{\rho}_j^2$ are 0.733, 0.641, 0.409, 0.155, 0.106, 0.062, and 0.005, which may provide a better feel for the possible rank. Formally, using the approximate LR test procedure mentioned in Section 7.6 gives the impression that the rank might need to be taken as large as $r = 4$ or $r = 5$. However, because of the conservative nature of this test procedure and the desire to explore a more parsimonious model, we further investigate the reduced-rank model with specification of lower ranks.

Thus, we examine ML fitting of models of various ranks. The procedure followed in calculation of the ML estimates for the model of each different rank is the partial least squares procedure described in Section 7.4. The BIC criterion is used to select a model of suitable rank, and the results for the BIC as well as the AIC criterion are presented in Table 7.2. It appears from these results that a model of rank 2 might be appropriate based on BIC, although the use of the AIC criterion would again require the use of model with rank as high as 5.

Table 7.2 Summary results on ML estimation of SUR models of various ranks fitted to the scanner data.

Rank	$\lvert\hat{\Sigma}_{\epsilon\epsilon}\rvert \times 10^8$	n^* = number of parameters	AIC	BIC
1	0.081145	14	−2.207	−1.823
2	0.031958	26	−2.878	−2.165
3	0.021850	36	−3.041	−2.054
4	0.014791	44	−3.257	−2.051
5	0.012370	50	−3.305	−1.935
6	0.011406	54	−3.299	−1.819
7	0.011406	56	−3.256	−1.721

The ML estimates of the component matrices B and A in the model of rank $r = 2$ are

$$\hat{B} = \begin{bmatrix} 1.0226 & -15.0856 & 18.5987 & -0.1566 & 4.7584 & -0.4678 & -0.7757 \\ 0.5840 & 8.3231 & -4.0433 & 5.0278 & 31.8771 & -0.4947 & -2.0161 \end{bmatrix},$$

$$\hat{A} = \begin{bmatrix} 0.1184 & 0.0945 \\ 0.0770 & 0.1166 \\ 0.0525 & 0.0736 \\ 0.1956 & 0.0008 \\ 0.2095 & 0.0740 \\ 0.1207 & 0.1504 \\ 0.1658 & 0.0510 \\ 0.1273 & 0.1859 \end{bmatrix}.$$

The resulting ML estimate $\hat{C} = \hat{A}\hat{B}$ is displayed in Table 7.3. Comparing elements of the rank 2 ML estimate of C with those of the full-rank estimate given in Table 7.1, we can observe that most of the coefficient estimates that were found to be significant under the full-rank estimation are well recovered within the rank 2 model specification, with the rank 2 model having less than one-half the number of parameters as the full-rank model. Also note that the price discount coefficient estimate for brand 3, which had the 'incorrect' sign in the full-rank estimation, now attains the correct sign in the model of rank 2.

Table 7.3 ML estimates of coefficient matrix $C = AB$ in reduced-rank SUR model of rank $r = 2$ for the scanner data.

Brand	Lag Sales (+)	Price (−)	Price Disc (+)	Prom (+)	Disp (+)	DD (−)	CPD (−)
1	0.176	−0.998	1.819	0.458	3.584	−0.102	−0.283
2	0.146	−0.191	0.960	0.574	4.084	−0.094	−0.295
3	0.096	−0.179	0.679	0.362	2.596	−0.061	−0.189
4	0.205	−2.883	3.605	0.010	1.188	−0.096	−0.168
5	0.257	−2.544	3.597	0.339	3.356	−0.135	−0.312
6	0.211	−0.569	1.637	0.737	5.368	−0.131	−0.397
7	0.199	−2.077	2.878	0.230	2.412	−0.103	−0.231
8	0.238	−0.374	1.616	0.915	6.532	−0.152	−0.474

From the ML estimate $\hat{B}$ it is interesting to note that the most important linear combination of the predictor variables (ignoring coefficients with small magnitudes) is $S_{t-1} - 15.1P_t + 18.6PD_t + 4.8D_t$. This linear combination might be interpreted to represent a market factor comprised of price and display variables. The price discount variable PD_t receives a larger weight than the display variable D_t. From the ML estimates in the first column of $\hat{A}$, we see that this linear combination has the highest impact on sales for brands 4, 5, and 7, medium impact for brands 1, 6, and 8, and smallest impact for brands 2 and 3. The second predictive linear combination includes a large contribution from the display variable D_t, and both the promotions variable PR_t and the competitors' price discount variable CPD_t enter more substantially than in the first linear combination. This predictive combination seems to have greater influence on brands 8, 6, and 2. To make more serious interpretations of the reduced-rank estimation results for these data, it would be necessary to delve into the extensive marketing literature. A goal here was to illustrate how certain canonical and reduced-rank procedures can be used for the SUR model setup. Further discussion related to interpretation of the estimation results will not be given here.

As a summary of the analysis results, we mention that the values of the ML residual variance estimates of log sales for the 8 brands under the full-rank SUR model fit were 0.0389, 0.1311, 0.0518, 0.0622, 0.2337, 0.1717, 0.0637, and 0.0625, whereas the corresponding values under the reduced-rank ($r = 2$) fit were 0.0520, 0.1318, 0.0620, 0.0674, 0.2670, 0.1720, 0.0593, and 0.0665. It is observed that the magnitudes of the residual variance estimates under the rank 2 model are comparable to those of the full-rank model. Taking into account the considerable reduction in the number of parameters that need to be estimated for the rank 2 model, slight increases in residual variances may be a small price to pay. An additional appeal of reduced-rank modeling of scanner data is that it provides a way to summarize, for managers, the large number of regression coefficient (elasticity) estimates through the 'canonical quantities' and these canonical quantities can also be used to keep track of market changes over time. Thus, the interpretation of large amounts of data can be simplified for managerial use.

8
Applications of Reduced-Rank Regression in Financial Economics

8.1 Introduction to Asset Pricing Models

In previous chapters, we have developed reduced-rank regression models of various forms, which have wide applications in a variety of contexts in the physical and social sciences. In this chapter we focus on the area of financial economics, where several applications of the reduced-rank regression models arise in a fairly natural way. Thus, the topics that will be examined in this chapter involve consideration of the contexts in which financial models arise from economic theories, the form of the models, and the empirical verification of these models through reduced-rank regression methods. In earlier chapters, the application of reduced-rank regression methods was mainly motivated from an empirical dimension-reduction aspect, whereas the use of reduced-rank models presented in this chapter results from a priori economic theory. With the high volume of financial data that are now routinely becoming available, these theories can be examined through tests of the models. We point out that, for the most part, only the basic reduced-rank methods developed in Chapter 2 are needed for the presentations considered here.

Of fundamental interest to financial economists is to examine the relationship between the risk of a financial security and its return. While it is obvious that risky assets can generally yield higher returns than risk-free assets, a quantification of the tradeoff between risk and expected return was made through the development of the Capital Asset Pricing Model (CAPM), for which the groundwork was laid by Markowitz (1959). A central feature of the CAPM is that the expected return is a linear function of the risk. The risk of an asset typically is measured by the covariability between its return and that of an appropriately defined 'market' portfolio. Examples of expected return-risk model relationships include the Sharpe (1964) and Lintner (1965) CAPM, the zero-beta CAPM of Black (1972), the arbitrage pricing theory (APT) due to Ross (1976), and the intertemporal asset pricing model by Merton (1973). The economy-wide models developed by Sharpe (1964), Lintner (1965), and Black (1972) are based on the work by Markowitz (1959) which assumes that investors would hold a mean-variance efficient portfolio. They showed that if investors have homogeneous expectations and hold efficient portfolios, then the market portfolio will itself be a mean-variance efficient portfolio. The CAPM equation (8.3), given below, is a direct implication of the mean-variance efficiency of the market portfolio. The main difference between the works of Sharpe (1964) and Lintner (1965) and the work of Black (1972) is that the former assume the existence of a risk-free lending and borrowing rate whereas the latter derived a more general version of the CAPM in the absence of a risk-free asset. Initially, we focus on the work of Black (1972) as it readily fits into the reduced-rank regression framework.

Gibbons (1982) developed the Black (1972) CAPM hypothesis through the multivariate regression model. Suppose there are m assets whose returns are observed over T time periods, and let r_{ik} denote the return on asset i in period k. The model is specified as follows:

$$r_{ik} = \alpha_i + \beta_i x_k + \epsilon_{ik}, \qquad i = 1, \ldots, m, \;\; k = 1, \ldots, T, \tag{8.1}$$

where x_k denotes the return on the market portfolio in period k. The coefficient β_i in (8.1) is the 'risk' of the ith asset, equal to

$$\beta_i = \mathrm{Cov}(r_{ik}, x_k)/\mathrm{Var}(x_k), \qquad i = 1, \ldots, m,$$

and the ϵ_{ik} are assumed to have zero mean and be uncorrelated over time, with the random vector $\epsilon_k = (\epsilon_{1k}, \ldots, \epsilon_{mk})'$ having positive-definite $m \times m$ covariance matrix $\Sigma_{\epsilon\epsilon} = \mathrm{Cov}(\epsilon_k)$. The above equations (8.1) can be written in the form of a multivariate regression model, as in Chapter 1, and represent "a statistical statement rather than one derived from financial theory". With the zero mean assumption on the errors in (8.1), we have

$$E(r_{ik}) = \alpha_i + \beta_i E(x_k), \qquad i = 1, \ldots, m. \tag{8.2}$$

The Black (1972) CAPM requires, for the mean-variance efficiency of the market portfolio, that the following relationship holds between the expected returns and the risks:

$$E(r_{ik}) = \gamma_0 + \beta_i [E(x_k) - \gamma_0], \qquad i = 1, \ldots, m, \tag{8.3}$$

where γ_0 is the expected return on the 'zero-beta' portfolio or any portfolio whose return is not correlated with the return on the market portfolio.

The zero-beta portfolio is defined to be the portfolio whose returns have the minimum variance of all portfolios whose returns are uncorrelated with the returns of the market portfolio. Comparing (8.2) with (8.3), we see that (8.3) implies that the intercepts of the model (8.1) must satisfy the constraints

$$\alpha_i = \gamma_0(1-\beta_i), \quad \text{for all} \quad i = 1, \ldots, m, \tag{8.4}$$

which is used as a basis for empirical testing of the CAPM model, that is, the efficiency of the market portfolio. A similar formulation of the problem of testing the efficiency of the market (or any specified) portfolio is obtained by Gibbons, Ross, and Shanken (1989), but based on the assumptions of the Sharpe (1964) and Lintner (1965) CAPM.

To connect the constraints (8.4) with a reduced-rank feature, we first observe that the model (8.1) can be written in a slightly different form as

$$y_{ik} \equiv r_{ik} - x_k = \alpha_i + (\beta_i - 1)x_k + \epsilon_{ik}.$$

We will then express this in the form of the multivariate regression model in (1.1). Define $Y_k = (r_{1k} - x_k, \ldots, r_{mk} - x_k)'$ and $X_k = (1, x_k)'$, where Y_k is the $m \times 1$ vector of 'excess' returns on the m assets. Further, write the $m \times 2$ coefficient matrix as

$$C = \begin{bmatrix} \alpha_1 & \alpha_2 & \cdots & \alpha_m \\ \beta_1 - 1 & \beta_2 - 1 & \cdots & \beta_m - 1 \end{bmatrix}' \equiv [\alpha, \, \beta - \mathbf{1}_m],$$

where $\mathbf{1}_m$ is an $m \times 1$ vector of ones. Now the model can be restated as the multivariate regression model

$$Y_k = CX_k + \epsilon_k, \qquad k = 1, \ldots, T, \tag{8.5}$$

where ϵ_k is the $m \times 1$ vector of random errors, with $E(\epsilon_k) = 0$ and $\text{Cov}(\epsilon_k) = \Sigma_{\epsilon\epsilon}$. Notice that under the constraints (8.4), the coefficient matrix C in (8.5) is expressible as

$$\begin{aligned} C = [\alpha, \, \beta - \mathbf{1}_m] &\equiv [\gamma_0(\mathbf{1}_m - \beta), \, \beta - \mathbf{1}_m] \\ &= [\beta - \mathbf{1}_m]\,[-\gamma_0, \, 1] = \beta^*\gamma', \end{aligned} \tag{8.6}$$

where $\gamma' = [-\gamma_0, \, 1]$ and $\beta^* = \beta - \mathbf{1}_m$. Thus, the matrix C is of reduced rank one, although there is some slight difference in the reduced-rank setup presented here compared to the approach for the model discussed in Chapter 2. While the form of (8.6) is similar to the form $C = AB$ in (2.4), the right-hand factor γ' in this setup (i.e., $C = \beta^*\gamma'$) is structured in that a normalization is directly incorporated into the second element of the vector γ. Recall, also, that the reduced-rank restriction on C implies that there is one constraint on the matrix C which can be stated as

$$C\gamma^* = 0 \tag{8.7}$$

where $\gamma^* = (1, \gamma_0)'$ is 'structured' as well. A main difference between (8.7) and (2.2) is that the constraints in (2.2) operate on the rows of the $m \times n$ matrix C whereas the above constraints operate on the columns of the $m \times 2$ matrix C. This difference in emphasis is due mainly to the convention adopted in Chapter 2, assumed for convenience of exposition, that $m \leq n$, while $m > n = 2$ in the present situation. For either case, methods developed in Chapters 1 and 2 are useful here.

8.2 Estimation and Testing in the Asset Pricing Model

Gibbons (1982) suggested the following iterative procedure for estimation of the coefficient matrix C in (8.5), subject to the constraints (8.4), which is very much in the spirit of the partial least squares method. Let $\boldsymbol{Y}$ and $\boldsymbol{X}$ denote the $m \times T$ and $2 \times T$ data matrices, respectively.
(i) In the first step, obtain the ordinary LS estimator $\tilde{C} = \boldsymbol{YX}'(\boldsymbol{XX}')^{-1}$ as given in (1.7), which is not constrained by (8.4). This yields the LS estimates $\tilde{\alpha}$ and $\tilde{\beta}$ of α and β. Then obtain $\tilde{\Sigma}_{\epsilon\epsilon} = (1/T)(\boldsymbol{Y} - \tilde{C}\boldsymbol{X})(\boldsymbol{Y} - \tilde{C}\boldsymbol{X})'$, the estimate of the contemporaneous error covariance matrix. (To avoid singularity, m must be greater than T, a condition that might not hold in the finance setting because of availability of data on a large number m of securities; one might typically have $m > T$.)
(ii) In step two, motivated by (8.4), compute the estimate $\hat{\gamma}_0 = \{\tilde{\alpha}'\tilde{\Sigma}_{\epsilon\epsilon}^{-1}(\mathbf{1}_m - \tilde{\beta})\}/(\mathbf{1}_m - \tilde{\beta})'\tilde{\Sigma}_{\epsilon\epsilon}^{-1}(\mathbf{1}_m - \tilde{\beta})$, a GLS estimate of γ_0 obtained by regressing the $\tilde{\alpha}_i$ on the $(1 - \tilde{\beta}_i)$, $i = 1, \ldots, m$, with the weights matrix $\tilde{\Sigma}_{\epsilon\epsilon}^{-1}$.
(iii) Using (8.3) and the estimate $\hat{\gamma}_0$ from (ii), perform m individual 'market model' regressions of $r_{ik} - \hat{\gamma}_0$ on $x_k - \hat{\gamma}_0$, without a constant term, to reestimate the β_i.
(iv) With the new estimates $\tilde{\beta}_i$, iterate the steps in (ii) and (iii) that yield $\hat{\gamma}_0$ and the new $\tilde{\beta}_i$ until convergence.

In several studies in finance (see Gibbons, Ross, and Shanken, 1989), following Sharpe (1964) and Lintner (1965), it is assumed that there exists a riskless asset which is usually taken as the U.S. Treasury bill rate. For this choice, the value of γ_0 is known and is set equal to the rate of return on the riskless asset. Then testing for the (known) constraints (8.4) can be done using the results in Chapter 1. For known γ_0, therefore known γ^*, the constraints (8.7) can be tested through the LR statistic in (1.16) and the approximate chi-squared distribution theory, or in this case that involves only a single constraint vector, through Hotelling's T^2-statistic given in (1.21). This follows from the same reasoning as in (1.21), and with $\overline{\Sigma}_{\epsilon\epsilon} = \{T/(T-2)\}\tilde{\Sigma}_{\epsilon\epsilon}$, the appropriate test statistic is

$$\mathcal{T}^2 = (\gamma^{*\prime}\tilde{C}'\overline{\Sigma}_{\epsilon\epsilon}^{-1}\tilde{C}\gamma^*)/\gamma^{*\prime}(\boldsymbol{XX}')^{-1}\gamma^*, \tag{8.8}$$

which has the Hotelling's T^2-distribution with $T-2$ degrees of freedom. Thus, $\mathcal{F}=\{(T-m-1)/m(T-2)\}\mathcal{T}^2$ has the F-distribution with m and T–m–1 degrees of freedom. This is the test discussed by Gibbons, Ross, and Shanken (1989) and the interpretation of the noncentrality parameter under the alternative that $C\gamma^* \neq 0$ in terms of the financial context has been of some interest in the finance literature. The constrained ML estimator of C under (8.7) is

$$\hat{C}=\tilde{C}\left[I_2-\frac{\gamma^*\gamma^{*\prime}(\boldsymbol{X}\boldsymbol{X}^{\prime})^{-1}}{\gamma^{*\prime}(\boldsymbol{X}\boldsymbol{X}^{\prime})^{-1}\gamma^*}\right], \tag{8.9}$$

which follows as a special case of the form given in (1.18) with $F_1=I_m$ and $G_2=\gamma^*$. Note that an alternate direct derivation of the test statistic (8.8) can be motivated as follows. With γ_0 known, model (8.1) can be written in the equivalent form as $(r_{ik}-\gamma_0)=\alpha_i^*+\beta_i(x_k-\gamma_0)+\epsilon_{ik}$, where $\alpha_i^*=\alpha_i-\gamma_0(1-\beta_i)\equiv 0$ under conditions (8.4). It follows that if the excess asset returns $r_{ik}-\gamma_0$ are regressed on the excess market return $x_k-\gamma_0$, the resulting intercepts are expected to be zero. The traditional multivariate T^2-test for the m intercepts to be zero yields the statistic in (8.8), noting that $\tilde{C}\gamma^*=\tilde{\alpha}-\gamma_0(\mathbf{1}_m-\tilde{\beta})\equiv\tilde{\alpha}^*$ is equal to the vector of LS estimates of the intercepts α_i^*.

When a riskless asset is not assumed to exist, so that γ_0 and hence γ^* is not known, we can obtain the constrained ML estimates and test the CAPM hypothesis of (8.4) using the reduced-rank regression methods developed in Chapter 2. Because of the rank one structure, and the simple nature of both γ in (8.6) and γ^* in (8.7), we can obtain the ML estimator of γ_0 in two ways. From general results in Section 2.3, the ML estimation criterion leads to minimizing (2.16) and the resulting solutions are given in (2.17). It follows, using the notation of the problem discussed here, that the ML estimates of the components β^* and γ in $C=\beta^*\gamma'$ are given by

$$\hat{\beta}^*=\tilde{\Sigma}_{\epsilon\epsilon}^{1/2}\hat{V}_1, \qquad \hat{\gamma}'=\hat{V}_1'\tilde{\Sigma}_{\epsilon\epsilon}^{-1/2}\tilde{C}, \tag{8.10}$$

where $\hat{V}_1$ is the (normalized) eigenvector that corresponds to the largest eigenvalue $\hat{\lambda}_1^2$ of the matrix $\tilde{\Sigma}_{\epsilon\epsilon}^{-1/2}\tilde{C}\hat{\Sigma}_{xx}\tilde{C}'\tilde{\Sigma}_{\epsilon\epsilon}^{-1/2}$. These estimates satisfy the normalization $\hat{\beta}^{*\prime}\tilde{\Sigma}_{\epsilon\epsilon}^{-1}\hat{\beta}^*=1$, while $\hat{\gamma}'\hat{\Sigma}_{xx}\hat{\gamma}=\hat{\lambda}_1^2$. Because of the disproportionate dimensions of Y_k and X_k involved (m is typically much larger than $n=2$), it would be practical to compute the equivalent ML estimates of β^* and γ based on the eigenvector of the 2×2 matrix $\hat{\Sigma}_{xx}^{1/2}\tilde{C}'\tilde{\Sigma}_{\epsilon\epsilon}^{-1}\tilde{C}\hat{\Sigma}_{xx}^{1/2}$. Let $\hat{V}_1^*$ be the (normalized) eigenvector that corresponds to the largest eigenvalue $\hat{\lambda}_1^2$ of the latter matrix. Then it follows from discussion given towards the end of Section 2.3 that an equivalent expression for the ML estimates is given by

$$\hat{\beta}^*=\tilde{C}\hat{\Sigma}_{xx}^{1/2}\hat{V}_1^*, \qquad \hat{\gamma}'=\hat{V}_1^{*\prime}\hat{\Sigma}_{xx}^{-1/2}, \tag{8.11}$$

which satisfy the normalization $\hat{\gamma}'\hat{\Sigma}_{xx}\hat{\gamma} = 1$, while $\hat{\beta}^{*\prime}\tilde{\Sigma}_{\epsilon\epsilon}^{-1}\hat{\beta}^{*} = \hat{\lambda}_1^2$. By rescaling the above ML estimate $\hat{\gamma}$ in (8.10) or (8.11) to have the form $[-\hat{\gamma}_0,\ 1]'$, we obtain the ML estimate $\hat{\gamma}_0$ of γ_0. Therefore, the iterative scheme suggested by Gibbons (1982) to obtain estimators of γ_0 and β can be replaced by a one-time eigenvalue and eigenvector calculation to obtain the ML estimates.

If the focus of estimation is on γ_0, this parameter can also be estimated through estimation of the constraints vector γ^* in (8.7). As discussed in Section 2.3, the ML estimate of the constraint corresponds to obtaining the eigenvector $\hat{V}_2$ (or $\hat{V}_2^*$) associated with the smallest eigenvalue $\hat{\lambda}_2^2$ of the matrix $\tilde{\Sigma}_{\epsilon\epsilon}^{-1/2}\tilde{C}\hat{\Sigma}_{xx}\tilde{C}'\tilde{\Sigma}_{\epsilon\epsilon}^{-1/2}$ (or of the matrix $\hat{\Sigma}_{xx}^{1/2}\tilde{C}'\tilde{\Sigma}_{\epsilon\epsilon}^{-1}\tilde{C}\hat{\Sigma}_{xx}^{1/2}$). This can also be motivated in this context through the T^2-statistic given in (8.8) (e.g., see Anderson, 1951, for a similar argument). For a given vector γ^*, the constrained ML estimate of the error covariance matrix $\Sigma_{\epsilon\epsilon}$ is easily obtained from the constrained ML estimate $\hat{C}$ given in (8.9) as

$$\hat{\Sigma}_{\epsilon\epsilon} = \tilde{\Sigma}_{\epsilon\epsilon} + \frac{1}{\gamma^{*\prime}\hat{\Sigma}_{xx}^{-1}\gamma^*}\tilde{C}\gamma^*\gamma^{*\prime}\tilde{C}' . \tag{8.12}$$

Since

$$\begin{aligned} |\hat{\Sigma}_{\epsilon\epsilon}| &= |\tilde{\Sigma}_{\epsilon\epsilon}|\,|I_m + (\gamma^{*\prime}\hat{\Sigma}_{xx}^{-1}\gamma^*)^{-1}\tilde{\Sigma}_{\epsilon\epsilon}^{-1/2}\tilde{C}\gamma^*\gamma^{*\prime}\tilde{C}'\tilde{\Sigma}_{\epsilon\epsilon}^{-1/2}| \\ &= |\tilde{\Sigma}_{\epsilon\epsilon}|\,(1 + (\gamma^{*\prime}\hat{\Sigma}_{xx}^{-1}\gamma^*)^{-1}\gamma^{*\prime}\tilde{C}'\tilde{\Sigma}_{\epsilon\epsilon}^{-1}\tilde{C}\gamma^*), \end{aligned}$$

minimizing $|\hat{\Sigma}_{\epsilon\epsilon}|$ with respect to the unknown constraint vector γ^* to arrive at the ML estimate is equivalent to minimizing the statistic $\mathcal{T}^2$. Therefore, it follows that the estimate $\hat{\gamma}^*$ can be obtained through the eigenvector that corresponds to the smallest eigenvalue $\hat{\lambda}_2^2$ of $\hat{\Sigma}_{xx}^{1/2}\tilde{C}'\tilde{\Sigma}_{\epsilon\epsilon}^{-1}\tilde{C}\hat{\Sigma}_{xx}^{1/2}$. Again, by appropriately rescaling the components of $\hat{\gamma}^*$, we obtain the ML estimate $\hat{\gamma}_0$ of γ_0. In either approach, the likelihood ratio statistic for testing H_0 : rank$(C) = 1$ is obtained from the results of Section 2.6 as $[T - (m+3)/2]\log(1 + \hat{\lambda}_2^2)$, and follows an asymptotic χ^2_{m-1} distribution under the null hypothesis.

Zhou (1991, 1995) formulated the testing of the asset pricing constraints in (8.4) as a problem in reduced-rank regression, and investigated small sample distributional properties of the likelihood ratio test (a test essentially based on $\hat{\lambda}_2^2$) for testing rank$(C) = 1$ versus rank$(C) = 2$. Shanken (1986) as well as Zhou (1991) derived the ML estimate of the parameter γ_0. The focus of Zhou's studies, however, was on testing for the rank of C rather than on estimates of γ_0 and β. Zhou (1991) comments that the accuracy of the ML estimates would be of interest to practitioners in the finance area, but observes that the standard error of the zero-beta rate estimate $\hat{\gamma}_0$ is not known. Because of the connections with reduced-rank estimation demonstrated here, we mention that the large sample distribution of the ML estimator $\hat{\gamma}_0$ can be derived using the results in Section 2.5.

The methods and results presented for the model (8.1) can easily be generalized to the multi-beta CAPM, which is often considered in the finance literature. The market model for the multi-beta CAPM can be written similar to (8.1) as

$$r_{ik} = \alpha_i + \sum_{j=1}^{n-1} \beta_{ij} x_{jk} + \epsilon_{ik}, \tag{8.13}$$

where $(x_{1k}, \ldots, x_{n-1,k})'$ is a $(n-1) \times 1$ vector of returns on the $(n-1)$ reference portfolios, and $\beta_i = (\beta_{i1}, \ldots, \beta_{i,n-1})'$ is the $(n-1) \times 1$ vector of risk measures. The constraints (8.4) generalize in the multi-beta CAPM to

$$\alpha_i = \gamma_0 \left(1 - \sum_{j=1}^{n-1} \beta_{ij} \right) \equiv \gamma_0 (1 - \beta_i' \mathbf{1}_{n-1}), \quad i = 1, \ldots, m. \tag{8.14}$$

Similar to the previous case, the model (8.13) can be reexpressed as

$$y_{ik} = r_{ik} - \frac{1}{n-1} \sum_{j=1}^{n-1} x_{jk} = \alpha_i + \sum_{j=1}^{n-1} \left(\beta_{ij} - \frac{1}{n-1} \right) x_{jk} + \epsilon_{ik},$$

and hence it can be written in the multivariate regression form as $Y_k = CX_k + \epsilon_k$, with $X_k = (1, x_{1k}, \ldots, x_{n-1,k})'$, and the $m \times n$ coefficient matrix C will be of reduced rank $n-1$ ($n < m$ is assumed) under the constraints of (8.14). The reduced-rank methods and results from Chapter 2 readily extend to this case. Some testing procedures related to (8.14) are discussed by Velu and Zhou (1997).

An alternate financial theory to the CAPM is the arbitrage pricing theory (APT) proposed by Ross (1976). A basic difference between the two theories stems from APT's treatment of the correlations among the asset returns. The APT assumes that returns on assets are generated by a number of industry-wide and market-wide factors; returns on any two securities are likely to be correlated if the returns are affected by the same factors. Although the CAPM allows for correlations among assets, it does not specify the underlying factors that induce the correlations. The APT where the factors are observable economic variables can also be formulated as a multivariate regression model with nonlinear (bilinear) parameter restrictions, that is, with reduced-rank structure. The details of the model as multivariate regression with nonlinear restrictions were first presented by McElroy and Burmeister (1988), and Bekker, Dobbelstein, and Wansbeek (1996) further identified the restrictions in the APT model as reduced-rank structure, and discussed the estimation of parameters, including the "undersized samples" problem when $m > T$. In the APT, it is assumed that the returns on the m assets at the kth time period are generated by a linear (observable) factor model, with n factors, as

$$r_k = E(r_k) + Bf_k + \epsilon_k, \quad k = 1, \ldots, T, \tag{8.15}$$

where r_k is an $m \times 1$ vector of returns, f_k is an $n \times 1$ vector of measured factors representing macroeconomic 'surprises', and B is an $m \times n$ matrix with typical element b_{ij} measuring the sensitivity of asset i to factor j. The random errors ϵ_k are assumed to follow the same assumptions as in model (8.5). The APT starts with model (8.15) and explains the cross-sectional variation in the expected returns $E(r_k)$ under the assumption that arbitrage profits are impossible without undertaking some risk and without making some net investment. The fundamental result of APT is that $E(r_k)$ is approximated by

$$E(r_k) = \delta_{0k} \mathbf{1}_m + B\delta,$$

where δ_{0k} is the (possibly time-varying) risk-free rate of return for period k, assumed to be known at the beginning of period k, and δ is an $n \times 1$ vector of premiums earned by an investor for assuming one unit of risk from each of the n factors. Combining the above two models we have

$$r_k = \delta_{0k} \mathbf{1}_m + B\delta + Bf_k + \epsilon_k. \tag{8.16}$$

Now, we let $Y_k = r_k - \delta_{0k} \mathbf{1}_m$, $X_k = [1,\ f_k']'$, and $C = [B\delta, B] \equiv B\,[\delta, I_n]$. Then the model (8.16) is in the same framework as (8.5), $Y_k = CX_k + \epsilon_k$, with the constraint on the coefficient matrix C given by

$$C \begin{bmatrix} 1 \\ -\delta \end{bmatrix} = 0.$$

Thus the $m \times (n+1)$ coefficient matrix C has a rank deficiency of one. The vector of risk premiums δ can be estimated from the eigenvector that corresponds to the smallest eigenvalue of $\hat{\Sigma}_{xx}^{1/2} \tilde{C}' \tilde{\Sigma}_{\epsilon\epsilon}^{-1} \tilde{C} \hat{\Sigma}_{xx}^{1/2}$. By normalizing the first element of this eigenvector as equal to -1, the remaining elements give the estimate $\hat{\delta}$ of δ. The associated estimator of B is then obtained as $\hat{B} = \boldsymbol{Y}\hat{\boldsymbol{X}}^{*\prime}(\hat{\boldsymbol{X}}^{*}\hat{\boldsymbol{X}}^{*\prime})^{-1}$, where $\hat{\boldsymbol{X}}^{*} = [\hat{X}_1^{*}, \ldots, \hat{X}_T^{*}]$ and the $\hat{X}_k^{*} = \hat{\delta} + f_k \equiv [\hat{\delta}, I_n]X_k$ are formed based on the estimator of δ.

McElroy and Burmeister (1988) suggested an iterative procedure to estimate the parameters in (8.16). Later, Bekker, Dobbelstein, and Wansbeek (1996) noted that the problem could be formulated as a reduced-rank regression setup as indicated above. Bekker et al. (1996) also derived the second-order properties of the estimators of B and δ and discussed the case of undersized samples, that is, when there are more assets than time periods or $m > T$. The asymptotic results can be derived directly through the approach of Silvey (1959) as demonstrated in Chapters 3 and 4. The case of undersized samples is important to consider in the finance context because we may often have $m > T$, and then the unconstrained estimate $\tilde{\Sigma}_{\epsilon\epsilon}$ of $\Sigma_{\epsilon\epsilon}$ will be singular. Two approaches were suggested for this case; one was simply to set $\Sigma_{\epsilon\epsilon} = I_m$, which leads to least squares reduced-rank estimates as discussed in Section 2.3. The second approach is to impose some factor-analysis structure on $\Sigma_{\epsilon\epsilon}$, thereby reducing the number of unknown parameters and resulting in a nonsingular estimate of $\Sigma_{\epsilon\epsilon}$. The unknown parameters in $\Sigma_{\epsilon\epsilon}$ can be estimated from $\tilde{\Sigma}_{\epsilon\epsilon}$.

8.3 Additional Applications of Reduced-Rank Regression in Finance

Additional applications of reduced-rank regression models occur in various areas of finance. In these contexts, the models typically arise as latent variables models, as discussed in Section 2.1, and they are described in Campbell, Lo, and MacKinlay (1997, pp. 446-448). Hansen and Hodrick (1983) and Gibbons and Ferson (1985) investigated the time-varying factor risk premiums; Campbell (1987) and Ferson and Harvey (1991) studied the areas of stock returns, forward currency premiums, international equity returns and capital integration; Stambaugh (1988) tested the theory of the term structure of interest rates. In all these studies, the models involved appear to have the reduced-rank structure. Because of the diverse nature of the various topics above, we refrain from discussing them except for the work of Gibbons and Ferson (1985), which follows somewhat from the models considered earlier.

The Black (1972) CAPM in (8.3) and other related models suffer from some methodological shortcomings; the expected returns $E(r_{ik})$ are assumed to be constant over a sufficient period of time and the market portfolio of risky assets, on which the return x_k is computed, must be observable. Gibbons and Ferson (1985) developed tests of asset pricing models that do not require the identification of the market portfolio or the assumption of constant risk premiums. These tests can be shown to be related to testing the rank of a regression coefficient matrix. We present only the simpler version of the model.

If γ_0 is assumed to be known in (8.3), then the returns r_{ik} and x_k can be adjusted by subtraction of this riskless rate of return γ_0 and this yields $E(r_{ik}^*) = \beta_i E(x_k^*)$, where $r_{ik}^* = r_{ik} - \gamma_0$ is the excess return on asset i and $x_k^* = x_k - \gamma_0$ is the excess return on the market portfolio. If the expected returns are changing over time and depend on the financial information prescribed by an n-dimensional vector X_k at time $k-1$, a generalization of the above model can be written as

$$E(r_{ik}^*|X_k) = \beta_i E(x_k^*|X_k), \qquad i = 1, \ldots, m, \tag{8.17}$$

where $r_{ik}^* = r_{ik} - \gamma_{0k}$ and $x_k^* = x_k - \gamma_{0k}$ are excess returns. The above holds for all assets, hence for $i = 1$ in particular. Therefore, if $\beta_1 \neq 0$,

$$E(r_{ik}^*|X_k) = (\beta_i/\beta_1) E(r_{1k}^*|X_k), \tag{8.18}$$

which does not involve the expected excess return on the market portfolio. Suppose the observed excess returns are assumed to be a linear function of the financial information X_k,

$$r_{ik}^{*} = \delta_i' X_k + \epsilon_{ik}, \tag{8.19}$$

where δ_i is an $n \times 1$ vector of regression coefficients, and suppose the assumptions on the ϵ_{ik} hold as stated earlier. Using $i = 1$ as a reference asset and substituting into both sides of (8.18) from (8.19) yields

$$\delta_i = (\beta_i/\beta_1)\delta_1, \qquad i \neq 1. \tag{8.20}$$

Observe that the choice of the reference asset is arbitrary and the coefficients β_i are not known. The conditions (8.20) lead to a reduced-rank regression setup as follows. The multivariate form of model (8.19) is given by

$$Y_k = CX_k + \epsilon_k, \qquad k = 1, \ldots, T, \tag{8.21}$$

where $Y_k = (r_{1k}^{*}, \ldots, r_{mk}^{*})'$ is the $m \times 1$ vector of excess returns and $C = [\delta_1, \ldots, \delta_m]'$ is an $m \times n$ regression coefficient matrix. The conditions (8.20) imply that $C' = \delta_1 [1, (\beta_2/\beta_1), \ldots, (\beta_m/\beta_1)]$ and hence that $\text{rank}(C) = 1$. It follows that the number of independent parameters to be estimated is $(m + n - 1)$ instead of mn. Gibbons and Ferson (1985) also present a more general version of model (8.17). In this, the excess return on the market portfolio is replaced by excess returns on r hedge portfolios, which are not necessarily observed, or by r reference assets. This leads to a multivariate regression model as in (8.21) and to the rank condition on the coefficient matrix that $\text{rank}(C) = r$. Therefore, all the methods that have been developed in Chapter 2 can be readily used in this setup.

8.4 Empirical Studies and Results on Asset Pricing Models

There exists an enormous amount of literature discussing empirical evidence on the CAPM. The early evidence was largely supportive of the CAPM but in the late 1970s less favorable evidence for the CAPM began to emerge. Because of the availability of voluminous financial data, the statistical tests of hypotheses related to the CAPM and other procedures discussed here can be carried out readily. We will briefly survey some empirical results that have been obtained related to models discussed in this chapter. For additional empirical results, readers may refer to Campbell, Lo, and MacKinlay (1997).

Gibbons (1982) used the monthly stock returns (r_k) provided by the Center for Research in Security Prices (CRSP) and the CRSP equal-weighted index was taken as the return on the market portfolio (x_k). The time period of 1926–1975 was divided into ten equal five-year subperiods. The number of stock returns was chosen to be $m = 40$. Within each subperiod it is assumed that the parameters of model (8.1) are stable. The general conclusion obtained from the study was that the restrictions (8.4) implied by the CAPM were rejected at 'reasonable significance levels in five out of ten subperiods'. Some diagnostic procedures were suggested to postulate alternative model specifications. Other studies also do not provide conclusive results. Therefore, it is suggested that in order to analyze financial data more thoroughly, new models need to be specified, or because the testing of restrictions is based on large sample theory, the tests must be evaluated through their small sample properties.

Zhou (1991) used the reduced-rank regression tools for estimation and studied the small sample distributions of the estimators and associated test statistics. In particular, Zhou (1991) obtained results on the small sample distribution of the smallest eigenvalue test statistic $\hat{\lambda}_2$ described in Section 8.2. He also considered the CRSP data, from 1926–1986, but instead of stock returns which may exhibit more volatility, he used 12 value-weighted industry portfolios. The analysis was again performed on a five-year subperiod basis. Out of 13 subperiods, the hypothesis (8.4) was rejected for all except four. He concluded that 'the rejection of efficiency is most likely caused by the fundamentally inefficient behavior of the (market) index'.

The widely quoted paper by Gibbons, Ross, and Shanken (1989) also considered empirical analyses of similar data. Recall that these authors considered the multivariate Hotelling's T^2-statistic (8.8) for testing for portfolio efficiency in the models discussed by Sharpe (1964) and Lintner (1965), where the risk-free rate of return, γ_0 in (8.4), is known. Although their focus was on studying the sensitivity of the test statistic (8.8), they also considered empirical analyses of monthly returns. The empirical examples that they presented illustrate the effect that different sets of assets can have on the outcome of the test. First they reported their analysis on portfolios that were sorted by their risks. Using monthly returns on $m = 10$ portfolios from January 1931 through December 1965, through the F-statistic indicated in Section 8.2, they confirm that the efficiency of the market portfolio cannot be rejected. For monthly returns data from 1926 through 1982, however, they constructed one set of $m = 12$ industry portfolios and another set of $m = 10$ portfolios based on the relative market value or size of the firms. For industry-based portfolios, the F-statistic rejected the efficiency hypothesis but for the size-based portfolios, it did not. Thus, the formation of portfolios could influence the outcome of the test procedure. Their overall conclusion is that 'the multivariate approach can lead to more appropriate conclusions', because in the case of the industry-based analysis, all 12 univariate t-statistics failed to reject the efficiency hypothesis.

Other empirical results reported by McElroy and Burmeister (1988), Gibbons and Ferson (1985), and Zhou (1995) use the tests for the rank of the coefficient matrix being equal to one in some form or another. McElroy and Burmeister (1988) considered monthly returns from CSRP over January 1972 to December 1982 for a sample of 70 firms. The predictor variables chosen for use in model (8.16) include four macroeconomic 'surprises' and the market return. The macro-factors include indices constructed from government and corporate bonds, deflation and treasury bills and others. These authors focus more on the interpretation of the regression coefficients of the model (8.16); tests for the rank of the resulting coefficient matrix were not reported. Gibbons and Ferson (1985) considered daily returns from 1962 to 1980 for several major companies. While the excess returns on these companies form the response variables, the two predictors used in model (8.19) are taken to be an indicator variable for Monday, when mean stock returns are observed to be lower typically than other days of the week, and a lagged return of a large portfolio on common stocks. The null hypothesis (8.20) is rejected in their analysis. Zhou (1995) applied reduced-rank methodology to examine the number of latent factors in the US equity market. For the 12 constructed industry portfolios, he examined monthly data from October 1941 to September 1986. Several predictors are included in the model: returns on equal-weighted index, treasury bill rate, yield on Moody's BAA-rated bonds, and so on. The general conclusion reached from his study was that the resulting regression matrix is of rank one consistent with the original CAPM hypothesis.

9
Alternate Procedures for Analysis of Multivariate Regression Models

In previous chapters, we have developed various reduced-rank multivariate regression models, and indicated their usefulness in different applications as dimension-reduction tools. We now briefly survey and discuss some other related multivariate regression modeling methodologies that have similar parameter reduction objectives as reduced-rank regression, such as multivariate ridge regression, partial least squares, joint continuum regression, and other shrinkage and regularization techniques. Some of these procedures are designed particularly for situations where there is a very large number n of predictor variables relative to the sample size T including, for example, $n > T$.

Multivariate Ridge Regression and Other Shrinkage Estimates

Consider the standard multivariate linear regression model as in (1.4), $\boldsymbol{Y} = C\boldsymbol{X} + \boldsymbol{\epsilon}$, where $\boldsymbol{Y} = [Y_1, \ldots, Y_T]$ is $m \times T$, $\boldsymbol{X} = [X_1, \ldots, X_T]$ is $n \times T$, $C = [C_{(1)}, \ldots, C_{(m)}]'$ is $m \times n$, $\boldsymbol{\epsilon} = [\epsilon_1, \ldots, \epsilon_T]$, and the ϵ_k are mutually independent with $E(\epsilon_k) = 0$ and $\text{Cov}(\epsilon_k) = \Sigma_{\epsilon\epsilon}$. The ordinary LS estimator of C is $\tilde{C} = \boldsymbol{Y}\boldsymbol{X}'(\boldsymbol{X}\boldsymbol{X}')^{-1}$, which also can be expressed in vector form as in (1.14),

$$\tilde{c} = \text{vec}(\tilde{C}') = (I_m \otimes (\boldsymbol{X}\boldsymbol{X}')^{-1}\boldsymbol{X})\,\text{vec}(\boldsymbol{Y}'). \tag{9.1}$$

As discussed in the introduction, Sclove (1971), among others, has argued that a deficiency of the LS estimator $\tilde{C}$ is that it takes no account of $\Sigma_{\epsilon\epsilon}$, the among regression equations error covariance matrix. To remedy this shortcoming and also to take advantage of possible improvements in the estimation of the individual $C_{(i)}$ even for the case where $\Sigma_{\epsilon\epsilon} = \sigma^2 I_m$ by using (univariate) ridge regression for that case, Brown and Zidek (1980, 1982), Haitovsky (1987), and others, suggested multivariate ridge regression estimators of $c = \text{vec}(C')$.

The ridge regression estimator of c proposed by Brown and Zidek (1980) is of the form

$$\begin{aligned}\hat{c}^*(K) &= (I_m \otimes \boldsymbol{XX}' + K \otimes I_n)^{-1}(I_m \otimes \boldsymbol{X})\text{vec}(\boldsymbol{Y}') \\ &= (I_m \otimes \boldsymbol{XX}' + K \otimes I_n)^{-1}(I_m \otimes \boldsymbol{XX}')\,\tilde{c}\,, \qquad (9.2)\end{aligned}$$

where K is the $m \times m$ ridge matrix. Brown and Zidek (1980) motivated (9.2) as a Bayes estimator, with prior distribution for c such that $c \sim N(0, V \otimes I_n)$, where V is an $m \times m$ covariance matrix, with $K = \Sigma_{\epsilon\epsilon} V^{-1}$ in (9.2) (and $\Sigma_{\epsilon\epsilon}$ is treated as known). By letting the choice of the ridge matrix K depend on the data $\boldsymbol{Y}$ in certain ways, adaptive multivariate ridge estimators are obtained. Mean square error properties of these adaptive and empirical Bayes estimators are compared, and their dominance over the ordinary LS estimator (i.e., $K = 0$) is considered by Brown and Zidek (1980).

Haitovsky (1987) considered a slightly more general class of multivariate ridge estimators, developed by using a more general prior distribution for the regression coefficients c of the form $c \sim N(\mathbf{1}_m \otimes \xi, V \otimes \Omega)$, where $V > 0$ is the $m \times m$ covariance matrix, except for scale factor, between any two columns of C, and $\Omega > 0$ is the $n \times n$ covariance matrix, except for scale factor, between any two rows (i.e., $C'_{(i)}$ and $C'_{(j)}$) in C. This leads to multivariate ridge estimators of the form

$$\begin{aligned}\hat{c}^* &= (I_m \otimes \boldsymbol{XX}' + K_w \otimes \Omega^{-1})^{-1}(I_m \otimes \boldsymbol{X})\text{vec}(\boldsymbol{Y}') \\ &= (I_m \otimes \boldsymbol{XX}' + K_w \otimes \Omega^{-1})^{-1}(I_m \otimes \boldsymbol{XX}')\,\tilde{c}\,, \qquad (9.3)\end{aligned}$$

where $K_w = \Sigma_{\epsilon\epsilon} V^{-1} W$ and $W = I_m - (\mathbf{1}'_m V^{-1} \mathbf{1}_m)^{-1} \mathbf{1}_m \mathbf{1}'_m V^{-1}$, and the error covariance matrix $\Sigma_{\epsilon\epsilon}$ is assumed known. Estimation for the more realistic situation where $\Sigma_{\epsilon\epsilon}$ (and the prior covariance matrices V and Ω) are unknown was also discussed by Haitovsky (1987), using multivariate mixed-effects and covariance components model ML estimation methods. (The situation of unknown covariance matrix was also considered by Brown and Zidek, 1982, for their class of ridge estimators.) Brown and Zidek (1980) and Haitovsky (1987) both noted particular multivariate shrinkage estimators, such as those of Efron and Morris (1972), result as special cases of their classes of multivariate ridge estimators. These estimators do not seem to have been used extensively in practical applications, although Brown and Payne (1975) used a specialized form of the estimators effectively in "election night forecasting".

Various other forms of estimators of the regression coefficient matrix C that 'shrink' the usual LS estimator $\tilde{C}$, in the spirit of Stein and Efron–Morris type shrinkage, were considered by Bilodeau and Kariya (1989),

Honda (1991), and Konno (1991), among others. These authors examined the risk properties of the proposed classes of shrinkage estimators, and for certain risk functions they presented sufficient conditions for these alternate estimators to have improved risk relative to the LS (minimax) estimator.

Canonical Analysis Multivariate Shrinkage Methods

Different shrinkage methods for the multivariate linear regression model based on classical canonical correlation analysis between Y_k and X_k were proposed by van der Merwe and Zidek (1980) and Breiman and Friedman (1997). In some respects, these methods may be viewed as generalizations of reduced-rank regression modeling. Recall, as shown in Sections 2.3 and 2.4, that the ML reduced-rank estimator of the regression coefficient matrix C, of rank $r < \min(m, n)$, can be expressed in terms of the canonical correlation analysis between Y_k and X_k. Specifically, the ML reduced-rank estimator can be written as

$$\hat{C}^{(r)} = \hat{A}_*^{(r)} \hat{B}_*^{(r)} = \hat{\Sigma}_{yy}^{1/2} \hat{V}_*^{(r)} \hat{V}_*^{(r)\prime} \hat{\Sigma}_{yy}^{-1/2} \tilde{C}, \tag{9.4}$$

where $\tilde{C} = \boldsymbol{Y}\boldsymbol{X}'(\boldsymbol{X}\boldsymbol{X}')^{-1} \equiv \hat{\Sigma}_{yx}\hat{\Sigma}_{xx}^{-1}$ is the ordinary LS estimator of C, $\hat{V}_*^{(r)} = [\hat{V}_1^*, \ldots, \hat{V}_r^*]$, and the $\hat{V}_j^*$ are normalized eigenvectors of the matrix $\hat{\Sigma}_{yy}^{-1/2}\hat{\Sigma}_{yx}\hat{\Sigma}_{xx}^{-1}\hat{\Sigma}_{xy}\hat{\Sigma}_{yy}^{-1/2}$ associated with its r largest eigenvalues $\hat{\rho}_j^2$, the squared sample canonical correlations between the Y_k and X_k. Letting $\hat{V}_* = [\hat{V}_1^*, \ldots, \hat{V}_m^*]$, an $m \times m$ orthogonal matrix, $\hat{H} = \hat{V}_*' \, \hat{\Sigma}_{yy}^{-1/2}$, and $\Delta_* = \mathrm{Diag}(1, \ldots, 1, 0, \ldots, 0)$, an $m \times m$ diagonal matrix with ones in the first r diagonals and zeros otherwise, the ML reduced-rank estimator in (9.4) can be expressed in the form

$$\hat{C}^{(r)} = \hat{H}^{-1} \Delta_* \hat{H} \, \tilde{C}. \tag{9.5}$$

As we have discussed previously in Sections 1.2 and 2.4, in practical application the predictor variables X_k and response variables Y_k will typically already be centered and be expressed in terms of deviations from sample mean vectors before the canonical correlation and least squares calculations are performed.

In the canonical analysis-based multivariate shrinkage methods, the estimator of C is formed as in (9.5) but with the diagonal matrix Δ_* replaced by a diagonal matrix $\hat{\Delta}$ of "shrinkage factors" $\{\hat{d}_j\}$. For example, in the generalized cross-validation (GCV) based approach of Breiman and Friedman (1997), the diagonal elements of $\hat{\Delta}$ used are given by

$$\hat{d}_j = \frac{(1-p)\,(\hat{\rho}_j^2 - p)}{(1-p)^2 \hat{\rho}_j^2 + p^2(1 - \hat{\rho}_j^2)}, \qquad j = 1, \ldots, m, \tag{9.6}$$

where $p = n/T$ is the ratio of the number n of predictor variables to the sample size T. In practice, the positive-part shrinkage rule is used in that $\hat{d}_j$

is replaced by 0 whenever $\hat{d}_j < 0$ (i.e., $\hat{\rho}_j^2 < n/T$) in (9.6). The form of the $\hat{d}_j$ in (9.6) is derived by Breiman and Friedman (1997) so as to minimize the generalized cross-validation prediction mean square error. The diagonal elements $\hat{d}_j$ also may be determined by a fully cross-validatory procedure. In the method proposed earlier by van der Merwe and Zidek (1980), the diagonal elements used for the matrix $\hat{\Delta}$ in place of Δ_* in (9.5) are

$$f_j = \left(\hat{\rho}_j^2 - \frac{n-m-1}{T}\right) / \hat{\rho}_j^2 \left(1 - \frac{n-m-1}{T}\right),$$

with the positive-part rule also employed. Breiman and Friedman (1997) also presented and discussed generalizations of their procedure, which combine the canonical analysis shrinkage methods with principal components regression and (univariate) ridge regression, especially useful for situations involving underdetermined ($n > T$) or poorly determined ($n < T$ but large relative to T, e.g., $n \approx T$) samples.

An alternate shrinkage procedure is also motivated by the extended reduced-rank regression model (3.3) of Section 3.1, and this suggested procedure was discussed explicitly by Reinsel (1998). The suggestion is to first estimate the regression of Y_k on a small number of the 'preferred' predictor variables, say Z_k, which are expected to yield canonical correlations that are a large portion of the total correlations due to all predictor variables, in the usual ordinary least squares method. Then use the previous canonical analysis multivariate shrinkage methods applied to the residuals, where the methods can be represented in terms of partial canonical correlation analysis. That is, suppose there exists a partition of the predictor variables as $(X_k', Z_k')'$, where X_k is $n_1 \times 1$, and Z_k is $n_2 \times 1$ with n_2 "small", such that Z_k contains predictor variables that are "known" or "expected" to give a large portion of the correlation with Y_k. Accordingly, write the multivariate regression model as $Y_k = CX_k + DZ_k + \epsilon_k$. Then, in the alternate suggested procedure, obtain the usual LS estimates in the regression of Y_k on Z_k only as $\tilde{D}^* = \hat{\Sigma}_{yz}\hat{\Sigma}_{zz}^{-1}$, and similarly perform the LS regression of X_k on Z_k to obtain the LS coefficient matrix $\tilde{D}_2 = \hat{\Sigma}_{xz}\hat{\Sigma}_{zz}^{-1}$. Then consider the shrinkage estimator of the regression coefficient matrix C of the form

$$\hat{C} = \hat{H}_1^{-1}\hat{\Delta}_1\hat{H}_1\,\tilde{C}, \tag{9.7}$$

where

$$\tilde{C} = \hat{\Sigma}_{yx.z}\hat{\Sigma}_{xx.z}^{-1} \equiv (\boldsymbol{Y} - \tilde{D}^*\boldsymbol{Z})(\boldsymbol{X} - \tilde{D}_2\boldsymbol{Z})'\{(\boldsymbol{X} - \tilde{D}_2\boldsymbol{Z})(\boldsymbol{X} - \tilde{D}_2\boldsymbol{Z})'\}^{-1}$$

is the ordinary LS estimate of C, with the corresponding estimate of D given by $\hat{D} = \tilde{D}^* - \hat{C}\tilde{D}_2$. In the same spirit as above, the shrinkage involved for $\hat{C}$ is developed and expressible in terms of the squared sample *partial* canonical correlations $\hat{\rho}_{1j}^2$ of the Y_k and X_k, after adjusting for Z_k. The rows of the matrix $\hat{H}_1$ in this case correspond to the sample partial canonical vectors for the variates Y_k, and $\hat{\Delta}_1$ in (9.7) is a diagonal matrix with elements $\hat{d}_{1j}$ of a similar form as in (9.6) but in terms of the $\hat{\rho}^2_{1j}$, that is,

$$\hat{d}_{1j} = \frac{(1-p_1)\,(\hat{\rho}^2_{1j} - p_1)}{(1-p_1)^2\hat{\rho}^2_{1j} + p_1^2(1-\hat{\rho}^2_{1j})}, \qquad j = 1, \ldots, m,$$

where $p_1 = n_1/(T - n_2)$. This form is derived by Reinsel (1998) based on consideration of the mean squared error matrix of prediction of Y_k using $(X_k', Z_k')'$, and bias-correction of the usual squared sample partial canonical correlation estimators.

Partial Least Squares

Partial least squares was introduced by Wold (1975) as an iterative computational algorithm for linear regression modeling, especially for situations where the number n of predictor variables is large. The multiple response ($m > 1$) version of partial least squares begins with a 'canonical covariance' analysis, a sequential procedure to construct predictor variables as linear combinations of the original n-dimensional set of predictor variables X_k. At the first stage, the initial predictor variable $x^*_{1k} = b_1' X_k$ is determined as the first 'canonical covariance' variate of the X_k, the linear combination of X_k that maximizes

$$\{(T-1) \times \text{sample covariance of } b'X_k \text{ and } \ell' Y_k\} \equiv b' \boldsymbol{X} \boldsymbol{Y}' \ell$$

over unit vectors b (and over all associated unit vectors ℓ). For given vector b, we know that $\ell = \boldsymbol{Y}\boldsymbol{X}'b/(b'\boldsymbol{X}\boldsymbol{Y}'\boldsymbol{Y}\boldsymbol{X}'b)^{1/2}$ is the maximizing choice of unit vector for ℓ, so that we then need to maximize $b'\boldsymbol{X}\boldsymbol{Y}'\boldsymbol{Y}\boldsymbol{X}'b$ with respect to unit vectors b. Thus, the first linear combination vector b_1 is the (normalized) eigenvector of $\boldsymbol{X}\boldsymbol{Y}'\boldsymbol{Y}\boldsymbol{X}'$ corresponding to its largest eigenvalue. (In the univariate case, $m = 1$ and $\boldsymbol{Y}$ is $1 \times T$, and so this reduces to $b_1 = \boldsymbol{X}\boldsymbol{Y}'/(\boldsymbol{Y}\boldsymbol{X}'\boldsymbol{X}\boldsymbol{Y}')^{1/2}$ with associated "eigenvalue" equal to $\boldsymbol{Y}\boldsymbol{X}'\boldsymbol{X}\boldsymbol{Y}'$.)

At any subsequent stage j of the procedure, there are j predictor variables $x^*_{ik} = b_i' X_k$, $i = 1, \ldots, j$, determined and the next predictor variable $x^*_{j+1,k} = b_{j+1}' X_k$, that is, the next unit vector b_{j+1} needs to be found. It is to be chosen as the vector b to maximize the sample covariance of $b'X_k$ and $\ell' Y_k$ (with unit vector ℓ also chosen to maximize the quantity), that is, we maximize $b'\boldsymbol{X}\boldsymbol{Y}'\boldsymbol{Y}\boldsymbol{X}'b$, subject to b_{j+1} being a unit vector and $x^*_{j+1,k} = b_{j+1}' X_k$ being orthogonal to the previous set of j predictor variables x^*_{ik}, $i = 1, \ldots, j$ (i.e., $b_{j+1}'\boldsymbol{X}\boldsymbol{X}'b_i = 0$, for $i = 1, \ldots, j$). The number w of predictors chosen until the sequential process is stopped is a regularization parameter of the procedure; its value is determined through cross-validation methods in terms of prediction mean square error. The final estimated regression equations are constructed by ordinary LS calculation of the response vectors Y_k on the first w 'canonical covariance' component predictor variables, $X_k^* = (x_{1k}^*, \ldots, x_{wk}^*)' \equiv \hat{B}X_k$, where $\hat{B} = [b_1, \ldots, b_w]'$, that is, the estimated regression equations are $\hat{Y}_k = \hat{A}X_k^* \equiv \hat{A}\hat{B}X_k$ with

$$\hat{A} = \boldsymbol{YX}^{*\prime}(\boldsymbol{X}^*\boldsymbol{X}^{*\prime})^{-1} \equiv \boldsymbol{YX}'\hat{B}'(\hat{B}\boldsymbol{XX}'\hat{B}')^{-1}$$

and $\boldsymbol{X}^* = \hat{B}\boldsymbol{X}$ is $w \times T$. In practice, the predictor variables X_k and response variables Y_k will typically already be expressed in terms of deviations from sample mean vectors, as we have discussed initially in Section 1.2.

Discussion of the sequential computational algorithms for the construction of the 'canonical covariance' components $x_{jk}^* = b_j'X_k$ required in the partial least squares regression procedure, and of various formulations and interpretations of the partial least squares method are given by Wold (1984), Helland (1988, 1990), Hoskuldsson (1988), Stone and Brooks (1990), Frank and Friedman (1993), and Garthwaite (1994). In particular, for univariate partial least squares, Garthwaite (1994) gives the interpretation of the linear combinations of predictor variables, $x_{jk}^* = b_j'X_k$, as weighted averages of predictors, where each individual predictor holds the residual information in an explanatory variable that is not contained in earlier linear combination components (x_{ik}^*, $i < j$), and the quantity to be predicted is the $1 \times T$ vector of residuals obtained from regressing y_k on the earlier components. Also, for the univariate case, Helland (1988, 1990) and Stone and Brooks (1990) showed that the linear combination vectors $\hat{B} = [b_1, \ldots, b_w]'$ are obtainable from nonsingular $(w \times w)$ linear transformation of $[\boldsymbol{XY}', (\boldsymbol{XX}')\boldsymbol{XY}', \ldots, (\boldsymbol{XX}')^{w-1}\boldsymbol{XY}']'$.

Joint Continuum Regression

Continuum regression, proposed by Stone and Brooks (1990) for the univariate response case and extended by Brooks and Stone (1994) to the multivariate case, may be viewed as a generalization of the method of partial least squares regression modeling and also of principal components regression. Like partial least squares, it is a sequential construction method, involving cross-validatory choice of certain control parameters. At a given stage j, j predictor variables $x_{ik}^* = b_i'X_k$, $i = 1, \ldots, j$, have been determined and the next predictor $x_{j+1,k}^* = b_{j+1}'X_k$, that is, the next linear combination vector b_{j+1} is chosen to maximize a certain criterion T_α (defined below) that expresses the potentiality of $x_{j+1,k}^*$ for predicting a similarly maximized linear combination of the response vector Y_k, subject to b_{j+1} being of unit length and $\{x_{j+1,k}^*\}$ being orthogonal to $\{x_{1k}^*, \ldots, x_{jk}^*\}$. The choice of the number w of predictor variables $x_{1k}^*, \ldots, x_{wk}^*$, and of the parameter α involved in the criterion T_α can be made on the basis of cross-validatory indices constructed from LS regression of Y_k on $X_k^* = (x_{1k}^*, \ldots, x_{wk}^*)'$. As in partial least squares, the predictor variables X_k and response variables Y_k may already be expressed in terms of deviations from sample mean vectors.

For $0 \leq \alpha < 1$, the form of the criterion T_α to determine the vector b at a given stage is

$$T_\alpha = \max_{||\ell||=1} \{(T-1) \times \text{sample covariance of } b'X_k \text{ and } \ell'Y_k\}^2 \times (b'\boldsymbol{X}\boldsymbol{X}'b)^{\{\alpha/(1-\alpha)\}-1}. \qquad (9.8)$$

The first factor in (9.8) is an indicator of the predictive ability of the linear combination $b'X_k$ for the Y_k, and the second factor is a power of the sum of squares of the $b'X_k$ relevant to principal components analysis of the X_k. The maximization (with respect to ℓ for given vector b) built into the criterion (9.8) is straightforward, with the factor in brackets in (9.8) equal to $\{b'\boldsymbol{X}\boldsymbol{Y}'\ell\}^2$, leading to the maximizing unit vector $\ell = \boldsymbol{Y}\boldsymbol{X}'b/(b'\boldsymbol{X}\boldsymbol{Y}'\boldsymbol{Y}\boldsymbol{X}'b)^{1/2}$. Thus, with $\gamma = \alpha/(1-\alpha)$, this gives the criterion as

$$T_\alpha = (b'\boldsymbol{X}\boldsymbol{Y}'\boldsymbol{Y}\boldsymbol{X}'b)\,(b'\boldsymbol{X}\boldsymbol{X}'b)^{\gamma-1}.$$

Numerical details of the method for determination of the vector b maximizing T_α (subject to the orthogonality constraints) are discussed by Stone and Brooks (1990) and Brooks and Stone (1994). As with partial least squares, the final estimated regression equations are obtained by ordinary LS calculation of the response vectors Y_k on the first w constructed predictor variables, $X_k^* = (x_{1k}^*, \ldots, x_{wk}^*)'$.

Brooks and Stone (1994) note that as $\alpha \to 1$ (so that $\gamma \to \infty$ and hence the criterion tends to be dominated by the second factor $(b'\boldsymbol{X}\boldsymbol{X}'b)^{\gamma-1}$), the predictors constructed by this method become the principal components of $\boldsymbol{X}$. For the case $\alpha = 0$, the criterion becomes $T_0 = b'\boldsymbol{X}\boldsymbol{Y}'\boldsymbol{Y}\boldsymbol{X}'b/b'\boldsymbol{X}\boldsymbol{X}'b$. For this case, no more than m predictor variables $x_{jk}^* = b_j'X_k$ are constructible; when $m \leq n < T$, it follows that the vectors b_j are directly related to the eigenvectors $\hat{V}_j$ of the matrix $\boldsymbol{Y}\boldsymbol{X}'(\boldsymbol{X}\boldsymbol{X}')^{-1}\boldsymbol{X}\boldsymbol{Y}'$ corresponding to the largest eigenvalues, with b_j' equal to a normalized version of $\hat{V}_j'\boldsymbol{Y}\boldsymbol{X}'(\boldsymbol{X}\boldsymbol{X}')^{-1}$. Hence, this case is equivalent to reduced-rank regression calculations with $\Sigma_{\epsilon\epsilon}$ taken to be of the form $\sigma^2 I_m$, the "least squares" method used by Davies and Tso (1982) as discussed in Section 2.3. In addition, the choice $\alpha = 1/2$ in the criterion gives simply $T_{1/2} = (b'\boldsymbol{X}\boldsymbol{Y}'\boldsymbol{Y}\boldsymbol{X}'b)$ and leads to a modification due to de Jong (1993) of a version of partial least squares regression determination. We note that for the univariate case, $m = 1$, Sundberg (1993) established a direct relation between first factor continuum regression (i.e., using only the first $w = 1$ predictor $x_{1k}^* = b_1'X_k$) and ridge regression, and discussed its implications.

In conclusion, we hope that the reduced-rank and alternate dimension-reduction tools for multivariate regression, as surveyed above, will develop further in interest and find broader use in the future as large volumes of data are routinely being collected for analysis in many areas of application.

Appendix

Table A.1 Biochemical data on urine samples of patients from a study by Smith et al. (1962).

y_1	y_2	y_3	y_4	y_5	x_1	x_2	x_3
17.60	1.50	1.50	1.88	7.50	0.98	2.05	2.40
13.40	1.65	1.32	2.24	7.10	0.98	1.60	3.20
20.30	0.90	0.89	1.28	2.30	1.15	4.80	1.70
22.30	1.75	1.50	2.24	4.00	1.28	2.30	3.00
18.50	1.20	1.03	1.84	2.00	1.28	2.15	2.70
12.10	1.90	1.87	2.40	16.80	1.22	2.15	2.50
10.10	2.30	2.08	2.68	0.90	1.22	1.90	2.80
14.70	2.35	2.55	3.00	2.00	1.30	1.75	2.40
14.40	2.50	2.38	2.84	3.80	1.30	1.55	2.70
18.10	1.50	1.20	2.60	14.50	1.35	2.20	3.10
16.90	1.40	1.15	1.72	8.00	1.35	3.05	3.20
23.70	1.65	1.58	1.60	4.90	1.35	2.75	2.00
18.00	1.60	1.68	2.00	3.60	1.35	2.10	2.30
14.80	2.45	2.15	3.12	12.00	1.40	1.70	3.10
15.60	1.65	1.42	2.56	5.20	1.40	2.35	2.80
16.20	1.65	1.62	2.04	10.20	1.41	1.85	2.10
17.50	1.05	1.56	1.48	9.60	1.41	2.65	1.50
14.10	2.70	2.77	2.56	6.90	1.62	3.05	2.60
22.50	0.85	1.65	1.20	3.50	1.62	4.30	1.60
17.00	0.70	0.97	1.24	1.90	2.05	3.50	1.80
12.50	0.80	0.80	0.64	0.70	2.05	4.75	1.00
21.50	1.80	1.77	2.60	8.30	2.30	1.95	3.30
13.00	2.20	1.85	3.84	13.00	2.30	1.60	3.50
13.00	3.55	3.18	3.48	18.30	2.15	2.40	3.30
12.00	3.65	2.40	3.00	14.50	2.15	2.70	3.40
22.80	0.55	1.00	1.14	3.30	2.30	4.75	1.60
18.40	1.05	1.17	1.36	4.90	2.30	4.90	2.80
8.70	4.25	3.62	3.84	19.50	2.62	1.15	2.50
9.40	3.85	3.36	5.12	1.30	2.62	0.97	2.80
15.00	2.45	2.38	2.40	20.00	2.55	3.25	2.70
12.90	1.70	1.74	2.48	1.00	2.55	3.10	2.30
12.10	1.80	2.00	2.24	5.00	2.70	2.45	2.50
11.50	2.25	2.25	3.12	5.10	2.70	2.20	3.40

Response variables are y_1 = pigment creatinine, y_2 = phosphate (mg/ml), y_3 = phosphorous (mg/ml), y_4 = creatinine (mg/ml), y_5 = choline (μg/ml); predictor variables are x_1 = weight (lbs/100), x_2 = volume (ml/100), x_3 = 100(specific gravity − 1).

Table A.2 Quarterly macroeconomic time series data for the United Kingdom, 1948–1956, from Klein et al. (1961).

y_1	y_2	y_3	y_4	y_5	x_1	x_2	x_3	x_4	x_5
95.43	99.61	89.74	99.3	91.3	99.0	98.7	97.5	97.8	99.7
103.11	102.69	95.67	100.0	99.3	99.8	99.7	100.2	99.3	102.2
100.27	98.79	108.76	100.7	102.0	100.4	100.3	100.7	101.1	101.6
101.59	98.91	105.83	100.0	107.4	100.8	101.3	101.6	101.9	102.2
101.03	100.11	105.54	100.7	113.3	100.6	102.2	101.5	102.6	101.2
108.71	103.09	98.67	107.4	106.5	100.9	102.5	100.8	102.6	102.9
105.97	99.79	97.46	109.7	103.7	101.1	103.2	99.1	102.6	102.9
107.89	101.51	104.13	107.9	115.3	101.7	103.2	102.9	103.3	102.9
110.03	103.21	107.24	104.6	123.2	101.7	104.1	105.4	105.9	104.6
115.11	104.89	102.37	112.9	119.3	101.8	104.1	108.4	107.7	106.8
112.57	102.09	105.36	104.5	125.4	102.1	104.1	108.2	109.5	105.9
116.59	105.51	99.03	104.2	134.4	102.6	106.7	116.5	113.4	106.3
113.73	106.51	87.34	112.1	122.7	102.6	109.8	125.0	119.5	109.5
122.41	106.39	73.07	120.1	130.6	102.7	112.0	134.5	127.7	114.6
116.27	102.79	79.36	125.8	123.2	103.3	113.9	134.8	134.7	116.9
116.59	104.51	93.43	120.9	125.9	103.3	117.7	133.7	137.3	117.3
116.03	103.51	119.64	119.0	131.9	103.2	120.9	134.8	137.7	119.3
115.41	106.29	151.67	113.2	115.0	102.9	121.8	132.3	138.1	121.9
108.97	103.59	142.96	100.2	106.7	102.8	123.0	128.6	136.0	121.5
115.59	104.81	131.53	106.2	118.9	102.8	125.9	125.0	133.8	122.0
116.73	105.31	127.74	113.3	119.3	102.6	127.5	123.5	132.5	123.1
121.71	110.79	111.97	123.5	121.7	102.9	127.8	121.1	130.8	124.9
117.97	105.49	106.86	118.7	122.3	103.4	129.0	119.1	131.2	123.6
125.89	109.41	104.03	121.0	134.2	103.8	130.0	117.9	130.3	122.8
126.73	106.11	108.44	120.4	131.0	103.9	131.6	116.4	129.9	124.6
131.41	111.59	93.47	121.5	130.7	104.3	134.1	118.4	129.9	126.3
126.27	108.09	89.06	120.0	129.0	104.9	135.1	119.9	129.9	127.0
134.29	114.71	84.23	121.8	129.9	105.3	136.0	120.7	129.9	126.2
134.73	107.81	80.64	138.2	143.3	105.2	139.4	123.8	130.8	128.3
139.41	113.19	76.87	125.8	125.1	105.6	143.9	122.3	131.6	130.2
130.97	112.59	77.06	137.4	138.4	106.2	144.8	121.5	133.4	130.7
141.29	114.91	71.43	137.1	147.6	106.6	145.5	123.5	134.2	131.7
135.73	112.01	75.74	136.9	144.1	106.5	150.2	125.0	136.0	134.6
140.41	114.59	78.77	136.6	151.2	106.7	154.9	125.5	136.4	137.6
130.97	110.99	95.26	128.9	135.8	107.0	155.9	122.6	138.1	136.3
137.89	119.21	88.93	132.4	152.7	107.0	156.5	127.4	139.0	136.1

Response variables are y_1 = industrial production, y_2 = consumption, y_3 = unemployment, y_4 = total imports, y_5 = total exports; predictor variables are x_1 = total labor force, x_2 = weekly wage rates, x_3 = price index of imports, x_4 = price index of exports, x_5 = price index of consumption.

Table A.3 Blood sugar concentrations ($y_0 - y_5$, in mg/100 ml) of 36 rabbits at 0, 1, 2, 3, 4, and 5 hours after administration of insulin dose, with two types of insulin preparation (standard, $x_1 = -1$, and test, $x_1 = +1$) and two dose levels (0.75 units, $x_2 = -1$, and 1.50 units, $x_2 = +1$), from a study by Volund (1980).

y_0	y_1	y_2	y_3	y_4	y_5	x_1	x_2
96	37	31	33	35	41	1	1
90	47	48	55	68	89	1	1
99	49	55	64	74	97	1	1
95	33	37	43	63	92	1	1
107	62	62	85	110	117	1	1
81	40	43	45	49	55	1	1
95	49	56	63	68	88	1	1
105	53	57	69	103	106	1	1
97	50	53	59	82	96	1	1
97	54	57	66	80	89	1	−1
105	66	83	95	97	100	1	−1
105	49	54	56	70	90	1	−1
106	79	92	95	99	100	1	−1
92	46	51	57	73	91	1	−1
91	61	64	71	80	90	1	−1
101	51	63	91	95	96	1	−1
87	53	55	57	78	89	1	−1
94	57	70	81	94	96	1	−1
98	48	55	71	91	96	−1	1
98	41	43	61	89	101	−1	1
103	60	56	61	76	97	−1	1
99	36	43	57	89	102	−1	1
97	44	51	58	85	105	−1	1
95	41	45	49	59	78	−1	1
109	65	62	72	93	104	−1	1
91	57	60	61	67	83	−1	1
99	43	48	52	61	86	−1	1
102	51	56	81	97	103	−1	−1
96	57	55	72	85	89	−1	−1
111	84	83	91	101	102	−1	−1
105	57	67	83	100	103	−1	−1
105	57	61	70	90	98	−1	−1
98	55	67	88	94	95	−1	−1
98	69	72	89	98	98	−1	−1
90	53	61	78	94	95	−1	−1
100	60	63	67	77	104	−1	−1

References

[1] Ahn, S. K. and Reinsel, G. C. (1988). Nested reduced-rank autoregressive models for multiple time series. *Journal of the American Statistical Association*, 83, 849–856.

[2] Ahn, S. K. and Reinsel, G. C. (1990). Estimation for partially nonstationary multivariate autoregressive models. *Journal of the American Statistical Association*, 85, 813–823.

[3] Aitchison, J. and Silvey, S. D. (1958). Maximum likelihood estimation of parameters subject to restraints. *Annals of Mathematical Statistics*, 29, 813–828.

[4] Akaike, H. (1974). A new look at the statistical model identification. *IEEE Transactions on Automatic Control*, AC-19, 716–723.

[5] Akaike, H. (1976). Canonical correlation analysis of time series and the use of an information criterion. In *Systems Identification: Advances and Case Studies*, Eds. R. K. Mehra and D. G. Lainiotis, pp. 27–96. New York: Academic Press.

[6] Albert, J. M. and Kshirsagar, A. M. (1993). The reduced-rank growth curve model for discriminant analysis of longitudinal data. *Australian Journal of Statistics*, 35, 345–357.

[7] Amemiya, Y. and Fuller, W. A. (1984). Estimation for the multivariate errors-in-variables model with estimated error covariance matrix. *Annals of Statistics*, 12, 497–509.

[8] Anderson, T. W. (1951). Estimating linear restrictions on regression coefficients for multivariate normal distributions. *Annals of Mathematical Statistics*, 22, 327–351.

[9] Anderson, T. W. (1963). The use of factor analysis in the statistical analysis of multiple time series. *Psychometrika*, 28, 1–25.

[10] Anderson, T. W. (1971). *The Statistical Analysis of Time Series.* New York: Wiley.

[11] Anderson, T. W. (1976). Estimation of linear functional relationships: approximate distributions and connections with simultaneous equations in econometrics (with discussion). *Journal of the Royal Statistical Society*, B 38, 1–36.

[12] Anderson, T. W. (1984a). *An Introduction to Multivariate Statistical Analysis*, Second Edition. New York: Wiley.

[13] Anderson, T. W. (1984b). Estimating linear statistical relationships. *Annals of Statistics*, 12, 1–45.

[14] Anderson, T. W. (1991). Trygve Haavelmo and simultaneous equations models. *Scandinavian Journal of Statistics*, 18, 1–19.

[15] Anderson, T. W. and Rubin, H. (1956). Statistical inference in factor analysis. In *Proceedings of the Third Berkeley Symposium on Mathematical Statistics and Probability*, 5, Ed. J. Neyman, pp. 111–150. Berkeley: University of California Press.

[16] Bartlett, M. S. (1938). Further aspects of the theory of multiple regression. *Proceedings of the Cambridge Philosophical Society*, 34, 33–40.

[17] Bartlett, M. S. (1947). Multivariate analysis. *Journal of the Royal Statistical Society*, B 9, 176–197.

[18] Bekker, P., Dobbelstein, P. and Wansbeek, T. (1996). The APT model as reduced-rank regression. *Journal of Business and Economic Statistics*, 14, 199–202.

[19] Bewley, R. and Yang, M. (1995). Tests for cointegration based on canonical correlation analysis. *Journal of the American Statistical Association*, 90, 990–996.

[20] Bhargava, A. K. (1979). Estimation of a linear transformation and an associated distributional problem. *Journal of Statistical Planning and Inference*, 3, 19–26.

[21] Bilodeau, M. and Kariya, T. (1989). Minimax estimators in the normal MANOVA model. *Journal of Multivariate Analysis*, 28, 260–270.

[22] Black, F. (1972). Capital market equilibrium with restricted borrowing. *Journal of Business*, 45, 444–454.

[23] Blattberg, R. C. and George E. I. (1991). Shrinkage estimation of price and promotional elasticities: Seemingly unrelated equations. *Journal of the American Statistical Association*, 86, 304–315.

[24] Blattberg, R. C., Kim, B. and Ye, J. (1994). Large scale databases: The new marketing challenge. In *The Marketing Information Revolution*, Eds. R. C. Blattberg, R. Glazer, and J. D. C. Little, Chap. 8, pp. 173–203. Boston: Harvard Business School Press.

[25] Blattberg, R. C. and Wisniewski, K. J. (1989). Price-induced patterns of competition. *Marketing Science*, 8, 291–309.

[26] Boot, J. C. G. and de Wit, G. M. (1960). Investment demand: An empirical contribution to the aggregation problem. *International Economic Review*, 1, 3–30.

[27] Bossaerts, P. (1988). Common nonstationary components of asset prices. *Journal of Economic Dynamics and Control*, 12, 347–364.

[28] Box, G. E. P. and Tiao, G. C. (1977). A canonical analysis of multiple time series. *Biometrika*, 64, 355–365.

[29] Breiman, L. and Friedman, J. H. (1997). Predicting multivariate responses in multiple linear regression (with discussion). *Journal of the Royal Statistical Society*, B 59, 3–54.

[30] Brillinger, D. R. (1969). The canonical analysis of stationary time series. In *Multivariate Analysis*, 2, Ed. P.R. Krishnaiah, pp. 331–350. New York: Academic Press.

[31] Brillinger, D. R. (1981). *Time Series: Data Analysis and Theory*, expanded edition. San Francisco: Holden-Day.

[32] Brooks, R. and Stone, M. (1994). Joint continuum regression for multiple predictands. *Journal of the American Statistical Association*, 89, 1374–1377.

[33] Brown, P. J. and Payne, C. (1975). Election night forecasting (with discussion). *Journal of the Royal Statistical Society*, A 138, 463–498.

[34] Brown, P. J. and Zidek, J. V. (1980). Adaptive multivariate ridge regression. *Annals of Statistics*, 8, 64–74.

[35] Brown, P. J. and Zidek, J. V. (1982). Multivariate regression shrinkage estimators with unknown covariance matrix. *Scandinavian Journal of Statistics*, 9, 209–215.

[36] Campbell, J. (1987). Stock returns and the term structure. *Journal of Financial Economics*, 18, 373–399.

[37] Campbell, J., Lo., A. and MacKinlay, A. C. (1997). *The Econometrics of Financial Markets.* Princeton, NJ: Princeton University Press.

[38] Campbell, N. (1984). Canonical variate analysis – a general formulation. *Australian Journal of Statistics*, 26, 86–96.

[39] Chi, E. M. and Reinsel, G. C. (1989). Models for longitudinal data with random effects and AR(1) errors. *Journal of the American Statistical Association*, 84, 452–459.

[40] Chinchilli, V. M. and Elswick, R. K. (1985). A mixture of the MANOVA and GMANOVA models. *Communications in Statistics, Theory and Methods*, 14, 3075–3089.

[41] Cragg, J. G. and Donald, S. G. (1996). On the asymptotic properties of LDU-based tests of the rank of a matrix. *Journal of the American Statistical Association*, 91, 1301–1309.

[42] Davies, P. T. and Tso, M. K. S. (1982). Procedures for reduced-rank regression. *Applied Statistics*, 31, 244–255.

[43] de Jong, S. (1993). SIMPLS: An alternative approach to partial least squares regression. *Chemometrics and Intelligent Laboratory Systems*, 18, 251–263.

[44] Diggle, P. J. (1988). An approach to the analysis of repeated measurements. *Biometrics*, 44, 959–971.

[45] Doan, T., Litterman, R. and Sims, C. (1984). Forecasting and conditional projection using realistic prior distributions. *Econometric Reviews*, 3, 1–100.

[46] Eckart, C. and Young, G. (1936). The approximation of one matrix by another of lower rank. *Psychometrika*, 1, 211–218.

[47] Efron, B. and Morris, C. (1972). Empirical Bayes on vector observations: An extension of Stein's method. *Biometrika*, 59, 335–347.

[48] Engle, R. F. and Granger, C. W. J. (1987). Co-integration and error correction: Representation, estimation, and testing. *Econometrica*, 55, 251–276.

[49] Ferson, W. and Harvey, C. (1991). The variation of economic risk premiums. *Journal of Political Economy*, 99, 385–415.

[50] Fisher, R. A. (1938). The statistical utilization of multiple measurements. *Annals of Eugenics*, 8, 376–386.

[51] Fortier, J. J. (1966). Simultaneous linear prediction. *Psychometrika*, 31, 369–381.

[52] Fountis, N. G. and Dickey, D. A. (1989). Testing for a unit root nonstationarity in multivariate autoregressive time series. *Annals of Statistics*, 17, 419–428.

[53] Frank, I. E. and Friedman, J. H. (1993). A statistical view of some chemometrics regression tools (with discussion). *Technometrics*, 35, 109–148.

[54] Gabriel, K. R. and Zamir, S. (1979). Lower rank approximation of matrices by least squares with any choice of weights. *Technometrics*, 21, 489–498.

[55] Gallant, A. R. and Goebel, J. J. (1976). Nonlinear regression with autocorrelated errors. *Journal of the American Statistical Association*, 71, 961–967.

[56] Garthwaite, P. H. (1994). An interpretation of partial least squares. *Journal of the American Statistical Association*, 89, 122–127.

[57] Geary, R. C. (1948). Studies in relations between economic time series. *Journal of the Royal Statistical Society*, B 10, 140–158.

[58] Geweke, J. F. (1996). Bayesian reduced rank regression in econometrics. *Journal of Econometrics*, 75, 121–146.

[59] Gibbons, M. R. (1982). Multivariate tests of financial models: a new approach. *Journal of Financial Economics*, 10, 3–27.

[60] Gibbons, M. and Ferson, W. (1985). Testing asset pricing models with changing expectations and an unobservable market portfolio. *Journal of Financial Economics*, 14, 217–236.

[61] Gibbons, M. R., Ross, S. A. and Shanken, J. (1989). A test of the efficiency of a given portfolio. *Econometrica*, 57, 1121–1152.

[62] Giles, D. E. A. and Hampton, P. (1984). Regional production relationships during the industrialization of New Zealand, 1935–1948. *Journal of Regional Science*, 24, 519–533.

[63] Glasbey, C. A. (1992). A reduced rank regression model for local variation in solar radiation. *Applied Statistics*, 41, 381–387.

[64] Gleser, L. J. (1981). Estimation in a multivariate "errors in variables" regression model: large sample results. *Annals of Statistics*, 9, 24–44.

[65] Gleser, L. J. and Olkin, I. (1966). A k-sample regression model with covariance. In *Multivariate Analysis*, pp. 59–72. New York: Academic Press.

[66] Gleser, L. J. and Olkin, I. (1970). Linear models in multivariate analysis. In *Essays in Probability and Statistics*, Ed. R. C. Bose, pp. 267–292. New York: Wiley.

[67] Gleser, L. J. and Watson, G. S. (1973). Estimation of a linear transformation. *Biometrika*, 60, 525–534.

[68] Goldberger, A. S. (1972). Maximum likelihood estimation of regressions containing unobservable independent variables. *International Economic Review*, 13, 1–15.

[69] Gonzalo, J. and Granger, C. W. J. (1995). Estimation of common long-memory components in cointegrated systems. *Journal of Business and Economic Statistics*, 13, 27–35.

[70] Grizzle, J. E. and Allen, D. M. (1969). Analysis of growth and dose response curves. *Biometrics*, 25, 357–381.

[71] Gudmundsson, G. (1977). Multivariate analysis of economic variables. *Applied Statistics*, 26, 48–59.

[72] Haitovsky, Y. (1987). On multivariate ridge regression. *Biometrika*, 74, 563–570.

[73] Hannan, E. J. and Kavalieris, L. (1984). Multivariate linear time series models. *Advances in Applied Probability*, 16, 492–561.

[74] Hansen, L. and Hodrick, R. (1983). Risk averse speculation in the forward foreign exchange market: An econometric analysis of linear models. In *Exchange Rates and International Macroeconomics*, Ed. J. Frenkel, pp. 113–152. Chicago: University of Chicago Press.

[75] Hastie, T. and Tibshirani, R. (1996). Discriminant analysis by Gaussian mixtures. *Journal of the Royal Statistical Society*, B 58, 155–176.

[76] Hatanaka, M. (1976). Several efficient two step estimators for the dynamic simultaneous equations model with autoregressive disturbances. *Journal of Econometrics*, 4, 189–204.

[77] Healy, J. D. (1980). Maximum likelihood estimation of a multivariate linear functional relationship. *Journal of Multivariate Analysis*, 10, 243–251.

[78] Heck, D. L. (1960). Charts of some upper percentage points of the distribution of the largest characteristic root. *Annals of Mathematical Statistics*, 31, 625–642.

[79] Helland, I. S. (1988). On the structure of partial least squares regression. *Communications in Statistics, Simulation and Computation*, B 17, 581–607.

[80] Helland, I. S. (1990). Partial least squares regression and statistical models. *Scandinavian Journal of Statistics*, 17, 97–114.

[81] Hendry, D. F. (1971). Maximum likelihood estimation of systems of simultaneous regression equations with errors generated by a vector autoregressive process. *International Economic Review*, 12, 257–272.

[82] Honda, T. (1991). Minimax estimators in the Manova model for arbitrary quadratic loss and unknown covariance matrix. *Journal of Multivariate Analysis*, 36, 113–120.

[83] Hoskuldsson, P. (1988). PLS regression methods. *Journal of Chemometrics*, 2, 211–228.

[84] Hotelling, H. (1933). Analysis of a complex of statistical variables into principal components. *Journal of Education Psychology*, 24, 417–441, 498–520.

[85] Hotelling, H. (1935). The most predictable criterion. *Journal of Education Psychology*, 26, 139–142.

[86] Hotelling, H. (1936). Relations between two sets of variables. *Biometrika*, 28, 321–377.

[87] Hsu, P. L. (1941). On the limiting distribution of roots of a determinantal equation. *Journal of London Mathematical Society*, 16, 183–194.

[88] Izenman, A. J. (1975). Reduced-rank regression for the multivariate linear model. *Journal of Multivariate Analysis*, 5, 248–264.

[89] Izenman, A. J. (1980). Assessing dimensionality in multivariate regression. In *Handbook of Statistics*, 1, Ed. P.R. Krishnaiah, pp. 571–592. New York: North Holland.

[90] Jenkins, G. M. and Alavi, A. S. (1981). Some aspects of modeling and forecasting multivariate time series. *Journal of Time Series Analysis*, 2, 1–47.

[91] Jennrich, R. I. and Schluchter, M. D. (1986). Unbalanced repeated measures models with structural covariance matrices. *Biometrics*, 42, 805–820.

[92] Johansen, S. (1988). Statistical analysis of cointegration vectors. *Journal of Economic Dynamics and Control*, 12, 231–254.

[93] Johansen, S. (1991). Estimation and hypothesis testing of cointegration vectors in Gaussian vector autoregressive models. *Econometrica*, 59, 1551–1580.

[94] Johansen, S. and Juselius, K. (1990). Maximum likelihood estimation and inference on cointegration–with applications to the demand for money. *Oxford Bulletin of Economics and Statistics*, 52, 169–210.

[95] Jones, R. H. (1986). Random effects and the Kalman filter. *Proceedings of the Business and Economic Statistics Section, American Statistical Association*, pp. 69–75.

[96] Jones, R. H. (1990). Serial correlation or random subject effects? *Communications in Statistics, Part B–Simulation and Computation*, 19, 1105–1123.

[97] Jöreskog, K. G. (1967). Some contributions to maximum likelihood factor analysis. *Psychometrika*, 32, 443–482.

[98] Jöreskog, K. G. (1969). A general approach to confirmatory maximum likelihood factor analysis. *Psychometrika*, 34, 183–202.

[99] Jöreskog, K. G. and Goldberger, A. S. (1975). Estimation of a model with multiple indicators and multiple causes of a single latent variable. *Journal of the American Statistical Association*, 70, 631–639.

[100] Kabe, D. G. (1975). Some results for the GMANOVA model. *Communications in Statistics*, 4, 813–820.

[101] Keller, W. J. and Wansbeek, T. (1983). Multivariate methods for quantitative and qualitative data. *Journal of Econometrics*, 22, 91–111.

[102] Khatri, C. G. (1966). A note on a MANOVA model applied to problems in growth curve. *Annals of the Institute of Statistical Mathematics*, 18, 75–86.

[103] Klein, L. R., Ball, R. J., Hazlewood, A. and Vandome, P. (1961). *An Econometric Model of the United Kingdom*. Oxford: Blackwell.

[104] Kohn, R. (1979). Asymptotic estimation and hypothesis testing results for vector linear time series models. *Econometrica*, 47, 1005–1030.

[105] Konno, Y. (1991). On estimation of a matrix of normal means with unknown covariance matrix. *Journal of Multivariate Analysis*, 36, 44–55.

[106] Kshirsagar, A. M. and Smith, W. B. (1995). *Growth Curves*. New York: Marcel Dekker.

[107] Laird, N. M. and Ware, J. H. (1982). Random-effects models for longitudinal data. *Biometrics*, 38, 963–974.

[108] Lawley, D. N. (1940). The estimation of factor loadings by the method of maximum likelihood. *Proceedings of the Royal Society*, A 60, 64–82.

[109] Lawley, D. N. (1943). The application of the maximum likelihood method to factor analysis. *British Journal of Psychology*, 33, 172–175.

[110] Lawley, D. N. (1953). A modified method of estimation in factor analysis and some large sample results. In *Uppsala Symposium on Psychological Factor Analysis*, pp. 35–42. Stockholm: Almqvist and Wicksell.

[111] Lee, Y. S. (1972). Some results on the distribution of Wilks' likelihood-ratio criterion. *Biometrika*, 59, 649–664.

[112] Lintner, J. (1965). The valuation of risk assets and the selection of risky investment in stock portfolios and capital budgets. *Review of Economics and Statistics*, 47, 13–37.

[113] Lütkepohl, H. (1993). *Introduction to Multiple Time Series Analysis*, Second Edition. Berlin: Springer-Verlag.

[114] Lyttkens, E. (1972). Regression aspects of canonical correlation. *Journal of Multivariate Analysis*, 2, 418–439.

[115] Markowitz, H. (1959). *Portfolio Selection: Efficient Diversification of Investments.* New York: Wiley.

[116] Marshall, A. S. and Olkin, I. (1979). *Inequalities: Theory of Majorization and Its Applications.* New York: Academic Press.

[117] McElroy, M. B. and Burmeister, E. (1988). Arbitrage pricing theory as a restricted nonlinear multivariate regression model: Iterated nonlinear seemingly unrelated regression models. *Journal of Business and Economic Statistics*, 6, 29–42.

[118] Merton, R. T. (1973). An intertemporal capital asset pricing model. *Econometrica*, 41, 867–887.

[119] Moran, P. A. P. (1971). Estimating structural and functional relationships. *Journal of Multivariate Analysis*, 1, 232–255.

[120] Muirhead, R. J. (1982). *Aspects of Multivariate Statistical Theory.* New York: Wiley.

[121] Neudecker, H. (1969). Some theorems on matrix differentiation with special reference to Kronecker matrix products. *Journal of the American Statistical Association*, 64, 953–963.

[122] Parzen, E. M. (1969). Multiple time series modeling. In *Multivariate Analysis*, 2, Ed. P. R. Krishnaiah, pp. 389–409. New York: Academic Press.

[123] Phillips, P. C. B. and Ouliaris, S. (1986). Testing for cointegration. Discussion Paper No. 809, Cowles Foundation for Research in Economics, Yale University.

[124] Pillai, K. C. S. (1967). Upper percentage points of the largest root of a matrix in multivariate analysis. *Biometrika*, 54, 189–194.

[125] Pillai, K. C. and Gupta, A. K. (1969). On the exact distribution of Wilks' criterion. *Biometrika*, 56, 109–118.

[126] Potthoff, R. F. and Roy, S. N. (1964). A generalized multivariate analysis of variance model useful especially for growth curve problems. *Biometrika*, 51, 313–326.

[127] Priestley, M. B., Rao, T. S. and Tong, H. (1974a). Identification of the structure of multivariable stochastic systems. In *Multivariate Analysis*, 3, Ed. P. R. Krishnaiah, pp. 351–368. New York: Academic Press.

[128] Priestley, M. B., Rao, T. S. and Tong, H. (1974b). Applications of principal component analysis and factor analysis in the identification of multivariable systems. *IEEE Transactions on Automatic Control*, 19, 730–734.

[129] Quenouille, M. H. (1968). *The Analysis of Multiple Time Series*. London: Griffin.

[130] Rao, C. R. (1955). Estimation and tests of significance in factor analysis. *Psychometrika*, 20, 93–111.

[131] Rao, C. R. (1964). The use and interpretation of principal component analysis in applied research. *Sankhya*, A, 26, 329–358.

[132] Rao, C. R. (1965). The theory of least squares when the parameters are stochastic and its application to the analysis of growth curves. *Biometrika*, 52, 447–458.

[133] Rao, C. R. (1966). Covariance adjustment and related problems in multivariate analysis. In *Multivariate Analysis*, 2, Ed. P. R. Krishnaiah, pp. 321–328. New York: Academic Press.

[134] Rao, C. R. (1967). Least squares theory using an estimated dispersion matrix and its application to measurement of signals. In *Proceedings of the 5th Berkeley Symposium on Mathematical Statistics and Probability*, 1, Ed. J. Neyman, pp. 355–372. Berkeley: University of California Press.

[135] Rao, C. R. (1973). *Linear Statistical Inference and Its Applications*, Second Edition. New York: Wiley.

[136] Rao, C. R. (1979). Separation theorems for singular values of matrices and their applications in multivariate analysis. *Journal of Multivariate Analysis*, 9, 362–377.

[137] Reiersøl, O. (1950). On the identifiability of parameters in Thurstone's multiple factor analysis. *Psychometrika*, 15, 121–149.

[138] Reinsel, G. (1979). FIML estimation of the dynamic simultaneous equations model with ARMA disturbances. *Journal of Econometrics*, 9, 263–281.

[139] Reinsel, G. (1983). Some results on multivariate autoregressive index models. *Biometrika*, 70, 145–156.

[140] Reinsel, G. C. (1985). Mean squared error properties of empirical Bayes estimators in a multivariate random effects general linear model. *Journal of the American Statistical Association*, 80, 642–650.

[141] Reinsel, G. C. (1997). *Elements of Multivariate Time Series Analysis*, Second Edition. New York: Springer-Verlag.

[142] Reinsel, G. C. (1998). On multivariate linear regression shrinkage and reduced-rank procedures. Technical Report, Department of Statistics, University of Wisconsin, Madison, WI.

[143] Reinsel, G. C. and Ahn, S. K. (1992). Vector autoregressive models with unit roots and reduced rank structure: Estimation, likelihood ratio test, and forecasting. *Journal of Time Series Analysis*, 13, 353–375.

[144] Reinsel, G. C., Tiao, G. C., Wang, M. N., Lewis, R. and Nychka, D. (1981). Statistical analysis of stratospheric ozone data for the detection of trends. *Atmospheric Environment*, 15, 1569–1577.

[145] Revankar, N. S. (1974). Some finite sample results in the context of two seemingly unrelated regression equations. *Journal of the American Statistical Association*, 69, 187–190.

[146] Robinson, P. M. (1972). Non-linear regression for multiple time-series. *Journal of Applied Probability*, 9, 758–768.

[147] Robinson, P. M. (1973). Generalized canonical analysis for time series. *Journal of Multivariate Analysis*, 3, 141–160.

[148] Robinson, P. M. (1974). Identification, estimation and large-sample theory for regressions containing unobservable variables. *International Economic Review*, 15, 680–692.

[149] Rochon, J. and Helms, R. W. (1989). Maximum likelihood estimation for incomplete repeated-measures experiments under an ARMA covariance structure. *Biometrics*, 45, 207–218.

[150] Ross, S. A. (1976). The arbitrage theory of capital asset pricing. *Journal of Economic Theory*, 13, 341–360.

[151] Roy, S. N. (1953). On a heuristic method of test construction and its use in multivariate analysis. *Annals of Mathematical Statistics*, 24, 220–238.

[152] Roy, S. N., Gnanadesikan, R. and Srivastava, J. N. (1971). *Analysis and Design of Certain Quantitative Multiresponse Experiments.* New York: Pergamon Press.

[153] Ryan, D. A. J., Hubert, J. J., Carter, E. M., Sprague, J. B. and Parrott, J. (1992). A reduced-rank multivariate regression approach to aquatic joint toxicity experiments. *Biometrics*, 48, 155–162.

[154] Sargent, T. J. and Sims, C. A. (1977). Business cycle modeling without pretending to have too much of a prior economic theory. In *New Methods in Business Cycle Research*, Ed. C.A. Sims, pp. 45–110, Federal Reserve Bank of Minneapolis, MN.

[155] Schatzoff, M. (1966). Exact distribution of Wilks' likelihood ratio criterion. *Biometrika*, 53, 347–358.

[156] Schott, J. R. (1997). *Matrix Analysis for Statistics.* New York: Wiley.

[157] Schwarz, G. (1978). Estimating the dimension of a model. *Annals of Statistics*, 6, 461–464.

[158] Sclove, S. L. (1971). Improved estimation of parameters in multivariate regression. *Sankhya* A, 33, 61–66.

[159] Shanken, J. (1986). Testing portfolio efficiency when the zero-beta rate is unknown: a note. *Journal of Finance*, 41, 269–276.

[160] Sharpe, W. F. (1964). Capital asset prices: a theory of market equilibrium under conditions of risk. *Journal of Finance*, 19, 425–442.

[161] Silvey, S. D. (1959). The Lagrangian multiplier test. *Annals of Mathematical Statistics*, 30, 389–407.

[162] Sims, C. A. (1981). An autoregressive index model for the U.S., 1948–1975. In *Large-Scale Macro-Econometric Models*, Eds. J. Kmenta and J. B. Ramsey, pp. 283–327. Amsterdam: North Holland.

[163] Skagerberg, B., MacGregor, J. and Kiparissides, C. (1992). Multivariate data analysis applied to low-density polyethylene reactors. *Chemometrics and Intelligent Laboratory Systems*, 14, 341–356.

[164] Smith, H., Gnanadesikan, R. and Hughes, J. B. (1962). Multivariate analysis of variance (MANOVA). *Biometrics*, 18, 22–41.

[165] Sprent, P. (1966). A generalized least-squares approach to linear functional relationships (with discussion). *Journal of the Royal Statistical Society*, B 28, 278–297.

[166] Srivastava, J. N. (1967). On the extension of Gauss–Markov theorem to complex multivariate linear models. *Annals of the Institute of Statistical Mathematics*, 19, 417–437.

[167] Srivastava, M. S. (1997). Reduced rank discrimination. *Scandinavian Journal of Statistics*, 24, 115–124.

[168] Srivastava, M. S. and Khatri, C. G. (1979). *An Introduction to Multivariate Statistics.* New York: North Holland.

[169] Srivastava, V. K. and Giles, D. E. A. (1987). *Seemingly Unrelated Regression Equations Models.* New York: Marcel Dekker.

[170] Stambaugh, R. (1988). The information in forward rates: Implications for models of the term structure. *Journal of Financial Economics*, 21, 41–70.

[171] Stanek, E. J. and Koch, G. G. (1985). The equivalence of parameter estimates from growth curve models and seemingly unrelated regression models. *American Statistician*, 39, 149–152.

[172] Stock, J. H. and Watson, M. W. (1988). Testing for common trends. *Journal of the American Statistical Association*, 83, 1097–1107.

[173] Stone, M. (1974). Cross-validatory choice and assessment of statistical predictions (with discussion). *Journal of the Royal Statistical Society*, B 36, 111–147.

[174] Stone, M. and Brooks, R. J. (1990). Continuum regression: Cross-validated sequentially constructed prediction embracing ordinary least squares, partial least squares and principal components regression (with discussion). *Journal of the Royal Statistical Society*, B 52, 237–269. Corrigendum (1992), 54, 906–907.

[175] Sundberg, R. (1993). Continuum regression and ridge regression. *Journal of the Royal Statistical Society*, B 55, 653–659.

[176] Takane, Y., Kiers, H. A. L. and de Leeuw, J. (1995). Component analysis with different sets of constraints on different dimensions. *Psychometrika*, 60, 259–280.

[177] ter Braak, C. J. F. (1990). Interpreting canonical correlation analysis through biplots of structure correlations and weights. *Psychometrika*, 55, 519–531.

[178] Theobald, C. M. (1975). An inequality with application to multivariate analysis. *Biometrika*, 62, 461–466.

[179] Theobald, C. M. (1978). Letter to the editors. *Applied Statistics*, 27, 79.

[180] Thurstone, L. L. (1947). *Multiple-Factor Analysis.* Chicago: University of Chicago Press.

[181] Tiao, G. C. and Box, G. E. P. (1981). Modeling multiple time series with applications. *Journal of the American Statistical Association*, 76, 802–816.

[182] Tiao, G. C. and Tsay, R. S. (1983). Multiple time series modeling and extended sample cross-correlations. *Journal of Business and Economic Statistics*, 1, 43–56.

[183] Tiao, G. C. and Tsay, R. S. (1989). Model specification in multivariate time series (with Discussion). *Journal of the Royal Statistical Society*, B 51, 157–213.

[184] Tintner, G. (1945). A note on rank, multicollinearity and multiple regression. *Annals of Mathematical Statistics*, 16, 304–308.

[185] Tintner, G. (1950). A test for linear relations between weighted regression coefficients. *Journal of the Royal Statistical Society*, B 12, 273–277.

[186] Tso, M. K.-S. (1981). Reduced-rank regression and canonical analysis. *Journal of the Royal Statistical Society*, B 43, 183–189.

[187] van den Wollenberg, A. L. (1977). Redundancy analysis: An alternative for canonical correlation analysis. *Psychometrika*, 42, 207–219.

[188] van der Leeden, R. (1990). *Reduced Rank Regression With Structured Residuals.* Leiden: DSWO Press.

[189] van der Merwe, A. and Zidek, J. V. (1980). Multivariate regression analysis and canonical variates. *Canadian Journal of Statistics*, 8, 27–39.

[190] Velu, R. P. (1991). Reduced rank models with two sets of regressors. *Applied Statistics*, 40, 159–170.

[191] Velu, R. P. and Reinsel, G. C. (1987). Reduced rank regression with autoregressive errors. *Journal of Econometrics*, 35, 317–335.

[192] Velu, R. P., Reinsel, G. C. and Wichern, D. W. (1986). Reduced rank models for multiple time series. *Biometrika*, 73, 105–118.

[193] Velu, R. P., Wichern, D. W. and Reinsel, G. C. (1987). A note on non-stationarity and canonical analysis of multiple time series models. *Journal of Time Series Analysis*, 8, 479–487.

[194] Velu, R. P. and Zhou, G. (1997). Testing multi-Beta asset pricing models. Working Paper (Olin-96-25), Washington University, St. Louis MO.

[195] Villegas, C. (1961). Maximum likelihood estimation of a linear functional relationship. *Annals of Mathematical Statistics*, 32, 1048–1062.

[196] Villegas, C. (1982). Maximum likelihood and least squares estimation in linear and affine functional models. *Annals of Statistics*, 10, 256–265.

[197] Volund, A. (1980). Multivariate bioassay. *Biometrics*, 36, 225–236.

[198] Wold, H. (1975). Soft modeling by latent variables: The nonlinear iterative partial least squares approach. In *Perspectives in Probability and Statistics: Papers in Honour of M. S. Bartlett*, Ed. J. Gani, pp. 117–142. New York: Academic Press.

[199] Wold, H. (1984). PLS regression. In *Encyclopedia of Statistical Sciences*, Eds. N. L. Johnson and S. Kotz, Vol. 6, pp. 581–591. New York: Wiley.

[200] Zellner, A. (1962). An efficient method of estimating seemingly unrelated regressions and tests for aggregation bias. *Journal of the American Statistical Association*, 57, 348–368.

[201] Zellner, A. (1963). Estimators for seemingly unrelated regression equations: Some exact finite sample results. *Journal of the American Statistical Association*, 58, 977–992.

[202] Zellner, A. (1970). Estimation of regression relationships containing unobservable variables. *International Economic Review*, 11, 441–454.

[203] Zhou, G. (1991). Small sample tests of portfolio efficiency. *Journal of Financial Economics*, 30, 165–191.

[204] Zhou, G. (1994). Analytical GMM tests: Asset pricing with time-varying risk premiums. *Review of Financial Studies*, 7, 687–709.

[205] Zhou, G. (1995). Small sample rank tests with applications to asset pricing. *Journal of Empirical Finance*, 2, 71–93.

Subject Index

Author Index

Lecture Notes in Statistics

For information about Volumes 1 to 62 please contact Springer-Verlag

Vol. 63: J. Pfanzagl, Estimation in Semiparametric Models: Some Recent Developments. iii, 112 pages, 1990.

Vol. 64: S. Gabler, Minimax Solutions in Sampling from Finite Populations. v, 132 pages, 1990.

Vol. 65: A. Janssen, D.M. Mason, Non-Standard Rank Tests. vi, 252 pages, 1990.

Vol 66: T. Wright, Exact Confidence Bounds when Sampling from Small Finite Universes. xvi, 431 pages, 1991.

Vol. 67: M.A. Tanner, Tools for Statistical Inference: Observed Data and Data Augmentation Methods. vi, 110 pages, 1991.

Vol. 68: M. Taniguchi, Higher Order Asymptotic Theory for Time Series Analysis. viii, 160 pages, 1991.

Vol. 69: N.J.D. Nagelkerke, Maximum Likelihood Estimation of Functional Relationships. V, 110 pages, 1992.

Vol. 70: K. Iida, Studies on the Optimal Search Plan. viii, 130 pages, 1992.

Vol. 71: E.M.R.A. Engel, A Road to Randomness in Physical Systems. ix, 155 pages, 1992.

Vol. 72: J.K. Lindsey, The Analysis of Stochastic Processes using GLIM. vi, 294 pages, 1992.

Vol. 73: B.C. Arnold, E. Castillo, J.-M. Sarabia, Conditionally Specified Distributions. xiii, 151 pages, 1992.

Vol. 74: P. Barone, A. Frigessi, M. Piccioni, Stochastic Models, Statistical Methods, and Algorithms in Image Analysis. vi, 258 pages, 1992.

Vol. 75: P.K. Goel, N.S. Iyengar (Eds.), Bayesian Analysis in Statistics and Econometrics. xi, 410 pages, 1992.

Vol. 76: L. Bondesson, Generalized Gamma Convolutions and Related Classes of Distributions and Densities. viii, 173 pages, 1992.

Vol. 77: E. Mammen, When Does Bootstrap Work? Asymptotic Results and Simulations. vi, 196 pages, 1992.

Vol. 78: L. Fahrmeir, B. Francis, R. Gilchrist, G. Tutz (Eds.), Advances in GLIM and Statistical Modelling: Proceedings of the GLIM92 Conference and the 7th International Workshop on Statistical Modelling, Munich, 13-17 July 1992. ix, 225 pages, 1992.

Vol. 79: N. Schmitz, Optimal Sequentially Planned Decision Procedures. xii, 209 pages, 1992.

Vol. 80: M. Fligner, J. Verducci (Eds.), Probability Models and Statistical Analyses for Ranking Data. xxii, 306 pages, 1992.

Vol. 81: P. Spirtes, C. Glymour, R. Scheines, Causation, Prediction, and Search. xxiii, 526 pages, 1993.

Vol. 82: A. Korostelev and A. Tsybakov, Minimax Theory of Image Reconstruction. xii, 268 pages, 1993.

Vol. 83: C. Gatsonis, J. Hodges, R. Kass, N. Singpurwalla (Editors), Case Studies in Bayesian Statistics. xii, 437 pages, 1993.

Vol. 84: S. Yamada, Pivotal Measures in Statistical Experiments and Sufficiency. vii, 129 pages, 1994.

Vol. 85: P. Doukhan, Mixing: Properties and Examples. xi, 142 pages, 1994.

Vol. 86: W. Vach, Logistic Regression with Missing Values in the Covariates. xi, 139 pages, 1994.

Vol. 87: J. Müller, Lectures on Random Voronoi Tessellations.vii, 134 pages, 1994.

Vol. 88: J. E. Kolassa, Series Approximation Methods in Statistics. Second Edition, ix, 183 pages, 1997.

Vol. 89: P. Cheeseman, R.W. Oldford (Editors), Selecting Models From Data: AI and Statistics IV. xii, 487 pages, 1994.

Vol. 90: A. Csenki, Dependability for Systems with a Partitioned State Space: Markov and Semi-Markov Theory and Computational Implementation. x, 241 pages, 1994.

Vol. 91: J.D. Malley, Statistical Applications of Jordan Algebras. viii, 101 pages, 1994.

Vol. 92: M. Eerola, Probabilistic Causality in Longitudinal Studies. vii, 133 pages, 1994.

Vol. 93: Bernard Van Cutsem (Editor), Classification and Dissimilarity Analysis. xiv, 238 pages, 1994.

Vol. 94: Jane F. Gentleman and G.A. Whitmore (Editors), Case Studies in Data Analysis. viii, 262 pages, 1994.

Vol. 95: Shelemyahu Zacks, Stochastic Visibility in Random Fields. x, 175 pages, 1994.

Vol. 96: Ibrahim Rahimov, Random Sums and Branching Stochastic Processes. viii, 195 pages, 1995.

Vol. 97: R. Szekli, Stochastic Ordering and Dependence in Applied Probability. viii, 194 pages, 1995.

Vol. 98: Philippe Barbe and Patrice Bertail, The Weighted Bootstrap. viii, 230 pages, 1995.

Vol. 99: C.C. Heyde (Editor), Branching Processes: Proceedings of the First World Congress. viii, 185 pages, 1995.

Vol. 100: Wlodzimierz Bryc, The Normal Distribution: Characterizations with Applications. viii, 139 pages, 1995.

Vol. 101: H.H. Andersen, M.Højbjerre, D. Sørensen, P.S.Eriksen, Linear and Graphical Models: for the Multivariate Complex Normal Distribution. x, 184 pages, 1995.

Vol. 102: A.M. Mathai, Serge B. Provost, Takesi Hayakawa, Bilinear Forms and Zonal Polynomials. x, 378 pages, 1995.

Vol. 103: Anestis Antoniadis and Georges Oppenheim (Editors), Wavelets and Statistics. vi, 411 pages, 1995.

Vol. 104: Gilg U.H. Seeber, Brian J. Francis, Reinhold Hatzinger, Gabriele Steckel-Berger (Editors), Statistical Modelling: 10th International Workshop, Innsbruck, July 10-14th 1995. x, 327 pages, 1995.

Vol. 105: Constantine Gatsonis, James S. Hodges, Robert E. Kass, Nozer D. Singpurwalla(Editors), Case Studies in Bayesian Statistics, Volume II. x, 354 pages, 1995.

Vol. 106: Harald Niederreiter, Peter Jau-Shyong Shiue (Editors), Monte Carlo and Quasi-Monte Carlo Methods in Scientific Computing. xiv, 372 pages, 1995.

Vol. 107: Masafumi Akahira, Kei Takeuchi, Non-Regular Statistical Estimation. vii, 183 pages, 1995.

Vol. 108: Wesley L. Schaible (Editor), Indirect Estimators in U.S. Federal Programs. viii, 195 pages, 1995.

Vol. 109: Helmut Rieder (Editor), Robust Statistics, Data Analysis, and Computer Intensive Methods. xiv, 427 pages, 1996.

Vol. 110: D. Bosq, Nonparametric Statistics for Stochastic Processes. xii, 169 pages, 1996.

Vol. 111: Leon Willenborg, Ton de Waal, Statistical Disclosure Control in Practice. xiv, 152 pages, 1996.

Vol. 112: Doug Fischer, Hans-J. Lenz (Editors), Learning from Data. xii, 450 pages, 1996.

Vol. 113: Rainer Schwabe, Optimum Designs for Multi-Factor Models. viii, 124 pages, 1996.

Vol. 114: C.C. Heyde, Yu. V. Prohorov, R. Pyke, and S. T. Rachev (Editors), Athens Conference on Applied Probability and Time Series Analysis Volume I: Applied Probability In Honor of J.M. Gani. viii, 424 pages, 1996.

Vol. 115: P.M. Robinson, M. Rosenblatt (Editors), Athens Conference on Applied Probability and Time Series Analysis Volume II: Time Series Analysis In Memory of E.J. Hannan. viii, 448 pages, 1996.

Vol. 116: Genshiro Kitagawa and Will Gersch, Smoothness Priors Analysis of Time Series. x, 261 pages, 1996.

Vol. 117: Paul Glasserman, Karl Sigman, David D. Yao (Editors), Stochastic Networks. xii, 298, 1996.

Vol. 118: Radford M. Neal, Bayesian Learning for Neural Networks. xv, 183, 1996.

Vol. 119: Masanao Aoki, Arthur M. Havenner, Applications of Computer Aided Time Series Modeling. ix, 329 pages, 1997.

Vol. 120: Maia Berkane, Latent Variable Modeling and Applications to Causality. vi, 288 pages, 1997.

Vol. 121: Constantine Gatsonis, James S. Hodges, Robert E. Kass, Robert McCulloch, Peter Rossi, Nozer D. Singpurwalla (Editors), Case Studies in Bayesian Statistics, Volume III. xvi, 487 pages, 1997.

Vol. 122: Timothy G. Gregoire, David R. Brillinger, Peter J. Diggle, Estelle Russek-Cohen, William G. Warren, Russell D. Wolfinger (Editors), Modeling Longitudinal and Spatially Correlated Data. x, 402 pages, 1997.

Vol. 123: D. Y. Lin and T. R. Fleming (Editors), Proceedings of the First Seattle Symposium in Biostatistics: Survival Analysis. xiii, 308 pages, 1997.

Vol. 124: Christine H. Müller, Robust Planning and Analysis of Experiments. x, 234 pages, 1997.

Vol. 125: Valerii V. Fedorov and Peter Hackl, Model-oriented Design of Experiments. viii, 117 pages, 1997.

Vol. 126: Geert Verbeke and Geert Molenberghs, Linear Mixed Models in Practice: A SAS-Oriented Approach. xiii, 306 pages, 1997.

Vol. 127: Harald Niederreiter, Peter Hellekalek, Gerhard Larcher, and Peter Zinterhof (Editors), Monte Carlo and Quasi-Monte Carlo Methods 1996, xii, 448 pp., 1997.

Vol. 128: L. Accardi and C.C. Heyde (Editors), Probability Towards 2000, x, 356 pp., 1998.

Vol. 129: Wolfgang Härdle, Gerard Kerkyacharian, Dominique Picard, and Alexander Tsybakov, Wavelets, Approximation, and Statistical Applications, xvi, 265 pp., 1998.

Vol. 130: Bo-Cheng Wei, Exponential Family Nonlinear Models, ix, 240 pp., 1998.

Vol. 131: Joel L. Horowitz, Semiparametric Methods in Econometrics, ix, 204 pp., 1998.

Vol. 132: Douglas Nychka, Walter W. Piegorsch, and Lawrence H. Cox (Editors), Case Studies in Environmental Statistics, viii, 200 pp., 1998.

Vol. 133: Dipak Dey, Peter Müller, and Debajyoti Sinha (Editors), Practical Nonparametric and Semiparametric Bayesian Statistics, xv, 408 pp., 1998.

Vol. 134: Yu. A. Kutoyants, Statistical Inference For Spatial Poisson Processes, vii, 284 pp., 1998.

Vol. 135: Christian P. Robert, Discretization and MCMC Convergence Assessment, x, 192 pp., 1998.

Vol. 136: Gregory C. Reinsel, Raja P. Velu, Multivariate Reduced-Rank Regression, xiii, 272 pp., 1998.

Printed in the United States
204666BV00002B/5/P